SOLIDWORKS 2018 For Novices

(Learn By Doing)

CADSoft Technologies

**SOLIDWORKS 2018 for Novices:
Learn By Doing**

CADSoft Technologies

CADSoft Technologies

ISBN: 978-1-64057-039-9

DEDICATION

To students, who are dedicated to learning new technologies

THANKS

To the faculty and employees of CADSoft Technologies for their cooperation

Training Programs offered by CADSoft Technologies

CADSoft Technologies provides quality training on various software packages including Computer Aided Design, Manufacturing and Engineering (CAD/CAM/CAE), Web and Programming Languages, animation, architecture, and GIS. The training is delivered at its training centers or online at any time, place, and pace to individuals, corporates as well as the students of colleges, universities, and CAD/CAM/CAE training centers. The main features of this program are:

Training in Classroom Settings and Online Training

A team of highly experienced instructors and qualified engineers at CADSoft Technologies conduct the classes in classroom environment at its centers. This team has authored several textbooks that are rated "one of the best" in their categories and are used in various colleges, universities, and training centers in North America, Europe, and in other parts of the world.

CADSoft Technologies with its cost effective and time saving initiative strives to deliver online training in the comfort of your home or work place, thereby relieving you from the hassles of traveling to training centers. CADSoft Technologies strives to be the best training institute in USA. CADSoft Technologies has developed career oriented programs by following state-of-the-art teaching and learning methodologies.

Courses Offered by CADSoft Technologies

CADSoft Technologies conducts classes on the following software packages:

CAD/CAM/CAE: *CATIA, Pro/ENGINEER Wildfire, Creo Parametric, Creo Direct, SOLIDWORKS, Autodesk Inventor, Solid Edge, NX, AutoCAD, AutoCAD LT, AutoCAD Plant 3D, Customizing AutoCAD, SolidCAM, NX CAM, NX Mold, Creo Mold, Alias Design, Alias Automotive, Ansys Fluent, and ANSYS*

Architecture and GIS: *Autodesk Revit Architecture, AutoCAD Civil 3D, Autodesk Revit Structure, AutoCAD Map 3D, Revit MEP, Navisworks, Primavera, and Bentley STAAD Pro, MS Project, MX Road, ArcGIS, and Raster Design*

Animation and VFX: *Autodesk 3ds Max, Chaosgroup VRay, Autodesk 3ds Max Design, Autodesk Maya, Blackmagic Design Fusion Studio, Adobe Premiere, Adobe Photoshop, Adobe Indesign, Adobe Illustrator, Corel Graphic Design CorelDraw, The Foundry NukeX, and MAXON CINEMA 4D*

Web and Programming: *C/C++, HTML5/CSS3, JavaScript, jQuery, Bootstrap, PHP, MySQL, Dreamweaver, VB.NET, Oracle, AJAX, and Java*

Personality Development*: Personality Development and Engineering Ethics/Soft Skills Course*

Table of Contents

Chapter 5: Advanced Dimensioning Techniques and Base Feature Options

Chapter 6: Creating Reference Geometries

Chapter 7: Advanced Modeling Tools-I

Chapter 8: Advanced Modeling Tools-II

Chapter 9: Editing Features

Chapter 10: Advanced Modeling Tools-III

Chapter 11: Advanced Modeling Tools-IV

Chapter 12: Assembly Modeling-I

Chapter 13: Assembly Modeling-II

Chapter 14: Working with Drawing Views-I

Chapter 15: Working with Drawing Views-II

Chapter 16: Surface Modeling

Chapter 17: Sheet Metal Design

This page is intentionally left blank

Preface

SOLIDWORKS 2018

SOLIDWORKS, originally developed by the SOLIDWORKS Corporation, USA, was acquired by Dassault Systemes, France, in 1997. Dassault Systemes is world's leading developer of product life cycle management (PLM) solutions. It is one of the fastest growing solid modeling software. It is a parametric, feature-based solid modeling tool that not only unites the three-dimensional (3D) parametric features with two-dimensional (2D) tools, but also addresses every design-through-manufacturing needs. SOLIDWORKS 2018 includes a number of customer requested enhancements, substantiating that it is completely tailored to address customers needs. Based mainly on the user feedback, this solid modeling tool is remarkably user-friendly and allows you to be productive from day one.

In SOLIDWORKS, you can easily generate the 2D drawing views of the components. The drawing views that can be generated include detailed, orthographic, isometric, auxiliary, section, and other views. You can use any predefined standard drawing document to generate the drawing views. Besides displaying the model dimensions in the drawing views or adding reference dimensions and other annotations, you can also add the parametric Bill of Materials (BOM) and balloons in the drawing view. If a component in the assembly is replaced, removed, or a new component is assembled, the modification will automatically reflect in the BOM placed in the drawing document. The bidirectional associative nature of this software ensures that any modification made in the model is automatically reflected in the drawing views and any modification made in the dimensions in the drawing views automatically updates the model.

In addition to creating solid models, assembly features, and drawing views, SOLIDWORKS enables you to effectively and easily create complex sheet metal components using a number of tools. Apart from modeling and detailing, you can print your solid models directly through 3D printers. You can also define position, orientation and other parameters of the model for 3D printing in SOLIDWORKS.

SOLIDWORKS 2018 for Novices: Learn By Doing textbook has been written to help the users who are interested in learning 3D design. Real-world mechanical engineering industry examples and tutorials have been used to ensure that the users can relate the knowledge of this textbook with the actual mechanical industry designs. In this edition, one chapter has been added on mold design. Some of the main features of the textbook are as follows:

- **Tutorial Approach**
  The author has adopted the tutorial point-of-view and the learn-by-doing approach throughout the textbook. This approach guides the users through the process of creating the models in the tutorials.

- **Real-world Mechanical Engineering Projects as Tutorials**
  The author has used the real-world mechanical engineering projects as tutorials in this textbook so that the readers can correlate the tutorials with the real-time models in the mechanical engineering industry.

- **Coverage of Major SOLIDWORKS Modes**
  All major modes of SOLIDWORKS are covered in this textbook. These include the **Part** mode, the **Assembly** mode, and the **Drawing** mode.

- **Tips and Notes**
  Additional information related to various topics is provided to the users in the form of tips and notes.

Symbols Used in this Textbook

Note

The author has provided additional information to the users about the topic being discussed in the form of Notes.

Tip

Special information on various techniques is provided in the form of Tips that will increase the efficiency of the users.

Formatting Conventions Used in the Textbook

Please refer to the following list for the formatting conventions used in this textbook.

- Names of tools, buttons, options, toolbars, and are written in boldface. Example: The **Extrude Boss/Base** tool, the **Mid-Plane** option, the **OK** button, the **Features** toolbar, and so on.

- Names of CommandManager, PropertyManager, rollouts, dialog box, drop-down lists, spinners, selection boxes, areas, edit boxes, check boxes, and radio buttons are written in boldface. Example: The **Features CommandManager**, the **Boss-Extrude PropertyManager**, the **Open** dialog box, the **End Condition** drop-down list, the **Depth** spinner, the **Direction of Extrusion** selection box, the **Draft outward** check box, and so on.

- Values entered in edit boxes are written in boldface. Example: Enter **5** in the **Radius** edit box.

- Names and paths of the files are written in italics. Example: *C:\Documents\SOLIDWORKS\c08\c08_tut01*

Naming Conventions Used in the Textbook

Tool

If you click on an item in a toolbar and a command is invoked to create/edit an object or perform some action, then that item is termed as tool.

For example:
To Create: **Line** tool, **Smart Dimension** tool, **Extruded Boss/Base** tool
To Modify: **Fillet** tool, **Draft** tool, **Trim Surface** tool
Action: **Zoom to Fit** tool, **Pan** tool, **Copy** tool

If you click on an item in a toolbar and a dialog box is invoked wherein you can set the properties to create/edit an object, then that item is also termed as tool, refer to Figure 1.

For example:
To Create: Extruded Boss/Base tool, **Mirror** tool, **Rib** tool
To Modify: Flex tool, **Deform** tool

In this textbook, the path to invoke a tool is given as:

CommandManager: Features > Extruded Boss/Base
SOLIDWORKS Menus: Insert > Boss/Base > Extrude

Flyout

A flyout is the one in which a set of tools are grouped together. You can identify a flyout with a down arrow on it. A flyout is given a name based on the types of tools grouped in it. For example, **Line** flyout, **View Settings** flyout, **Fillet** flyout, and so on; refer to Figure 1.

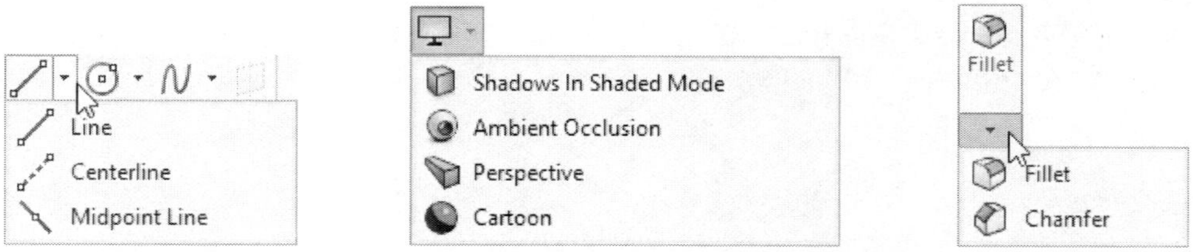

Figure 1 *The* ***Line****,* ***View Settings****, and* ***Fillet*** *flyouts*

PropertyManager

The naming conventions for the components in a PropertyManager are mentioned in Figure 2.

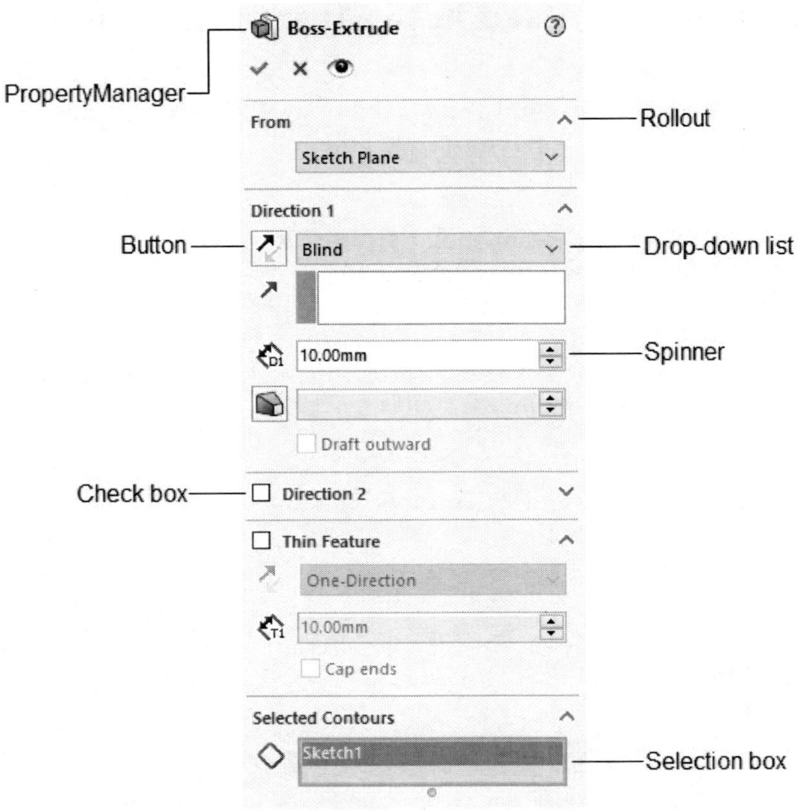

*Figure 2 The **Boss-Extrude PropertyManager***

Button

The items in a dialog box that has a 3D shape like a button is termed as **Button**. For example, **OK** button, **Cancel** button, and so on.

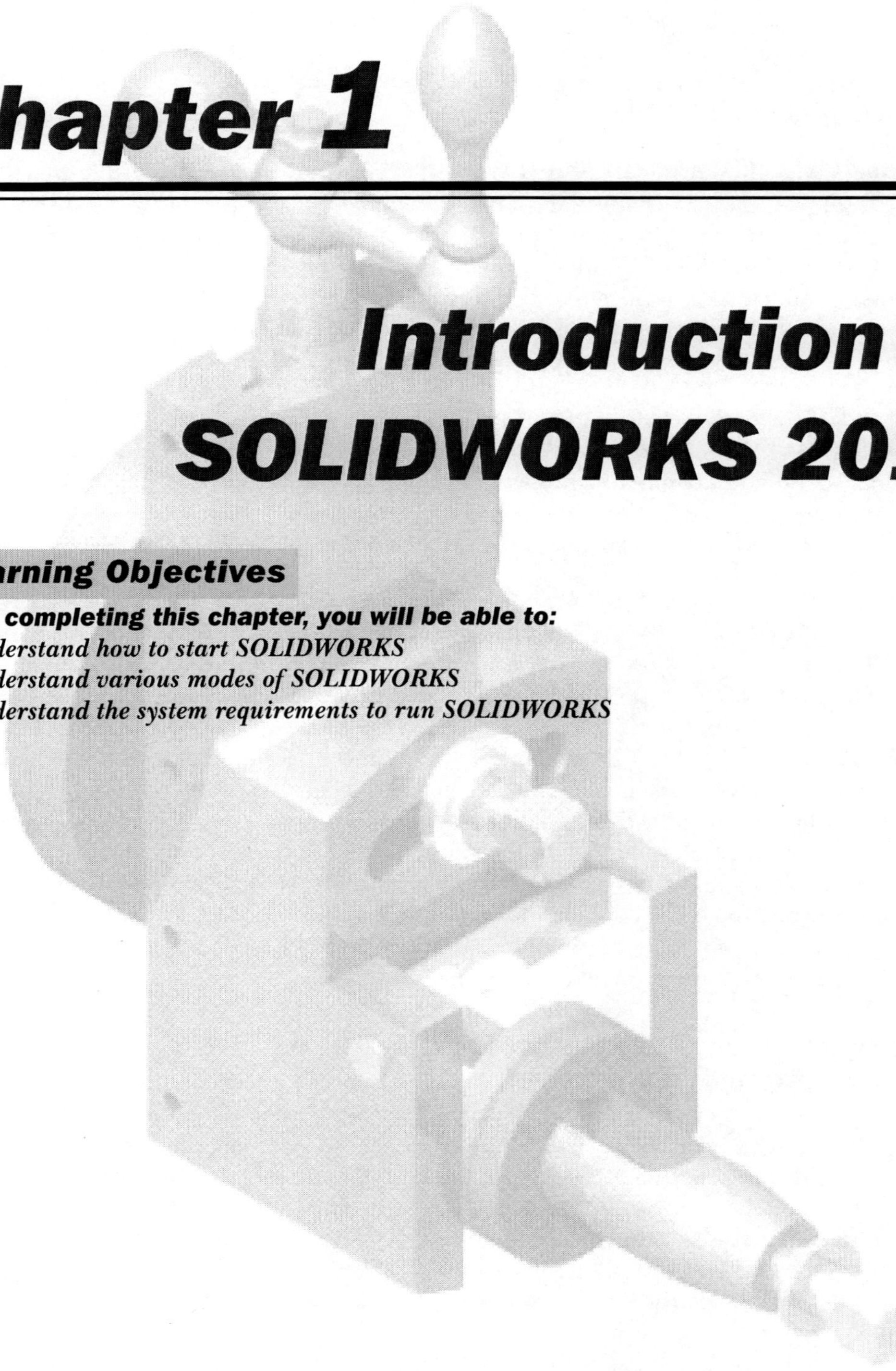

Chapter 1

Introduction to SOLIDWORKS 2018

Learning Objectives

After completing this chapter, you will be able to:

- *Understand how to start SOLIDWORKS*
- *Understand various modes of SOLIDWORKS*
- *Understand the system requirements to run SOLIDWORKS*

INTRODUCTION TO SOLIDWORKS 2018

Welcome to the world of Computer Aided Design (CAD) with SOLIDWORKS. If you are a new user of this software package, you will be joining hands with thousands of users of this parametric, feature-based, and one of the most user-friendly software packages. If you are familiar with the previous releases of this software, you will be able to upgrade your designing skills with this improved release of SOLIDWORKS.

SOLIDWORKS, developed by the SOLIDWORKS Corporation, USA, is a feature-based, parametric solid-modeling mechanical design and automation software. SOLIDWORKS is the first CAD package to use the Microsoft Windows graphical user interface. The use of the drag and drop (DD) functionality of Windows makes this CAD package extremely easy to learn. The Windows graphic user interface makes it possible for the mechanical design engineers to innovate their ideas and implement them in the form of virtual prototypes or solid models, large assemblies, subassemblies, and detailing and drafting.

SOLIDWORKS is one of the products of SOLIDWORKS Corporation, which is a part of Dassault Systemes. SOLIDWORKS also works as platform software for a number of software. This implies that you can also use other compatible software within the SOLIDWORKS window. There are a number of software provided by the SOLIDWORKS Corporation, which can be used as add-ins with SOLIDWORKS. Some of the software that can be used on SOLIDWORKS's work platform are listed below:

SOLIDWORKS Motion	SOLIDWORKS Routing	ScanTo3D	eDrawings
SOLIDWORKS Simulation	SOLIDWORKS Toolbox	PhotoView 360	CircuitWorks
SOLIDWORKS Plastics	SOLIDWORKS Inspection	TolAnalyst	

As mentioned earlier, SOLIDWORKS is a parametric, feature-based, and easy-to-use mechanical design automation software. It enables you to convert the basic 2D sketch into a solid model by using simple but highly effective modeling tools. It also enables you to create the virtual prototype of a sheet metal component and the flat pattern of the component. This helps you in the complete process planning for designing and creating a press tool. SOLIDWORKS helps you to extract the core and the cavity of a model that has to be molded or cast. With SOLIDWORKS, you can also create complex parametric shapes in the form of surfaces. Some of the important modes of SOLIDWORKS are discussed next.

Part Mode

The **Part** mode of SOLIDWORKS is a feature-based parametric environment in which you can create solid models. In this mode, you are provided with three default planes named as **Front Plane**, **Top Plane**, and **Right Plane**. First, you need to select a sketching plane to create a sketch for the base feature. On selecting a sketching plane, you enter the sketching environment. The sketches for the model are drawn in the sketching environment using easy-to-use tools. After drawing the sketches, you can dimension them and apply the required relations in the same sketching environment. The design intent is captured easily by adding relations and equations and using the design table in the design. You are provided with the standard hole library known as the **Hole Wizard** in the **Part** mode. You can create simple holes, tapped holes, counterbore holes, countersink holes, and so on by using this wizard. The holes can be of any standard such as ISO, ANSI, JIS, and so on. You can also create complicated surfaces by using the surface modeling tools available in the **Part** mode. Annotations such as weld symbols, geometric tolerance, datum references, and surface finish symbols can be added to the model within the **Part** mode. The standard features that are used frequently can be saved as library features and retrieved when needed. The palette feature library of SOLIDWORKS contains a number of standard mechanical parts and features. You can also create the sheet metal components in this mode of SOLIDWORKS by using the related tools. Besides this, you can also analyze the part model for various stresses applied to it in the real physical conditions by using an easy and user-friendly tool called SimulationXpress. It helps you reduce the cost and time in physically testing your design in real testing conditions (destructive tests). You can also analyze the component during modeling in the SOLIDWORKS windows. In addition, you can work with the weld modeling within the **Part** mode of SOLIDWORKS by creating steel structures and adding weld beads. All standard weld types and welding conditions are available for your reference. You can extract the core and the cavity in the **Part** mode by using the mold design tools.

Assembly Mode

In the **Assembly** mode, you can assemble components of the assembly with the help of the required tools. There are two methods of assembling the components:

1. Bottom-up assembly
2. Top-down assembly

In the bottom-up assembly method, the assembly is created by assembling the components created earlier and maintaining their design intent. In the top-down method, the components are created in the assembly mode. You may begin with some ready-made parts and then create other components in the context of the assembly. You can refer to the features of some components of the assembly to drive the dimensions of other components. You can assemble all components of an assembly by using a single tool, the **Mate** tool. While assembling the components of an assembly, you can also animate the assembly by dragging. Besides this, you can also check the working of your assembly. Collision detection is one of the important features in this mode. Using this feature, you can rotate and move components as well as detect the interference and collision between the assembled components. You can see the realistic motion of the assembly by using physical dynamics. Physical simulation is used to simulate the assembly with the effects of motors, springs, and gravity on the assemblies.

Drawing Mode

The **Drawing** mode is used for the documentation of the parts or the assemblies created earlier in the form of drawing views. The procedure for creating drawing views is called drafting. There are two types of drafting done in SOLIDWORKS:

1. Generative drafting
2. Interactive drafting

Generative drafting is a process of generating drawing views of a part or an assembly created earlier. The parametric dimensions and the annotations added to the component in the **Part** mode can be generated in the drawing views. Generative drafting is bidirectionally associative in nature. Automatic BOMs and balloons can be added to an assembly while generating the drawing views of it.

In interactive drafting, you have to create the drawing views by sketching them using normal sketching tools and then add dimensions to them.

SYSTEM REQUIREMENTS

The system requirements to ensure the smooth functioning of SOLIDWORKS on your system are as follows:

- Microsoft Windows 10, Windows 8.1 or Windows 8 (64 bit only) or Windows 7 (SP1 required).
- Intel or AMD Processor with SSE2 support.
- 2 GB RAM minimum (8 GB recommended).
- Hard disk space 5 GB minimum (10 GB recommended).
- A certified graphics card and driver.
- Microsoft Office 2007 or later.
- Adobe Acrobat higher than 8.0.7.
- DVD drive and Mouse or any other compatible pointing device.
- Internet Explorer version 8 or higher.

GETTING STARTED WITH SOLIDWORKS

Install SOLIDWORKS 2018 on your system; the shortcut icon of SOLIDWORKS 2018 will automatically be created on the desktop. Double-click on this icon; the system will prepare to start SOLIDWORKS and after sometime, the SOLIDWORKS window will be displayed on the screen. On opening SOLIDWORKS for the first time, the **SOLIDWORKS License Agreement** dialog box will be displayed, as shown in Figure 1-1. Choose the **Accept** button in this dialog box; the **SOLIDWORKS 2018** window will open and the **SOLIDWORKS Resources** task pane will be displayed on the right. Also, the **Welcome - SOLIDWORKS 2018** dialog box is invoked simultaneously, as shown in Figure 1-2. This window can be used to open a new file or an existing file.

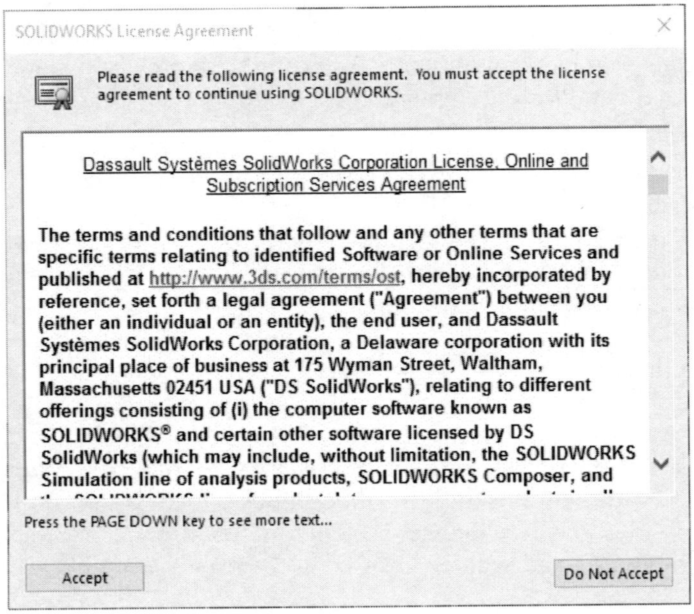

*Figure 1-1 The **SOLIDWORKS License Agreement** dialog box*

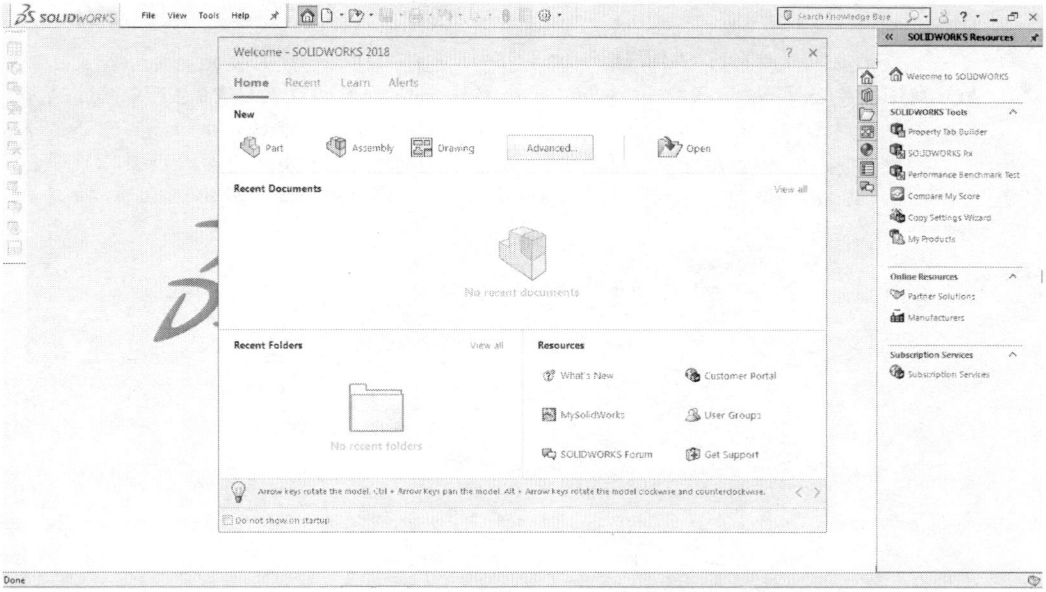

***Figure 1-2** The SOLIDWORKS 2018 window and the **SOLIDWORKS Resources** task pane*

If the **SOLIDWORKS Resources** task pane is not displayed or expanded, choose the **SOLIDWORKS Resources** button located on the right side of the window to display it. This task pane can be used to open online tutorials and to visit the website of SOLIDWORKS partners. Choose the **Part** button from the **Welcome - SOLIDWORKS 2018** dialog box or the **New** button from the Menu Bar to create a new document. If you start a new document using the **New** button from the Menu Bar then the **New SOLIDWORKS Document** dialog box will be displayed, as shown in Figure 1-3.

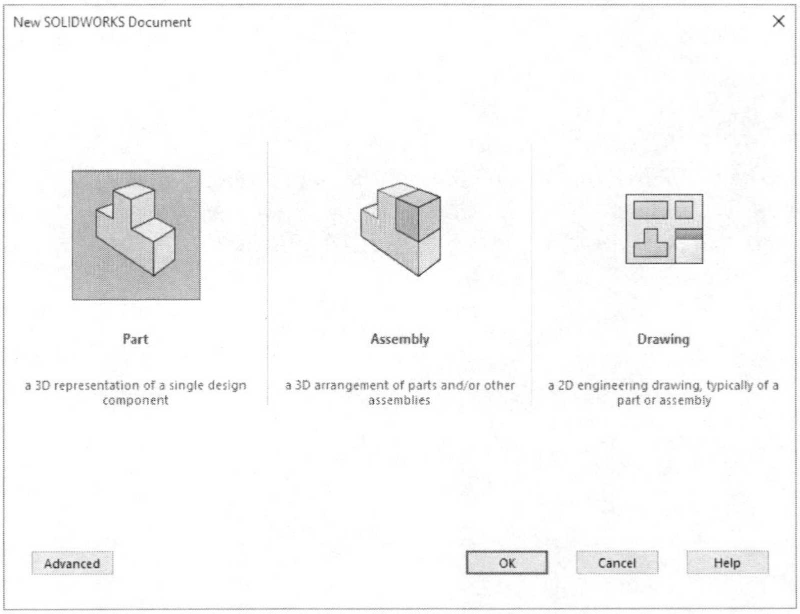

***Figure 1-3** The **New SOLIDWORKS Document** dialog box*

Note

If you are starting SOLIDWORKS 2018 for the first time, then on invoking the ***New SOLIDWORKS Document*** *dialog box or the* ***Welcome - SOLIDWORKS 2018*** *dialog box, the* ***Units and Dimension Standard*** *dialog box will be displayed, as shown in Figure 1-4. Using this dialog box, you can specify the default units and dimension standards for SOLIDWORKS. In this book, the unit system used is MMGS (millimeter, gram, second) and the dimension standard used is ISO.*

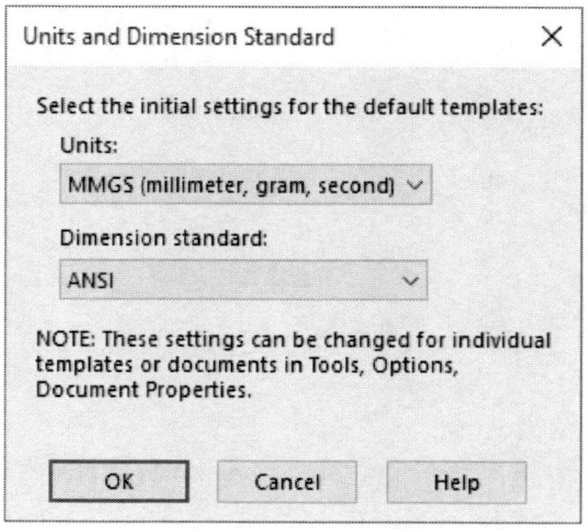

Figure 1-4 *The* ***Units and Dimension Standard*** *dialog box*

Choose the **Part** button to create a part model and then choose **OK** from the **New SOLIDWORKS Document** dialog box to enter the **Part** mode of SOLIDWORKS. Hover the cursor over the SOLIDWORKS logo; the SOLIDWORKS Menus will be displayed on the right of the logo. Note that the task pane is automatically closed once you start a new file and click in the drawing area. The initial screen display on starting a new part file of SOLIDWORKS using the **New** button in the Menu Bar is shown in Figure 1-5.

Tip

In SOLIDWORKS, the tip of the day will be displayed at the bottom of the ***Welcome - SOLIDWORKS 2018*** *dialog box. You can click on the arrows to view additional tips. These tips help you use SOLIDWORKS efficiently. It is recommended that you view at least 2 or 3 tips every time you start a new session of SOLIDWORKS 2018.*

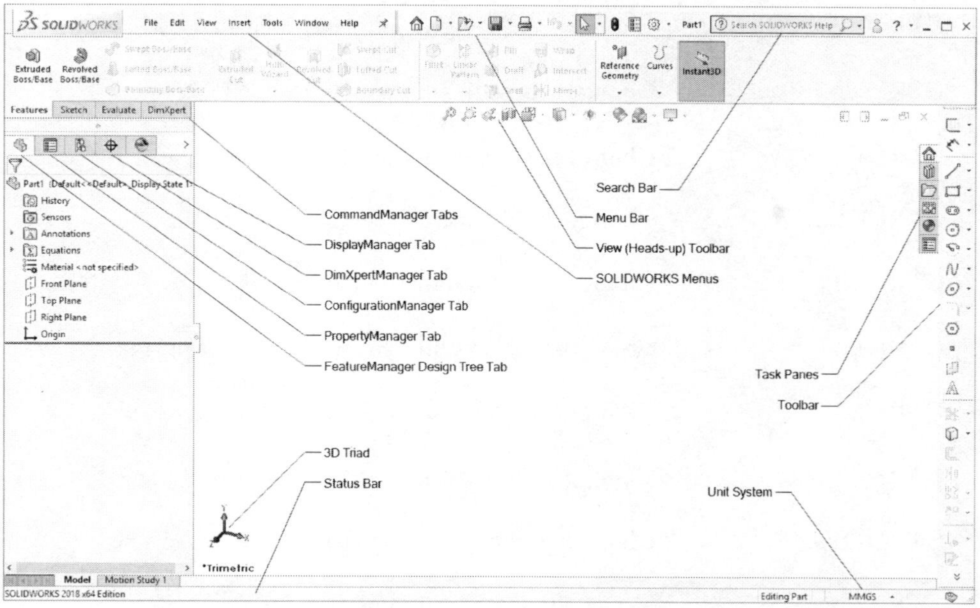

Figure 1-5 *The components of a new part document*

It is evident from the screen that SOLIDWORKS is a very user-friendly solid modeling software. Apart from the default CommandManager shown in Figure 1-5, you can also invoke other CommandManagers. To do so, move the cursor on a CommandManager tab and right-click; a shortcut menu will be displayed. Choose the required CommandManager from the shortcut menu; it will be added. Besides the existing CommandManager, you can also create a new CommandManager.

HOT KEYS

SOLIDWORKS is more popularly known for its mouse gesture functionality. However, you can also use the keys of the keyboard to invoke some tools, windows, dialog boxes, and so on. These keys are known as hot keys. Some hot keys along with their functions are given next.

Hot Key	**Function**
F11	Full screen
S	Invokes the shortcut bar
R	Invokes the recent documents
F	Fits the object in the drawing over the screen
Z	Zooms out
SPACE BAR	Invokes the **Orientation** menu

CTRL+1	Changes the current view to the Front View
CTRL+2	Changes the current view to the Back View
CTRL+3	Changes the current view to the Left View
CTRL+4	Changes the current view to the Right View
CTRL+5	Changes the current view to the Top View
CTRL+6	Changes the current view to the Bottom View
CTRL+7	Changes the current view to the Isometric View
CTRL+8	Changes the current view to the Normal View
CTRL+SHIFT+Z	Changes the current view to the Previous View
CTRL+Arrows	Moves the feature along the arrows direction
SHIFT+Arrows	Rotates the feature along the arrows direction
CTRL+B	Rebuilds the model
CTRL+Z	Invokes the **Undo** tool
CTRL+N	Invokes the **New SOLIDWORKS Document** dialog box
CTRL+O	Invokes the **Open** window
CTRL+S	Saves the document
CTRL+P	Prints the document
CTRL+A	Selects all the parts in the document
CTRL+C	Copies the selected feature
CTRL+V	Pastes the selected feature
CTRL+X	Cuts the selected feature
ALT+F	Opens the **File** menu
ALT+E	Opens the **Edit** menu
ALT+V	Opens the **View** menu

ALT+I	Opens the **Insert** menu
ALT+T	Opens the **Tool** menu
ALT+W	Opens the **Window** menu
ALT+H	Opens the **Help** menu
CTRL+W	Closes the current document

This page is intentionally left blank

Chapter 2

Drawing Sketches for Solid Models

Learning Objectives

After completing this chapter, you will be able to:

- *Understand the sketching environment*
- *Start a new document*
- *Set the document options*
- *Learn the sketcher terms*
- *Use various sketching tools*
- *Use the drawing display tools*

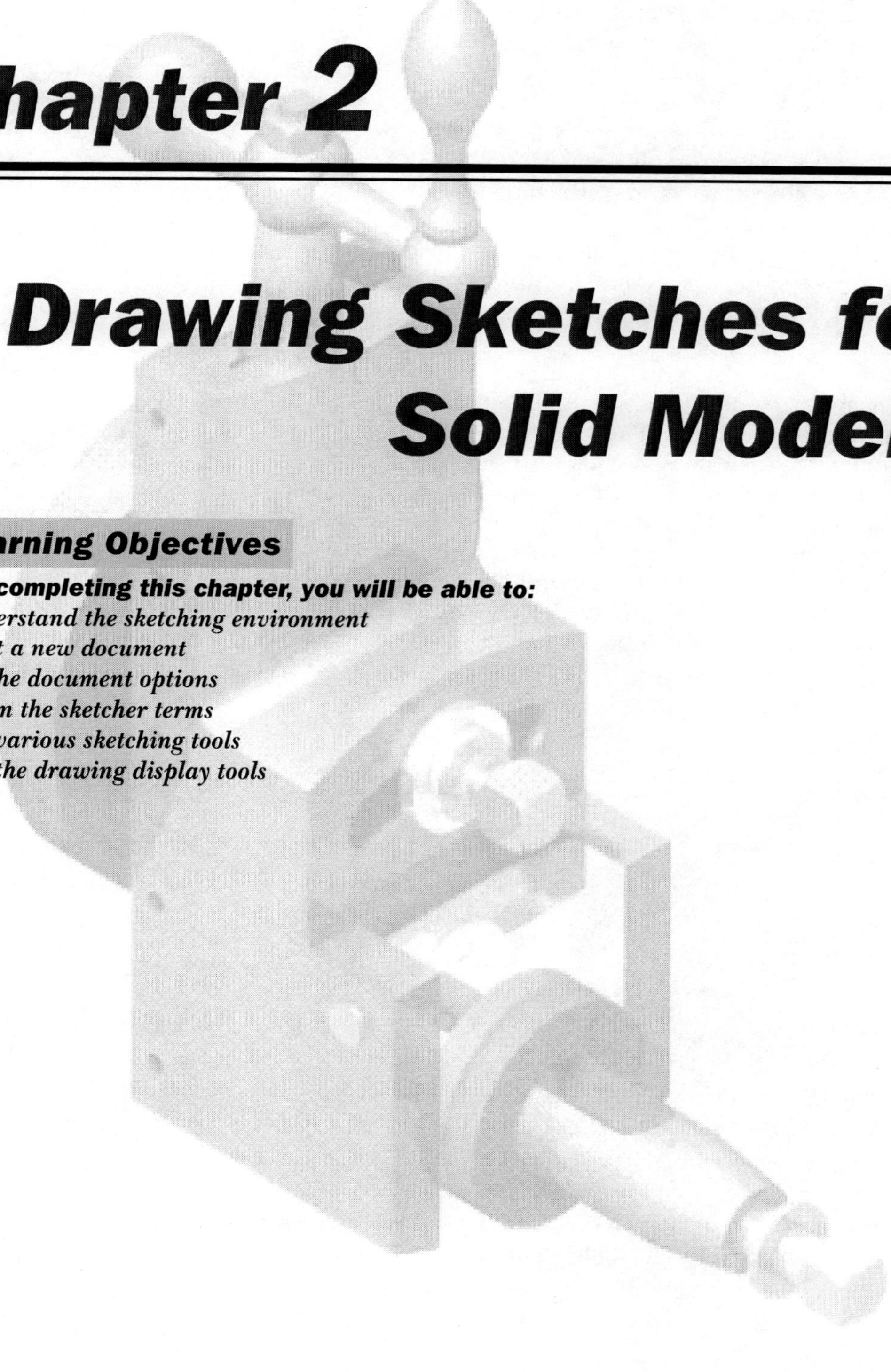

THE SKETCHING ENVIRONMENT

Most of the products designed by using SOLIDWORKS are a combination of sketched, placed, and derived features. The placed and derived features are created without drawing a sketch, but the sketched features require a sketch to be drawn first. Generally, the base feature of any design is a sketched feature and is created using the sketch. Therefore, while creating any design, the first and foremost requirement is to draw a sketch for the base feature. Once you have drawn the sketch, you can convert it into the base feature and then add the other sketched, placed, and derived features to complete the design. In this chapter, you will learn to create the sketch for the base feature using various sketching tools.

In general terms, a sketch is defined as the basic contour for a feature. For example, consider the solid model of a spanner shown in Figure 2-1.

This spanner consists of a base feature, cut feature, mirror feature (cut on the back face), fillets, and an extruded text feature. The base feature of this spanner is shown in Figure 2-2. It is created using a single sketch drawn on the **Front Plane**, refer to Figure 2-3. This sketch is drawn in the sketching environment using various sketching tools. Therefore, to draw the sketch of the base feature, you first need to invoke the sketching environment where you will draw the sketch.

Figure 2-1 *Solid model of a spanner*

Figure 2-2 *Base feature of the spanner*

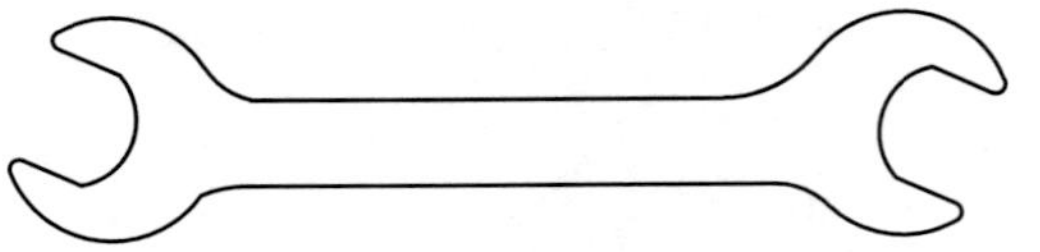

Figure 2-3 *Sketch for the base feature of the spanner*

Note

Once you are familiar with various options of SOLIDWORKS, you can also use a derived feature or a derived part as the base feature.

The sketching environment of SOLIDWORKS can be invoked anytime in the **Part** or **Assembly** mode. You just have to specify that you need to draw the sketch of a feature and then select the plane on which you need to draw the sketch.

STARTING A NEW SESSION OF SOLIDWORKS 2018

Double-click on the **SOLIDWORKS 2018** icon; the SOLIDWORKS 2018 window will be displayed. If you are starting the SOLIDWORKS application for the first time after installing it, the **SOLIDWORKS License Agreement** dialog box will be displayed. Choose **Accept** from this dialog box; the **SOLIDWORKS** interface window along with the **Welcome - SOLIDWORKS 2018** dialog box will be displayed, as shown in Figure 2-4. Click anywhere in SOLIDWORKS 2018 window to collapse the **Welcome - SOLIDWORKS 2018** dialog box.

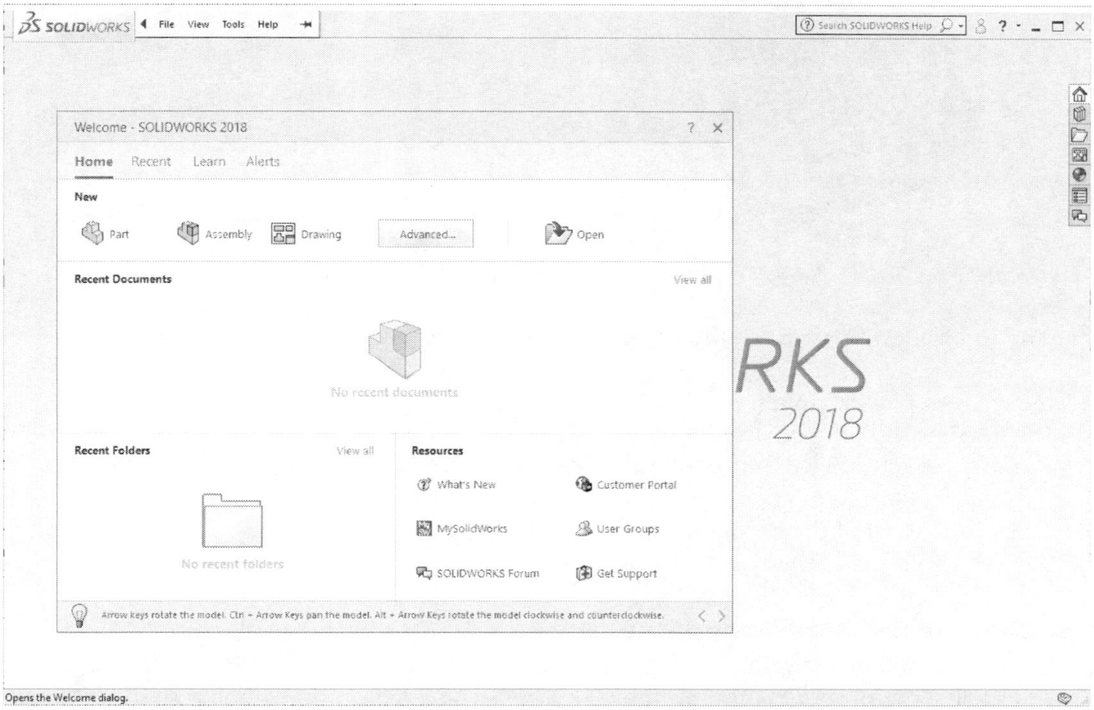

*Figure 2-4 The **SOLIDWORKS** window*

TUTORIALS

Tutorial 1

In this tutorial, you will draw the basic sketch of the revolved solid model shown in Figure 2-5. The sketch of this model is shown in Figure 2-6. Do not dimension the sketch as the solid model and its dimensions are given for your reference only. **(Expected time: 30 min)**

Figure 2-5 Revolved solid model for Tutorial 1

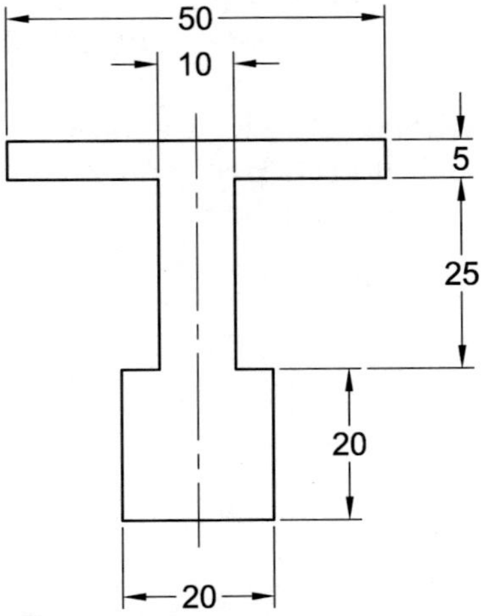

Figure 2-6 Sketch of the revolved solid model

The following steps are required to complete this tutorial:

a. Start a new part document.
b. Invoke the sketching environment.
c. Modify the settings of the snap and grid so that the cursor jumps through a distance of 5 mm.
d. Draw the sketch of the model using the **Line** tool.
e. Save the sketch and then close the document.

Opening a New Part Document

1. Start SOLIDWORKS by double-clicking on the shortcut icon of SOLIDWORKS 2018 available on the desktop of your computer; the SOLIDWORKS 2018 window along with the **Welcome - SOLIDWORKS 2018** dialog box will be displayed.

2. Choose the **Part** button available in the **New** area of the **Home** tab in the **Welcome - SOLIDWORKS 2018** dialog box. A new SOLIDWORKS part document is invoked. You can also select the **New** button from the Menu Bar; the **New SOLIDWORKS Document** dialog box is displayed, as shown in Figure 2-7.

3. In the **New SOLIDWORKS Document** dialog box, the **Part** button is chosen by default. Therefore, choose the **OK** button; a new SOLIDWORKS part document starts.

   You need to invoke the sketching environment to draw the sketch.

4. Choose the **Sketch** tab from the **CommandManager** and then select the **Sketch** tool from the **Sketch CommandManager**; the **Edit Sketch PropertyManager** is displayed and you are prompted to select a plane on which you want to draw the sketch.

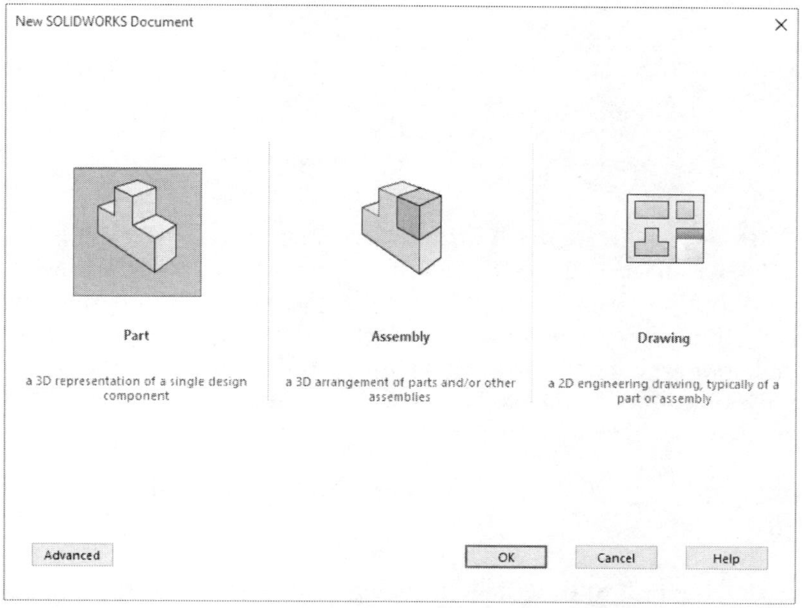

Figure 2-7 *The* ***New SOLIDWORKS Document*** *dialog box*

5. Select the **Front Plane** from the drawing area; the sketching environment is invoked and the plane gets oriented normal to the view. You will notice that red colored arrows are displayed at the center of the screen, indicating that you are in the sketching environment. Also, the confirmation corner with the **Exit Sketch** and **Cancel** options is displayed on the upper right corner in the graphics area. The screen display in the sketching environment of SOLIDWORKS 2018 is shown in Figure 2-8.

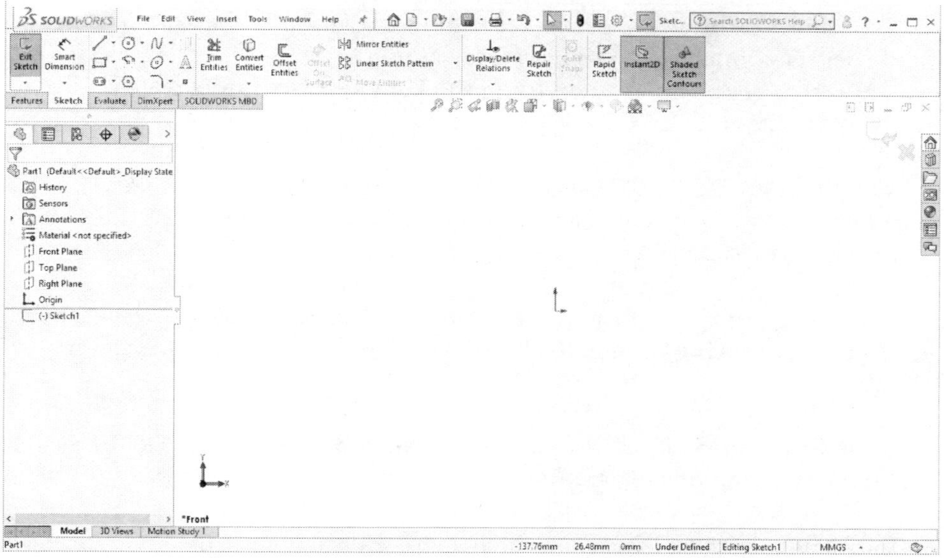

Figure 2-8 *Screen display in the sketching environment*

Tip

In SOLIDWORKS 2018, you can set the orientation of sketching plane automatically parallel to screen whenever you start a new sketch or edit an existing sketch. This setting can toggled by selecting the ***Auto-rotate view normal to sketch plane on sketch creation and sketch edit*** *check box in* ***Sketch*** *area in* ***System Options - General*** *dialog box.*

Modifying the Snap, Grid, and Dimensioning Unit Settings

It is assumed that while installing SOLIDWORKS, you have selected the **MMGS (millimeters, gram, second)** option for measuring the length. Therefore, the length of an entity will be measured in millimeters in the current file. But, if you have selected some other unit at the time of installation, you need to change the linear and angular unit settings before drawing the sketch. For this tutorial, you need to modify the grid and snap settings so that the cursor jumps through a distance of 5 mm.

1. Choose the **Options** button from the Menu Bar; the **System Options - General** dialog box is displayed.

2. Choose the **Document Properties** tab; the name of the dialog box changes to the **Document Properties - Drafting Standard**.

Note

If you have selected millimeters as the unit of measurement while installing SOLIDWORKS, skip steps 3 and 4 in this section.

3. Select the **Units** option from the area on the left to display the options related to the linear and angular units.

4. Select the **MMGS (millimeter, gram, second)** radio button in the **Unit system** area. Also, select the **degrees** option in the **Units** column as unit of angle, if it is not selected by default. You can also change the unit system using the **Unit system** button, which is located at the right-side of the status bar.

   It is evident from Figure 2-2 that the dimensions in the sketch are multiples of 5. Therefore, you need to modify the grid and snap settings so that the cursor jumps through a distance of 5 mm instead of 10 mm.

5. Select the **Grid/Snap** option from the area on the left to display grid options. Set the value in the **Major grid spacing** spinner to **50** and the value in the **Minor-lines per major** spinner to **10**.

6. Select the **Display grid** check box if it is cleared. Next, choose the **Go To System Snaps** button; the system options related to relations and snaps are displayed.

7. Select the **Grid** check box from the **Sketch snaps** area and clear the **Snap only when grid is displayed** check box. Choose **OK** to exit the dialog box.

Note that in the sketching environment, the lower right corner of the drawing area displays the information about the status of sketch and location of the cursor in the X, Y, and Z coordinates. You will use the coordinates displayed to draw the sketch of the model. These coordinates will be modified as you move the cursor around the drawing area. If you move the cursor after initial settings, the coordinates will show an increment of 5 mm instead of the default increment of 10 mm.

Drawing the Sketch

It is evident from Figure 2-2 that the sketch will be drawn using the **Line** tool. Therefore, you need to start drawing the sketch from the lower left corner of the sketch.

1. Choose the **Line** tool from the **Sketch CommandManager**; the arrow cursor is replaced by the line cursor.

2. Move the line cursor to the origin.

3. Left-click at this point and move the cursor horizontally toward the right. You will notice that the symbol of the **Horizontal** relation is displayed below the line cursor and the length and angle of the line are displayed above the line cursor.

4. Left-click again when the length of the line above the line cursor shows 20.

   The first horizontal line is drawn. As you are drawing continuous lines, the endpoint of the line drawn is automatically selected as the start point of the next line.

5. Move the line cursor vertically upward. The symbol of the **Vertical** relation is displayed on the right of the line cursor and the length of the line is displayed above the line cursor. Click when the length of the line on the line cursor is displayed as 20.

6. Move the cursor horizontally toward the left and click when the length of the line on the line cursor is displayed as 5.

7. Move the line cursor vertically upward and press the left mouse button when the length of the line on the line cursor is displayed as 25.

8. Move the line cursor horizontally toward the right and click when the length of the line on the line cursor is displayed as 20.

9. Move the line cursor vertically upward and click when the length of the line on the line cursor is displayed as 5.

10. Press F on the keyboard to fit the sketch on the screen.

11. Move the line cursor horizontally toward the left and click when the length of the line on the line cursor is displayed as 50.

12. Move the line cursor vertically downward and click when the length of the line on the line cursor is displayed as 5.

13. Move the line cursor horizontally toward the right and press the left mouse button when the length of the line on the line cursor is displayed as 20.

14. Move the line cursor vertically downward and click when the length of the line on the line cursor is displayed as 25.

15. Move the line cursor horizontally toward the left and click when the length of the line on the line cursor is displayed as 5.

16. Move the line cursor vertically downward to the start point of the first line. Click when an orange circle is displayed; the final sketch for Tutorial 1 is created, as shown in Figure 2-9. In this figure, the grid display and **Shaded Sketch Contours** settings are turned off for clarity.

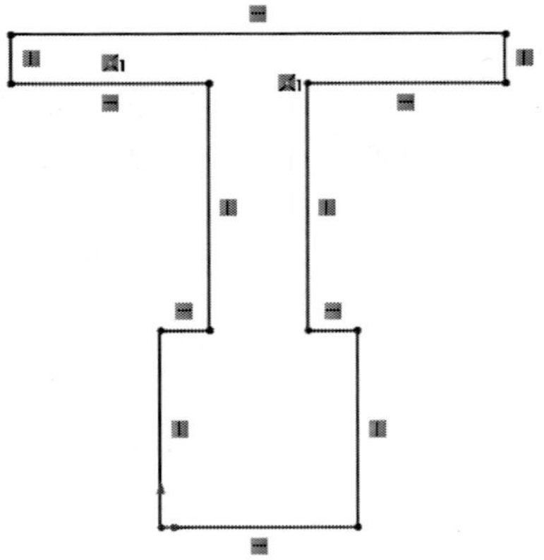

Figure 2-9 *Final sketch for Tutorial 1*

17. Right-click and then choose the **Select** option from the shortcut menu to exit the **Line** tool.

Note

In Figure 2-9, the display of relations is turned on. To turn off the display of relations, choose the ***View Sketch Relations*** *button from the* ***Hide/Show Items*** *flyout in the* ***View (Head-Up)*** *toolbar.*

Tip

To turn off the grid display, right-click in the drawing area to display a shortcut menu. Choose the ***Display Grid*** *button to turn off the grid display. This is a toggle button.*

Saving the Sketch

It is recommended that you create a separate folder for saving the tutorial files of this book. Next, you can save the tutorials of a chapter in the folder of that chapter.

1. Choose the **Save** button from the Menu Bar to invoke the **Save As** dialog box. Create the *SOLIDWORKS* folder inside the *\Documents* folder and then create the *c02* folder inside the *SOLIDWORKS* folder.

2. Enter **c02_tut01** as the name of the document in the **File name** edit box and choose the **Save** button. The document is saved at the location *\Documents\SOLIDWORKS\c02*.

3. Close the document by choosing **File** > **Close** from the SOLIDWORKS menus.

Tutorial 2

In this tutorial, you will draw the sketch of the solid model shown in Figure 2-10. The sketch of the model is shown in Figure 2-11. Do not dimension the sketch as the solid model and the dimensions are given for your reference only. **(Expected time: 30 min)**

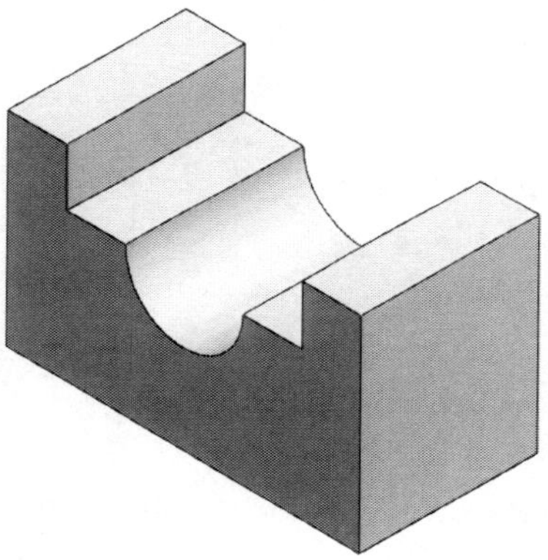

Figure 2-10 *Solid model for Tutorial 2*

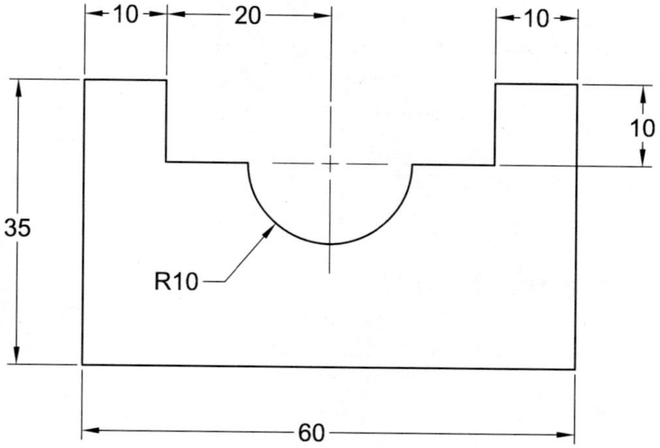

Figure 2-11 *Sketch for Tutorial 2*

The following steps are required to complete this tutorial:

a. Start SOLIDWORKS and then start a new part document.
b. Invoke the sketching environment.
c. Modify the snap and grid settings so that the cursor jumps through a distance of 5 mm.
d. Draw the sketch using the **Line** tool, refer to Figure 2-12.
e. Save the sketch and then close the file.

Opening a New File

1. Choose the **New** button from the Menu Bar to invoke the **New SOLIDWORKS Document** dialog box.

Make sure that the **Part** button is chosen in the **New SOLIDWORKS Document** dialog box.

2. Choose the **OK** button from the dialog box.

You need to invoke the sketching environment to draw the sketch of the model first.

3. Choose the **Sketch** tool from the **Sketch CommandManager** and select the **Front Plane** to invoke the sketching environment.

Modifying the Snap, Grid, Dimensioning Unit Settings

It is evident from Figure 2-11 that the dimensions in the sketch are multiples of 5. Therefore, you need to modify the grid and snap settings so that the cursor jumps through a distance of 5 mm.

1. Choose the **Options** button from the Menu Bar to invoke the **System Options - General** dialog box. In this dialog box, choose the **Document Properties** tab.

2. Select the **Grid/Snap** option from the area on the left and select the **Display Grid** check box, if it is cleared. Set the value in the **Major grid spacing** spinner to **50** and in the **Minor-lines per major** spinner to **10**.

3. Choose the **Go To System Snaps** button to display the **System Options - Relations/Snaps** dialog box. Make sure that the **Grid** check box is selected in the **System Options - Relations/Snaps** dialog box. Next, choose the **OK** button.

   When you move the cursor, the coordinates displayed close to the lower right corner of the drawing area show an increment of 5 mm.

Drawing the Sketch

The sketch will be drawn using the **Line** tool. The arc in the sketch will also be drawn using the same tool. You need to start the drawing from the lower left corner of the sketch.

1. Invoke the **Line** tool by pressing the L key; the arrow cursor is replaced by the line cursor.

2. Move the line cursor to origin.

3. Click at this point and move the cursor horizontally toward the right. Click again when the length of the line above the line cursor displays 60; a horizontal line of 60 mm length is drawn.

4. Move the line cursor vertically upward and click when the length above the line cursor is displayed as 35.

5. Choose the **Zoom to Fit** button from the **View (Heads-Up)** toolbar to fit the sketch into the screen.

   As mentioned earlier, you can invoke the drawing display tools while some other tools are still active. After modifying the drawing display area, the **Line** tool that was active before invoking the drawing display tool will be restored and you can continue drawing lines using the **Line** tool.

6. Move the line cursor horizontally toward the left and click the left mouse button when the length of the line above the line cursor shows the value 10.

7. Move the line cursor vertically downward and click when the length of the line above the line cursor is displayed as 10.

8. Move the line cursor horizontally toward the left and click when the length of the line above the line cursor is displayed as 10.

   Next, you need to draw an arc normal to the last line using the **Line** tool. It is recommended to use the **Line** tool when you need to draw a sketch that is a combination of lines and arcs. This increases productivity by reducing the time taken in invoking tools for drawing an arc and a line.

9. Move the line cursor away from the endpoint of the last line and then move it back close to the endpoint; the arc mode is invoked.

10. Move the arc cursor vertically downward up to the next grid point.

11. Move the arc cursor toward the left.

    You will notice that a normal arc is being drawn and the angle and radius of the arc are displayed above the line cursor.

12. Move the cursor to the left and click when the angle value on the arc cursor is displayed as 180 and the radius value is displayed as 10; an arc normal to the last line is drawn and the line mode is invoked.

13. Move the line cursor horizontally toward the left and click when the length of the line on the line cursor is displayed as 10.

14. Move the line cursor vertically upward and click when the length of the line on the line cursor is displayed as 10.

15. Move the line cursor horizontally toward the left and click when the length of the line on the line cursor is displayed as 10.

16. Move the line cursor to the start point of the first line and click when an orange circle is displayed.

17. Press the ESC key to exit the **Line** tool.

    This completes the sketch. However, you need to modify the drawing display area such that the sketch fits the screen.

18. Press the F key to modify the drawing display area. The final sketch for Tutorial 2 with the grid display and the **Shaded Sketch Contours** settings turned off is shown in Figure 2-12.

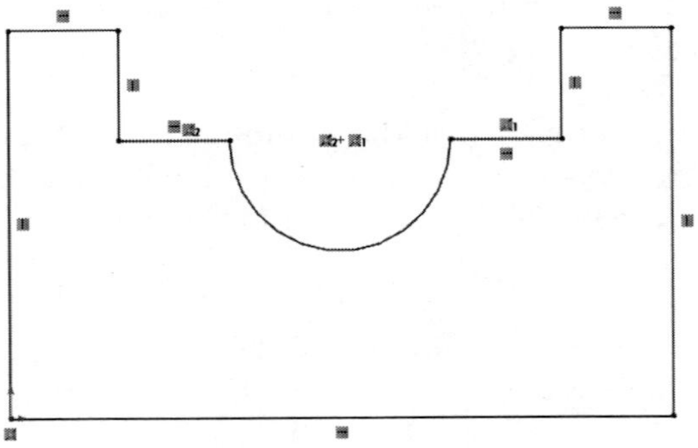

Figure 2-12 *Final sketch for Tutorial 2*

Saving the Sketch

1. Choose the **Save** button from the Menu Bar to invoke the **Save As** dialog box.

2. Enter **c02_tut02** as the name of the document in the **File name** edit box. Choose the **Save** button and then save the file at the location *\Documents\SOLIDWORKS\c02*.

3. Close the document by choosing **File > Close** from the SOLIDWORKS menus.

EXERCISES

Exercise 1

Draw the sketch of the model shown in Figure 2-13. The sketch to be drawn is shown in Figure 2-14. Do not dimension the sketch. The solid model and its dimensions are given for your reference only. **(Expected time: 30 min)**

Figure 2-13 *Solid model for Exercise 1*

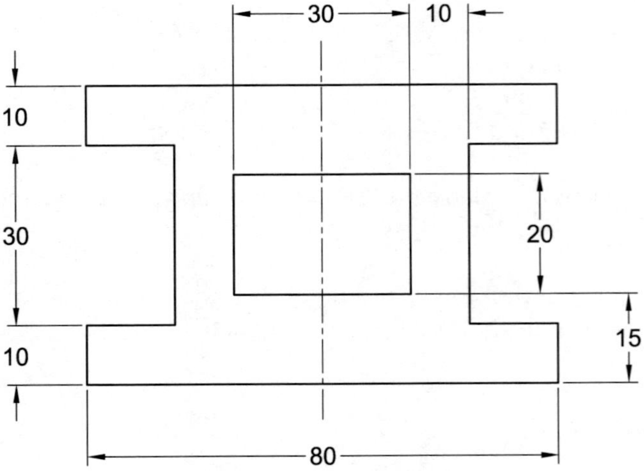

Figure 2-14 *Sketch for Exercise 1*

Exercise 2

Draw the sketch of the model shown in Figure 2-15. The sketch to be drawn is shown in Figure 2-16. Do not dimension the sketch. The solid model and its dimensions are given for your reference only. **(Expected time: 30 min)**

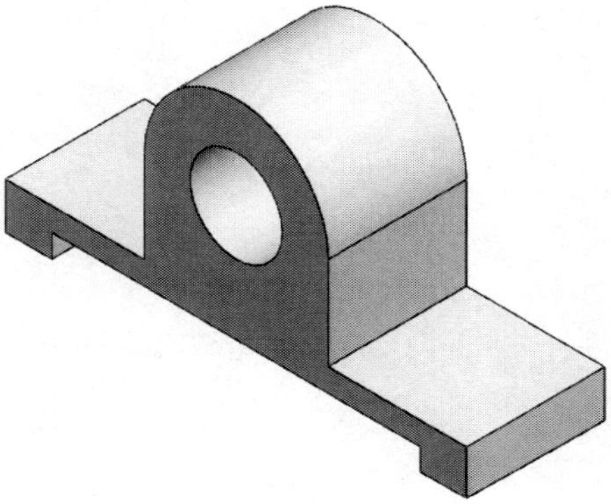

Figure 2-15 *Solid model for Exercise 2*

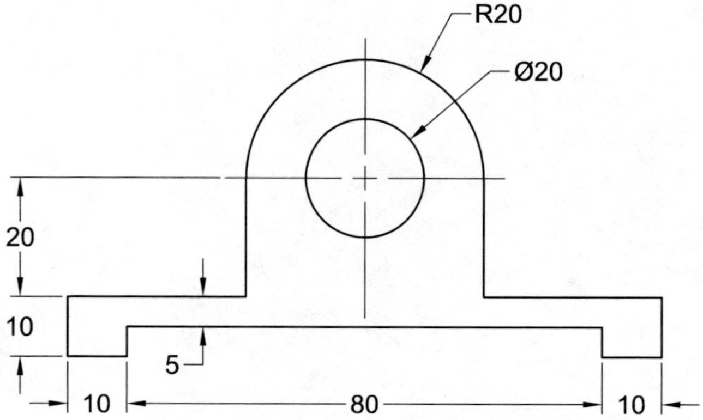

Figure 2-16 *Sketch for Exercise 2*

This page is intentionally left blank

Chapter 3

Editing and Modifying Sketches

Learning Objectives

After completing this chapter, you will be able to:

- *Edit sketches using various editing tools*
- *Create circular patterns of sketched entities*
- *Modify sketched entities*
- *Modify sketches by dynamically dragging sketched entities*

EDITING SKETCHED ENTITIES

In SOLIDWORKS, there are various tools that can be used to edit the sketched entities. These tools are used to trim, extend, offset, or mirror the sketched entities. You can also perform various other editing operations by using these tools. Various editing operations and the tools used to perform them are:

a. Trim entities
b. Extend entities
c. Convert entities
d. Intersection curve
e. Sketch fillet
f. Sketch chamfer
g. Offset entities
h. Offset on surface
i. Mirror entities
j. Move entities
k. Rotate entities
l. Scale entities
m. Stretch entities
n. Copy entities

CREATING PATTERNS

Sometimes, while creating a base feature, you may need to place the sketched entities in a particular arrangement such as along linear edges or around a circle. Figures 3-1 and 3-2 show the base features with slots. These slots are created with the help of linear and circular patterns of the sketched entities.

Figure 3-1 *Base feature with slots created along the linear edges*

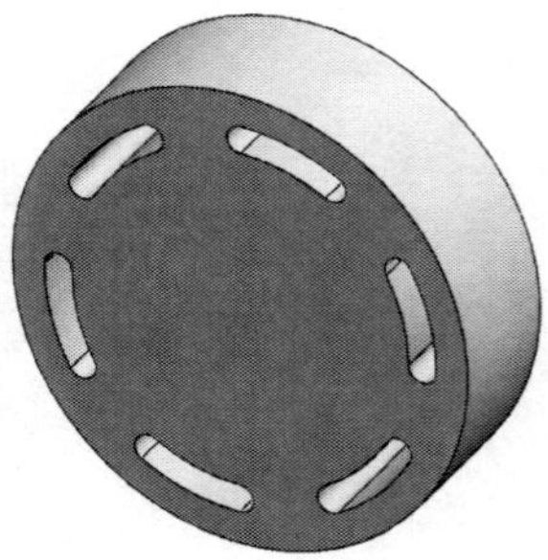

Figure 3-2 *Base feature with slots created in circular pattern*

MODIFYING SKETCHED ENTITIES

Most of the sketches require modification at some stages of design. Therefore, it is important for a designer to understand the process of modification in SOLIDWORKS.

TUTORIALS

Tutorial 1

In this tutorial, you will draw the base sketch of the model shown in Figure 3-3. The sketch of the model is shown in Figure 3-4. You will draw the sketch with a mirror line and a mirror tool. After drawing the sketch, you will modify it by dragging the sketched entities.

(Expected time: 30 min)

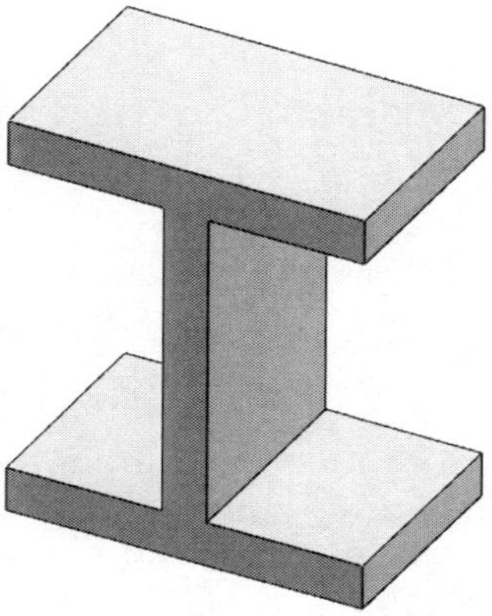

Figure 3-3 Solid model for Tutorial 1

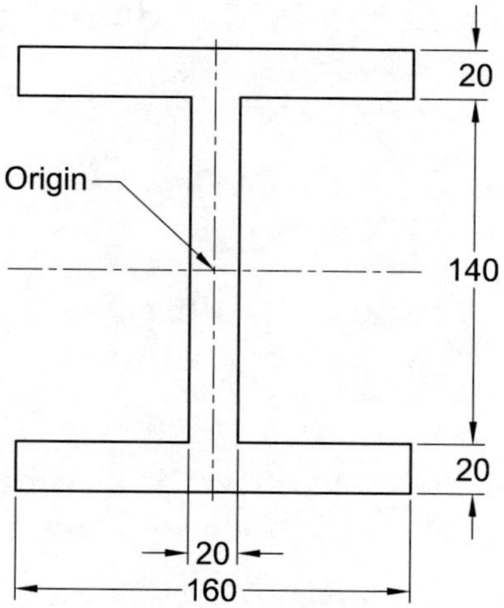

Figure 3-4 Sketch for Tutorial 1

The following steps are required to complete this tutorial:

a. Start SOLIDWORKS and then start a new part document.
b. Invoke the sketching environment.
c. Create centerlines and convert one of the centerlines to mirror line using the **Dynamic Mirror Entities** tool.
d. Create the sketch in the third quadrant; the sketch will automatically be mirrored on the other side of the mirror line, refer to Figure 3-5.
e. Mirror the entire sketch along the second mirror line, refer to Figure 3-6.
f. Modify the sketch by dragging the sketched entities, refer to Figure 3-7.

Starting SOLIDWORKS and then a New SOLIDWORKS Document

1. Start SOLIDWORKS by double-clicking on the shortcut icon of SOLIDWORKS 2018 on the desktop of your computer; the SOLIDWORKS 2018 window along with the **Welcome - SOLIDWORKS 2018** dialog box will be displayed.

2. Choose the **Part** button available in the **New** area of the **Home** tab in the **Welcome - SOLIDWORKS 2018** dialog box. A new **SOLIDWORKS** part document starts.

Next, you need to invoke the sketching environment.

3. Choose the **Sketch** tab from the **CommandManager**, if it is not selected by default. Next, choose the **Sketch** button from the **Sketch CommandManager**; the **Edit Sketch PropertyManager** is invoked and you are prompted to select the plane to create the sketch.

4. Select the **Front Plane**; the sketching environment is invoked and the plane is oriented normal to the view.

Modifying the Grid and Snap Settings and the Dimensioning Units

Before drawing the sketch, you need to modify the grid and snap settings to make the cursor jump through a distance of 10 mm.

1. Choose the **Options** button from the Menu Bar; the **System Option - General** dialog box is displayed.

While installing SOLIDWORKS, if you have selected a unit other than millimeter to measure the length then you need to select millimeter as the unit for the current drawing by following the next two steps.

2. Choose the **Document Properties** tab and select the **Units** option from the area on the left-side of the **Document Properties - Drafting Standard** dialog box.

3. Next, select the **MMGS (millimeter, gram, second)** radio button in the **Unit system** area and select **degrees** from the drop-down list in the cell corresponding to the **Angle** row and the **Unit** column, if it is not selected by default.

4. Next, click on the **Grid/Snap** option from the area on the left in the dialog box. Set the values **100** and **10** in the **Major grid spacing** spinner and the **Minor-lines per major** spinner, respectively.

On invoking the sketching environment, if the grid is displayed by default, you can turn off its display by clearing the **Display grid** check box in the **Grid** area.

5. Choose the **Go To System Snaps** button and select the **Grid** check box. Next, clear the **Snap only when grid is displayed** check box, if it is selected.

6. After making the necessary settings, choose the **OK** button.

The coordinates displayed close to the lower right corner of the SOLIDWORKS window show an increment of 10 mm when you move the cursor in the drawing area after exiting the dialog box.

Drawing the Centerlines and Converting Them into Mirror Lines

It is recommended that you draw symmetrical sketches about an axis by using any of the mirroring tools. In this tutorial, you will draw the sketch using the **Line**, **Dynamic Mirror**, and **Mirror Entities** tools. However, you can also complete this tutorial by using the **Line** and **Dynamic Mirror** tools.

1. Choose the **Centerline** button from the **Line** flyout available in the **Sketch CommandManager**.

2. Move the line cursor to a location whose coordinates are 0 mm, 100 mm, and 0 mm. You may need to zoom in the drawing to locate that point.

3. Specify the start point of the centerline at this point, move the cursor vertically downward and draw a line of 200 mm length. You may need to zoom and pan the drawing to draw a line of this length.

4. Now, double-click anywhere on the screen to end the current chain or right-click anywhere in the drawing area to invoke the shortcut menu and choose the **End chain** option from it.

5. Move the line cursor to a location whose coordinates are -100 mm, 0 mm, and 0 mm.

6. Specify the start point of the centerline and move the cursor horizontally toward the right to draw a line of 200 mm length. Pan the drawing, if required.

7. Press F to fit the drawing on the screen. Right-click and choose the **Select** option from the shortcut menu; the line cursor is replaced by the select cursor.

8. Select the vertical centerline.

9. Choose **Tools > Sketch Tools > Dynamic Mirror** tool from the SOLIDWORKS menus; the vertical centerline is converted into a mirror line and the dynamic mirror option is activated.

   You can confirm the creation of the mirror line and the activation of the dynamic mirror option by observing the symmetrical symbol displayed on both ends of the centerline.

Drawing the Sketch

Next, you need to draw the sketch of the base feature. You will draw the sketch in the third quadrant and the same sketch will be created automatically on the other side of the mirror line. The symmetrical relation is applied between the parent entity and the mirrored entity.

1. Press the L key on the keyboard; the **Line** tool is invoked.

2. Move the line cursor to a location whose coordinates are 0 mm, -100 mm, and 0 mm.

3. Specify this point as the start point of the line and move the cursor horizontally toward the left.

4. Specify the endpoint of the line when 80 is displayed above the line cursor.

   You will notice that as soon as you specify the endpoint of the line, a mirror image is automatically created on the other side of the mirror line. The line drawn as the mirrored entity gets merged with the line drawn on the left. Therefore, the entire line becomes a single entity. Remember that the lines will merge only if one of the endpoints of the line drawn is coincident with the mirror line.

5. Move the cursor vertically upward and click to specify the endpoint of the line when the length of the line above the line cursor shows the value 30.

   You will notice that as soon as you specify the endpoint of the line, a mirror image is created automatically on the other side of the mirror line.

6. Move the line cursor toward the right and click to specify the endpoint of the line when the length of the line above the line cursor shows the value 30; a mirror image is created automatically on the other side of the mirror line.

7. Move the line cursor vertically upward and click to specify the endpoint of the line when the line cursor snaps the horizontal centerline. Exit the **Line** tool. The sketch after drawing the lines is shown in Figure 3-5.

Mirroring the Entire Sketch

After creating one-half of the sketch, you need to mirror the entire sketch about the horizontal centerline. But, first you need to disable automatic mirroring.

1. Right-click and choose **Recent Commands** > **Dynamic Mirror Entities** from the shortcut menu to disable dynamic mirroring.

2. Use the box selection method to select all the lines sketched earlier and also the horizontal centerline. Make sure you do not select the vertical centerline.

3. Choose the **Mirror Entities** tool from the **Sketch CommandManager**; the entire sketch is mirrored about the horizontal centerline. The sketch after mirroring the sketched entities is shown in Figure 3-6.

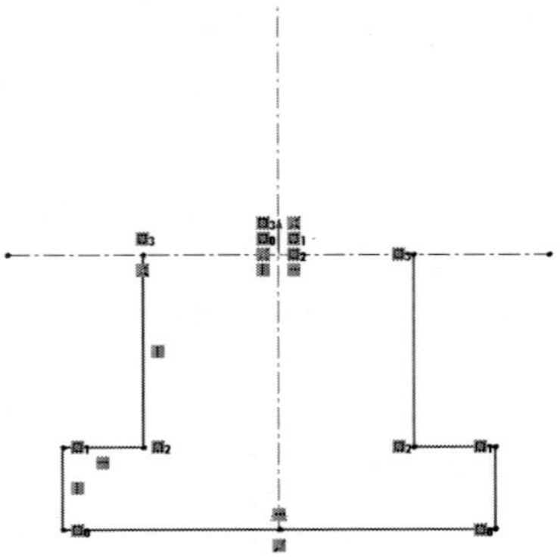

Figure 3-5 Sketch after drawing the lines

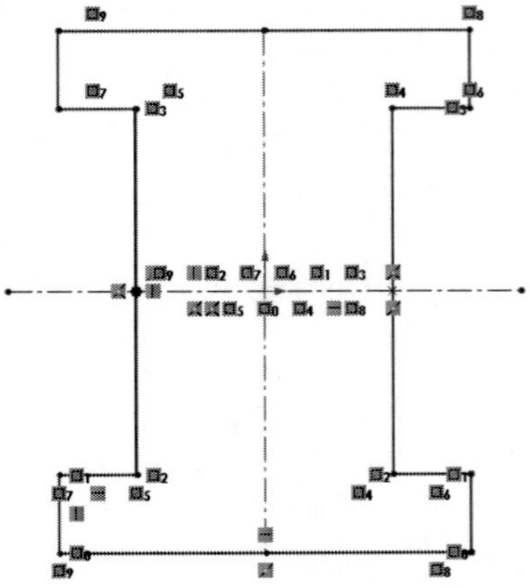

Figure 3-6 Sketch after mirroring sketched entities

Modifying the Sketch by Dragging Entities

Next, you need to modify the sketch by dragging. While dragging the entities, you will observe that the corresponding mirrored entity is also being modified.

1. Select the lower right vertical line; the **Line Properties PropertyManager** is displayed on the left of the drawing area. You will notice that the value in the **Length** spinner is 30.

2. As the required length of this line is 20, set the value in the **Length** spinner to **20** in the **Parameters** rollout. You will observe that all the dependent lines are also modified. This is because they are created as mirror images.

3. Choose the **Close Dialog** button or click once in the drawing area.

4. Select the midpoint of the right vertical line that passes through the horizontal centerline; the **Point PropertyManager** is displayed. You will notice that the coordinates of this point are 50,0.

5. Press and hold the left mouse button and drag the point toward the origin; the value in the **X Coordinate** spinner of the **Parameters** rollout changes dynamically.

6. Release the left mouse button when the **X Coordinate** spinner shows the value 10. The final sketch after modifying the sketched entities by dragging them is shown in Figure 3-7.

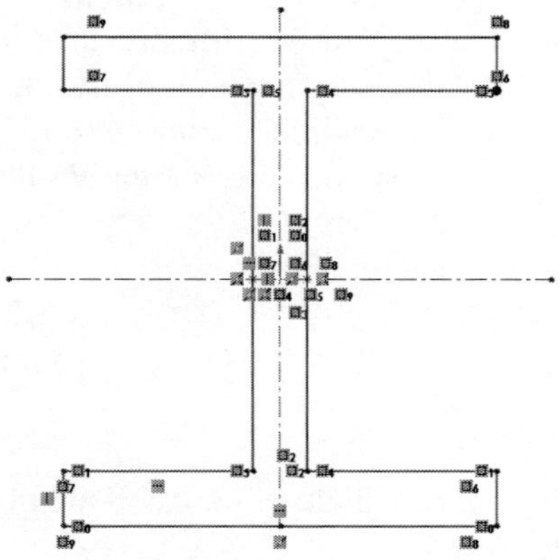

Figure 3-7 *Final sketch for Tutorial 1*

Saving the Sketch

1. Choose the **Save** button from the Menu Bar to invoke the **Save As** dialog box.

2. Choose the **New Folder** button from the **Save As** dialog box. Enter the name of the folder as **c03** and press ENTER. Enter the name of the document as **c03_tut01** in the **File name** edit box and choose the **Save** button.

3. Press CTRL+W to close the file.

Tutorial 2

In this tutorial, you will create the base sketch of the model shown in Figure 3-8. The sketch of the model is shown in Figure 3-9. You will create the sketch of the base feature by using the sketch tools. Also, you will modify and edit the sketch using various modifying options. Do not create the center marks and centerlines as they are for your reference only.

(Expected time: 30 min)

Figure 3-8 Solid Model for Tutorial 2

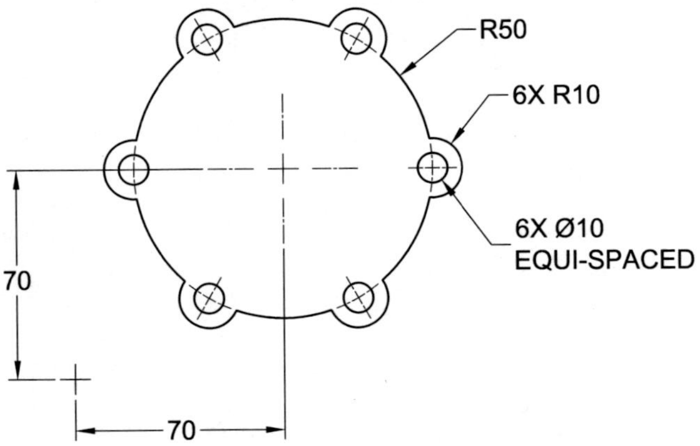

Figure 3-9 Sketch for Tutorial 2

The following steps are required to complete this tutorial:

a. Start a new part document.
b. Invoke the sketching environment.
c. Draw the outer loop of the sketch, refer to Figures 3-10 and 3-11.
d. Create a circle to define the hole in the outer loop.
e. Use the **Circular Sketch Pattern** tool to create a circular pattern of circles in the outer loop, refer to Figure 3-12.

Starting a New Document

1. Choose the **New** button from the Menu Bar to invoke the **New SOLIDWORKS Document** dialog box.

2. In this dialog box, the **Part** button is chosen by default. Choose the **OK** button.

   Next, you need to invoke the sketching environment.

3. Choose the **Sketch** button from the **Sketch CommandManager**; the **Edit Sketch PropertyManager** is invoked and you are prompted to select the plane to create the sketch.

4. Select the **Front Plane** as the sketching plane; the sketching environment is invoked and the plane is oriented normal to the view.

5. Choose the **Options** button from Menu Bar; the **System Option - General** dialog box is displayed.

6. Choose the **Document Properties** tab from this dialog box and select the **Units** option from the area on the left. Set the units to millimeters by selecting the **MMGS (millimeter, gram, second)** option from the **Units system** area.

7. Next, choose the **Grid/Snap** option from the area on the left. Set the value of the **Minor-lines per major** spinner to **20** and the **Major grid spacing** spinner to **100**.

8. Choose the **OK** button after making the necessary settings.

Drawing the Outer Loop of the Sketch

It is evident from Figure 3-9 that the sketch consists of the outer loop and inner cavities. It is recommended that you create the outer loop of the sketch first and then the inner cavities.

The origin of the sketching environment is in the middle of the drawing area and you need to create the sketch in the first quadrant. Therefore, it is recommended that you modify the drawing area by relocating the origin.

1. Press the CTRL key and the middle mouse button and then drag the cursor in such a way that the origin of the sketch is moved near the lower left corner of the drawing area.

2. Press and hold the right mouse button and drag the cursor to the right; the sketching tools are displayed in the mouse gesture. Move the cursor over the **Circle** tool; the **Circle** tool is invoked and the **Circle PropertyManager** is displayed.

3. Make sure that the **Circle** button is chosen in the **Circle Type** rollout of the **Circle PropertyManager**.

4. Move the cursor to a location whose coordinates are 70 mm, 70 mm, and 0 mm.

5. Click to specify the center point of the circle at this location and move the cursor horizontally toward the right. Click when the radius above the circle cursor shows the value **50**.

6. Choose the **Zoom to Area** button from the **View (Heads-Up)** toolbar. Press and hold the left mouse button and drag the cursor to define a window such that the sketched circle and the origin are placed in the window.

7. Release the left mouse button; the display area of the sketch is increased.

8. Invoke the **Circle** tool by using the mouse gesture and move the cursor to the right quadrant of the circle. All the quadrants of the circle are displayed and the right quadrant on which you have placed the cursor is displayed in orange. Also, the symbol of coincident constraint is displayed below the cursor.

9. Specify the center point of the circle at this location and move the cursor horizontally toward the right. When the value of the radius above the circle cursor shows 10, press the left mouse button.

10. Press the S key from the keyboard; a toolbar is displayed. Choose the **Trim Entities** button; the **Trim PropertyManager** is displayed. Next, choose the **Trim to closest** button from this PropertyManager.

11. Trim the sketch such that it looks similar to the one shown in Figure 3-10. Exit the **Trim** tool.

12. Select the trimmed circle of radius **10** mm and choose the **Circular Sketch Pattern** button from the **Pattern** flyout in the **Sketch CommandManager**; the **Circular Pattern PropertyManager** is displayed and the preview of the circular pattern with the default setting is displayed in the drawing area.

    On doing this, you will notice that the center of the circular pattern is placed at the origin and an arrow is displayed, indicating that the origin is the center of the circular pattern. But, in this sketch, the center of the circular pattern is not at the origin, so you need to modify it. This can be done by setting the coordinates of the point in the **Center X** and **Center Y** spinners in the **Parameters** rollout of this PropertyManager. However, it is recommended that you drag the arrow displayed at the center of the pattern to the required location.

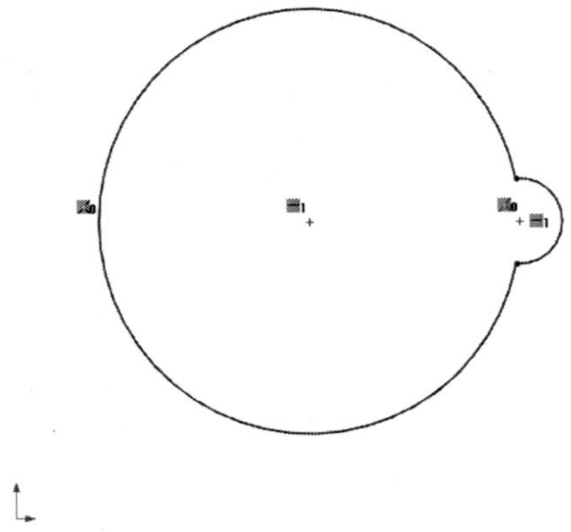

Figure 3-10 *Sketch after trimming the unwanted entities*

13. Move the circular pattern cursor to the control point available at the end of the arrow.

14. Press and hold the left mouse button at the control point and drag it to the center of the 100 mm diameter circle. Release the left mouse button when an orange circle is displayed at the center point of the circle.

    You will notice that both the **Center X** and **Center Y** spinners display the value **70** mm. This is because the center of the 100 mm diameter circle is located at a distance of 70 mm along the X and Y axis directions.

15. Set **6** in the **Number of Instances** spinner.

16. Clear the **Dimension angular spacing** check box if it is selected.

17. Select the **Display instance count** check box. Accept the remaining default values and choose the **OK** button to create the pattern.

18. Trim the unwanted portion of the 100 mm diameter circle using the **Trim Entities** tool. You need to use the **Trim to closest** button for this trimming.

19. Choose **OK** from the **Trim PropertyManager** after trimming. The outer loop of the sketch is created, as shown in Figure 3-11.

Sketching the Holes

Next, you need to draw the sketch of the holes. It is evident from Figure 3-9 that you need to create six circles. After drawing the first circle, you need to create the other five circles by creating a circular pattern of the parent circle.

1. Invoke the **Circle** tool by using the mouse gesture.

2. Select the center point of the **10** mm radius arc on the left quadrant of the larger circle as the center point of the new circle.

3. Press and hold the CTRL key and then draw a circle of diameter close to **10**. Make sure you press the CTRL key so that the cursor does not snap to the points or grid.

4. Set **5** in the **Radius** spinner in the **Circle PropertyManager**.

Creating the Circular Pattern of the Holes

1. Choose the **Circular Sketch Pattern** button from the **Pattern** flyout in the **Sketch CommandManager**; the **Circular Pattern PropertyManager** is displayed and the preview of the circular pattern is displayed with an arrow at the center.

2. Move the circular pattern cursor to the control point that is available at the end of the arrow head indicating the center point of the pattern. Press and hold the left mouse button at the control point and drag it to the center of the **100** mm diameter circle. Release the left mouse button when an orange circle is displayed at the center point of the circle.

3. Set **6** in the **Number of Instances** spinner.

4. Clear the **Dimension angular spacing** check box.

5. Select the **Display instance count** check box. Accept the remaining default values and choose the **OK** button to create the pattern.

   The final sketch of the model is shown in Figure 3-12. In this figure, all the relations are hidden for clarity.

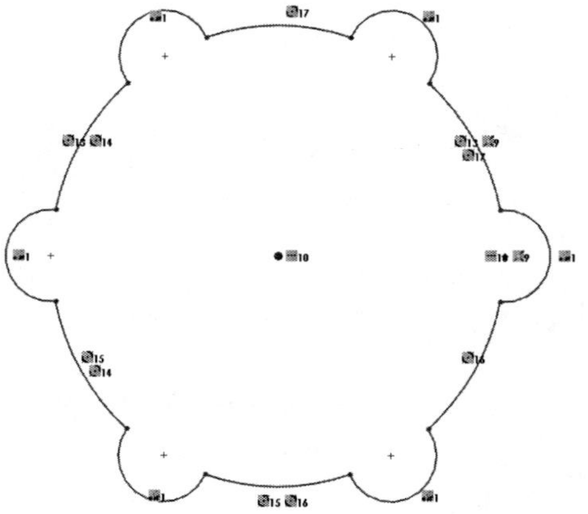

Figure 3-11 *Outer loop of the sketch*

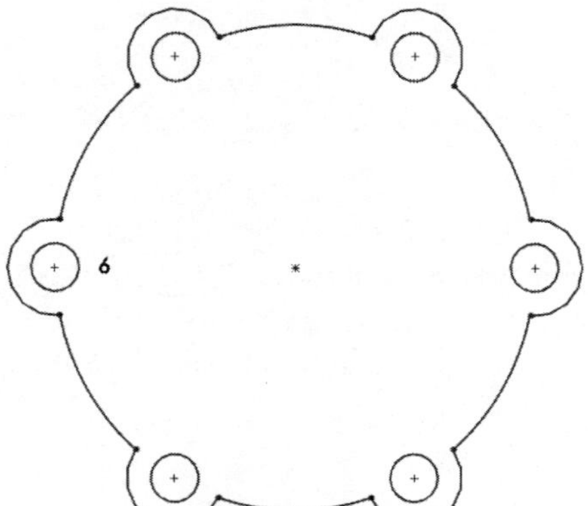

Figure 3-12 *Final sketch for Tutorial 2*

Saving the Sketch

1. Choose the **Save** button from the **Menu Bar** to invoke the **Save As** dialog box.

2. Enter the name of the document as **c03_tut02** in the **File name** edit box and choose the **Save** button.

The document is saved at the location */Documents/SOLIDWORKS/c03*.

3. Close the file by choosing **File > Close** from the SOLIDWORKS menus.

EXERCISES

Exercise 1

Create the sketch of the model shown in Figure 3-13. The sketch of the model is shown in Figure 3-14. The solid model and dimensions are given for reference only.

(Expected time: 30 min)

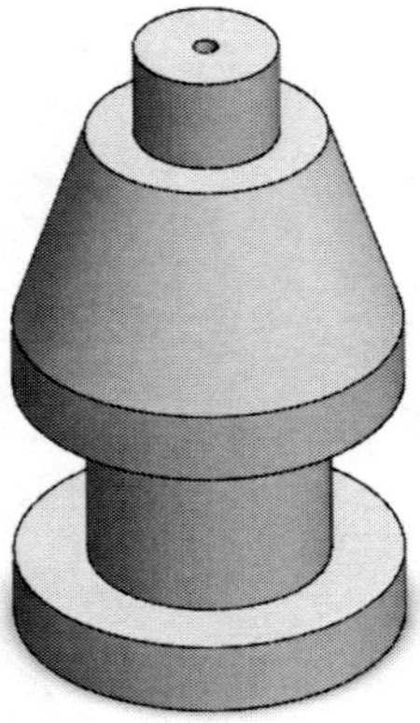

Figure 3-13 Solid model for Exercise 1

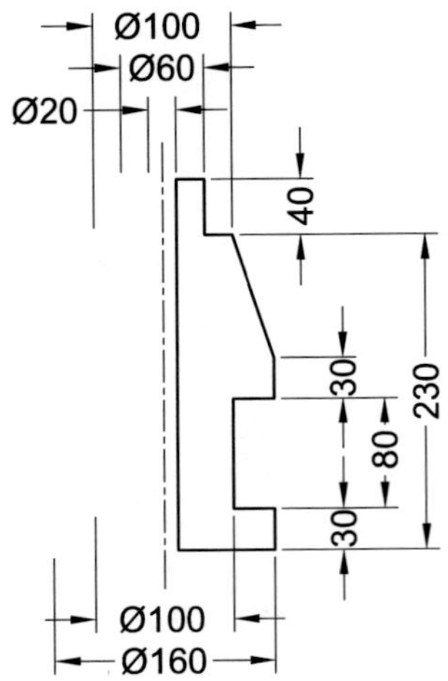

Figure 3-14 Sketch for Exercise 1

Exercise 2

Create the sketch of the model shown in Figure 3-15. The sketch of the model is shown in Figure 3-16. The solid model and its dimensions are given only for reference. Create the sketch on one side and then mirror it on the other side. Make sure you do not use the **Dynamic Mirror** tool to draw this sketch. This is because if you draw the sketch using this tool, some relations are applied to the sketch. These relations interfere while creating fillets.

(Expected time: 30 min)

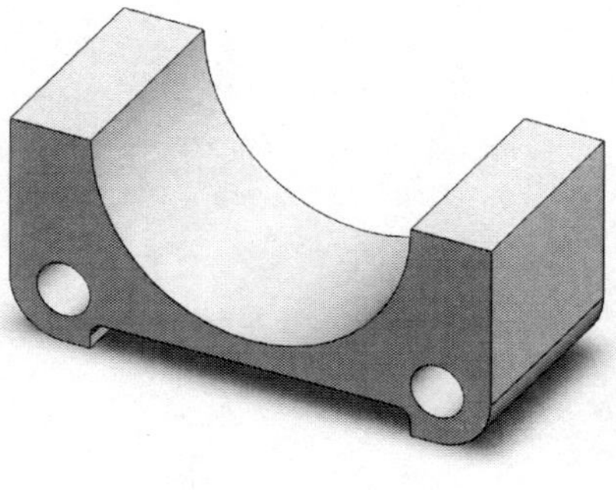

Figure 3-15 Solid model for Exercise 2

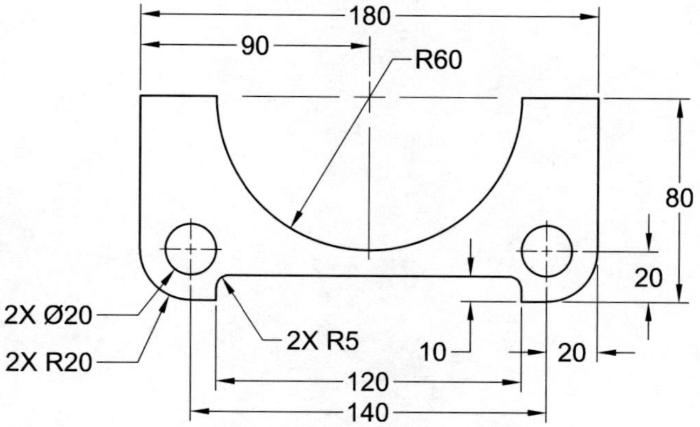

Figure 3-16 Sketch for Exercise 2

This page is intentionally left blank

Chapter 4

Adding Relations and Dimensions to Sketches

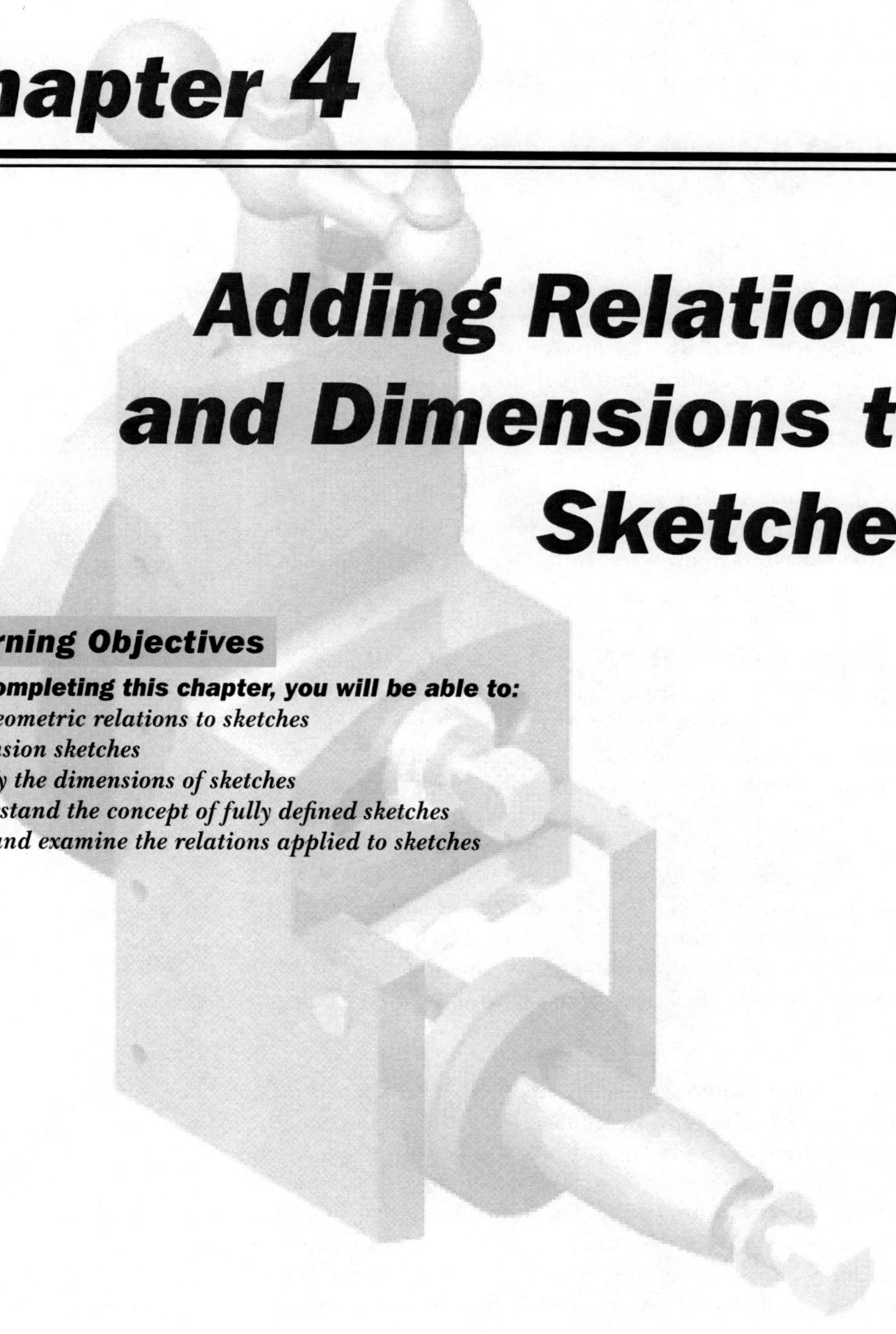

Learning Objectives

After completing this chapter, you will be able to:

- *Add geometric relations to sketches*
- *Dimension sketches*
- *Modify the dimensions of sketches*
- *Understand the concept of fully defined sketches*
- *View and examine the relations applied to sketches*

APPLYING GEOMETRIC RELATIONS TO SKETCHES

Geometric relations are the logical operations that are performed to add relationships (such as tangent, perpendicular, or parallel) between the sketched entities, planes, axes, edges, or vertices. The relations applied to the sketched entities are used to capture the design intent. Geometric relations constrain the degree of freedom of the sketched entities. You can apply relations to a sketch by using the **Add Relations PropertyManager** and Automatic Relations.

Applying Relations using the Add Relations PropertyManager

CommandManager: Sketch > Display/Delete Relations flyout > Add Relation

The **Add Relations PropertyManager** is widely used to apply relations to a sketch in the sketching environment of SOLIDWORKS. To invoke this PropertyManager, choose the **Add Relation** button from the **Display/Delete Relations** flyout in the **Sketch CommandManager**, as shown in Figure 4-1. Alternatively, right-click on an entity in the drawing area and then choose the **Add Relation** option from the **Sketch Tools** cascade menu displayed; the **Add Relations PropertyManager** will be displayed, refer to Figure 4-2. Also, the confirmation corner will be displayed at the top right corner of the drawing area.

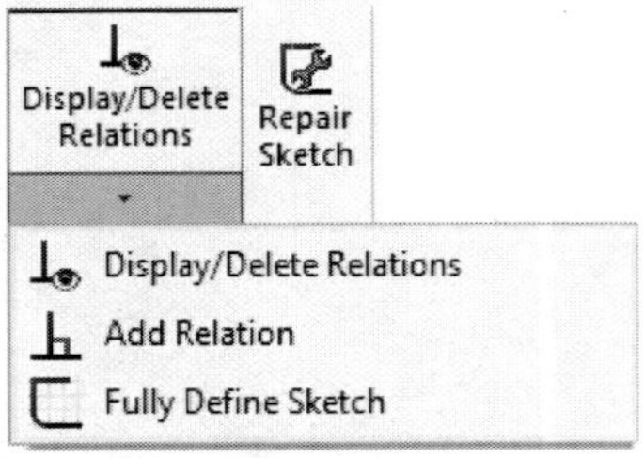

***Figure 4-1** Tools in the **Display/Delete Relations** flyout*

Automatic Relations

Automatic relations are applied automatically to a sketch while drawing. For example, you will notice that when you specify the start point of a line and move the cursor horizontally toward the right or left, the horizontal line symbol is displayed below the line cursor. This is the symbol of the **Horizontal** relation that is applied to the line while drawing. If you move the cursor vertically downward or upward, the vertical line symbol for the **Vertical** relation will be displayed below the line cursor. If you move the cursor to the intersection of two or more sketched entities, the intersection symbol will appear below the cursor. Similarly, other relations are also applied automatically to the sketch when you draw it.

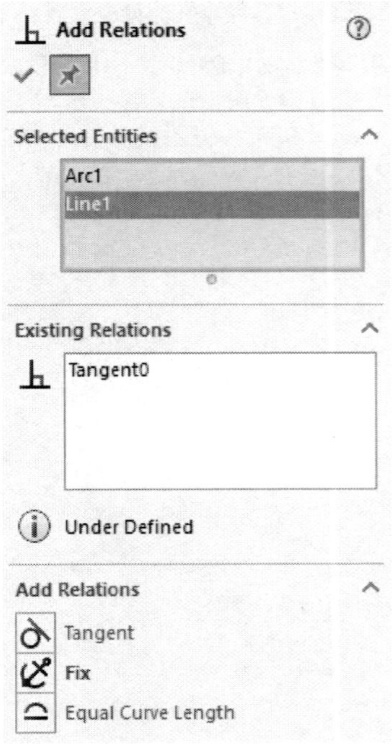

Figure 4-2** The **Add Relations PropertyManager

You can activate the automatic relations option if it is not activated. To do so, choose the **Options** button from the Menu Bar; the **System Options - General** dialog box will be displayed. Select the **Relations/Snaps** option from the area on the left, then select the **Automatic relations** check box from this dialog box, and finally choose the **OK** button. You can also activate or deactivate the Automatic Relations option from the SOLIDWORKS menus. To activate it, choose **Tools > Sketch Settings > Automatic**

Relations from the SOLIDWORKS menus. This is a toggle option. So, you can use the same option to deactivate it.

The Horizontal, Vertical, Coincident, Midpoint, Intersection, Tangent, Perpendicular, and Coradial relations can be applied automatically to a sketch while drawing.

> **Tip**
> *While drawing a sketch, you will observe that two types of inferencing lines are displayed: one in blue and the other in yellow. The yellow inferencing line indicates that a relation has been applied automatically to the sketch whereas the blue inferencing line indicates that no automatic relation has been applied.*

DIMENSIONING A SKETCH

After drawing sketches and adding relations, the most important step in creating a design is dimensioning. SOLIDWORKS being a parametric software, the entity on dimensioning is driven by the specified value irrespective of the original size. Therefore, when you apply and modify the dimension of an entity, it is forced to change its size in accordance with the specified dimension value.

You can dimension any kind of entity by using the **Smart Dimension** tool, which is available in the **Sketch CommandManager**. If you use the **Smart Dimension** tool, the type of dimension to be applied will depend on the type of entity selected. For example, if you select a line, then a horizontal, vertical, or aligned dimension will be applied. If you select a circle, a diametric dimension will be applied. Similarly, if you select an arc, a radial dimension will be applied. However, if you want to apply a particular type of dimension, then choose the required tool from the **Smart Dimension** flyout. You can also invoke these tools from the **Dimensions/Relations** toolbar.

As soon as you place the dimension, the **Modify** dialog box will be displayed, as shown in Figure 4-3. You can modify the default dimension value by using the spinner or by entering a new value in the edit box available in the **Modify** dialog box. You can also drag the thumbwheel provided below the spinner to the right to increase and to the left to decrease the value.

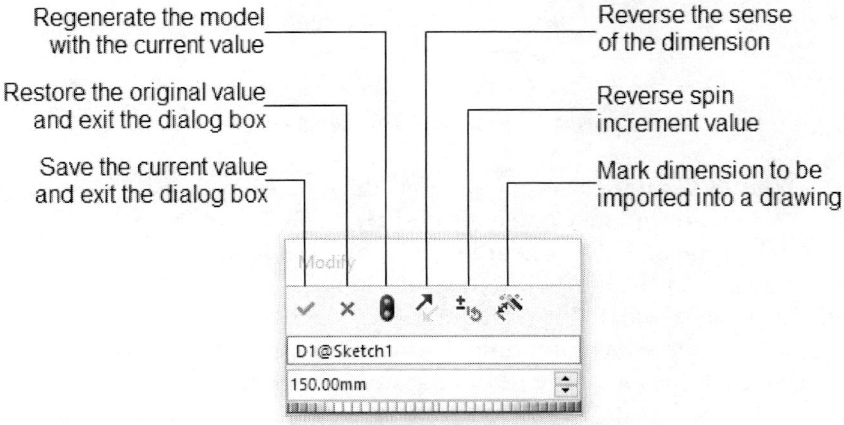

Figure 4-3 *The **Modify** dialog box*

Tip
*If the **Modify** dialog box is not displayed when you place the dimension, you need to set its preference manually. To do this, invoke the **System Options - General** dialog box and select the **Input dimension value** check box.*

CONCEPT OF A FULLY DEFINED SKETCH

It is important for you to understand the concept of fully defined sketches. While creating a model, you first need to draw the sketch of the base feature and then proceed further to create other features. After creating the sketches, you need to add the required relations and dimensions to constrain the sketch with respect to the surroundings. After adding the required relations and dimensions, the sketch may exist in any of the six states discussed next:

TUTORIALS

Tutorial 1

In this tutorial, you will draw the sketch of the model shown in Figure 4-4. You will draw the sketch using the mirror line and then add the required relations and dimensions to it. The sketch of the model is shown in Figure 4-5. The solid model is given for reference only. (**Expected time: 30 min**)

Figure 4-4 *Solid Model for Tutorial 1*

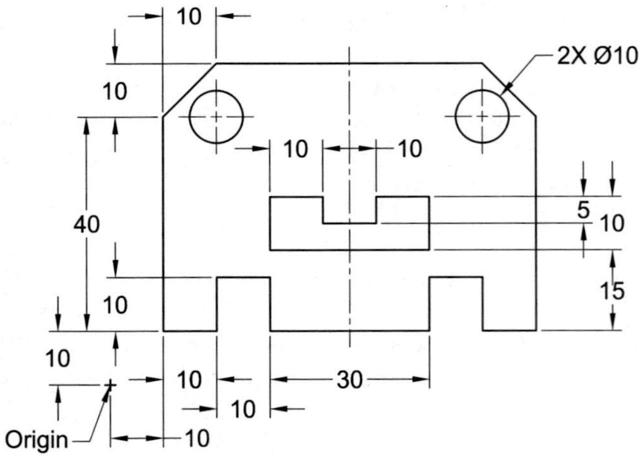

Figure 4-5 *Sketch of the model*

The following steps are required to complete this tutorial:

a. Start SOLIDWORKS and then start a new part document.
b. Create a mirror line using the **Centerline** and **Dynamic Mirror** tools.
c. Draw the sketch of the model on one side of the mirror line so that it is automatically drawn on the other side, refer to Figures 4-6 through 4-11.
d. Add relations to the sketch, refer to Figure 4-13.
e. Add dimensions to the sketch and fully define the sketch, refer to Figures 4-14 and 4-15.
f. Save the sketch and then close the document.

Starting SOLIDWORKS and a New Part Document

1. Start SOLIDWORKS by double-clicking on the shortcut icon of SOLIDWORKS 2018 available on the desktop of your computer, the SOLIDWORKS 2018 window along with the **WELCOME- SOLIDWORKS 2018** dialog box will be displayed.

2. Choose the **New** button from the Menu Bar; the **New SOLIDWORKS Document** dialog box with the **Part** button chosen is displayed.

3. Choose the **OK** button from the **New SOLIDWORKS Document** dialog box; a new SOLIDWORKS part document is started.

4. Choose the **Sketch** button from the **Sketch CommandManager** and then select the **Front Plane** to invoke the sketching environment.

   In the previous chapters, you used the grid and snap settings to create sketches. From this chapter, you are recommended not to use those settings. By doing so, you can draw sketches at arbitrary locations and then apply dimensions and relations to them.

5. If the grid is displayed, invoke the **System Options - Relations/Snaps** dialog box and then clear the **Grid** check box from the **Sketch snaps** area to hide the grid.

6. Set the unit for measuring linear dimensions to millimeters and the unit for angular dimensions to degree using the **Document Properties - Units** dialog box and choose **OK**. However, if you have selected millimeters as unit while installing SOLIDWORKS, you can skip this step.

Drawing the Mirror Line

In this tutorial, you will draw the sketch of the given model with the help of the **Dynamic Mirror** tool. So, when you draw an entity on one side of the centerline, the same entity will be drawn automatically on its other side. Also, the symmetric relation will be applied to the entities on both sides of the centerline. Therefore, if you modify an entity on one side of the centerline, the same modification will be reflected in the mirrored entity and vice-versa. However, to follow these steps, first you need to draw the centerline to proceed to the next steps.

The origin of the sketching environment is placed at the center of the drawing area and you need to create the sketch in the first quadrant. Therefore, it is recommended that you modify the drawing area such that the area in the first quadrant is increased. This can be done by using the **Pan** tool.

1. Press the CTRL key and the middle mouse button and then drag the cursor toward the bottom left corner of the drawing area.

2. Choose the **Centerline** button from the **Line** flyout in the **Sketch CommandManager**; the **Insert Line PropertyManager** is displayed. 

3. Move the line cursor to a location whose coordinates are close to 45 mm, 70 mm, and 0 mm.

4. Click to specify the start point of the centerline and move the line cursor vertically downward to draw a line of length close to 80 mm. As soon as you specify the endpoint of the centerline, a rubber-band line is attached to the line cursor.

5. Right-click to display the shortcut menu and then choose the **Select** option from it to exit the **Centerline** tool.

6. Press the F key; the sketch is zoomed and it fits on the screen.

7. Select the centerline and choose **Tools > Sketch Tools > Dynamic Mirror** from the SOLIDWORKS menus to convert the centerline into a mirror line.

Drawing the Sketch

You will draw the sketch on the right of the mirror line and the same sketch will automatically be drawn on the other side of the mirror line.

1. Choose the **Line** button from the **Sketch CommandManager**; the arrow cursor is replaced by the line cursor.

2. Move the line cursor close to the lower side of the centerline; the cursor snaps to the mirror line and a coincident symbol is displayed below the cursor.

3. Click to specify the start point of the line at this point and move the cursor horizontally toward the right. Specify the endpoint of the line when its length shows a value close to 15. As soon as you click the left mouse button to specify the endpoint of the line, a line of the same length is drawn automatically on the other side of the mirror line. Figure 4-6 shows the mirrored entity created automatically on the other side of the mirror line. The display of relations in this figure can be turned off by choosing the **View Sketch Relations** button in the **Hide/Show Items** flyout in the **View (Heads-Up)** toolbar.

   Note that the mirrored entity that is automatically created on the left of the mirror line is merged with the line drawn on the right. Therefore, the entire line becomes a single entity. The mirror image of the line will merge with the line that you draw only if one of the endpoints of the line is coincident with the mirror line.

4. Move the cursor vertically upward. Specify the endpoint of the line when the length of the line displays a value close to 10. Figure 4-7 shows the sketch after drawing the vertical lines.

5. Move the line cursor horizontally toward the right. Specify the endpoint of the line when the length of the line displays a value close to 10.

6. Move the line cursor vertically downward. Specify the endpoint when the length of the line displays a value close to 10.

7. Move the line cursor horizontally toward the right. Specify the endpoint when the length of the line displays a value close to 10.

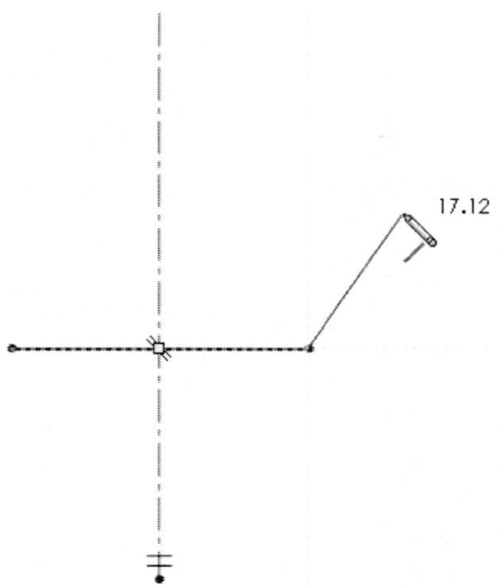

Figure 4-6 *Sketching using automatic mirroring*

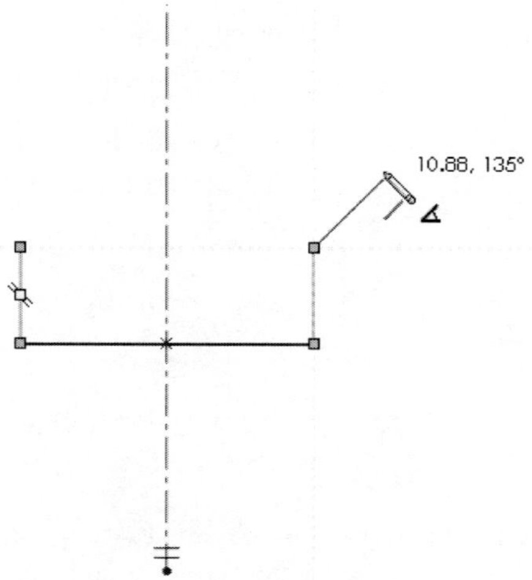

Figure 4-7 *Sketch after drawing the vertical lines*

8. Move the line cursor vertically upward. Specify the endpoint when the length of the line displays a value close to 40.

9. Move the line cursor at an angle close to 135-degree from the horizontal reference and to a length close to 14, specify the endpoint of the line. Figure 4-8 shows the sketch after drawing the inclined line.

10. Move the line cursor horizontally toward the left. Specify the endpoint when the cursor snaps to the mirror line and the mirror line is highlighted. Double-click anywhere in the drawing area to end the creation of line. The sketch after completing the outer profile is shown in Figure 4-9.

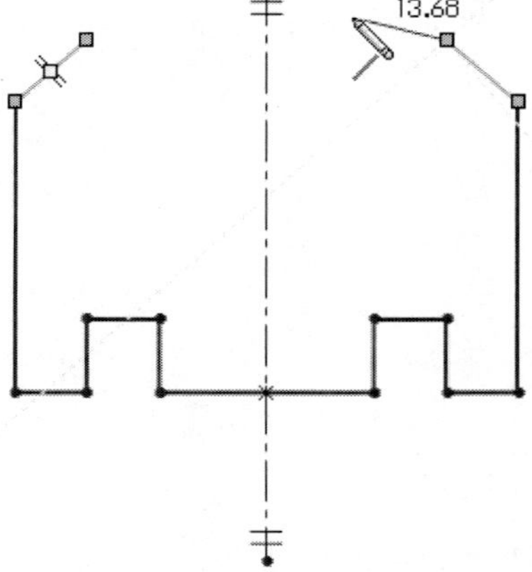

Figure 4-8 *Sketch after drawing the inclined line*

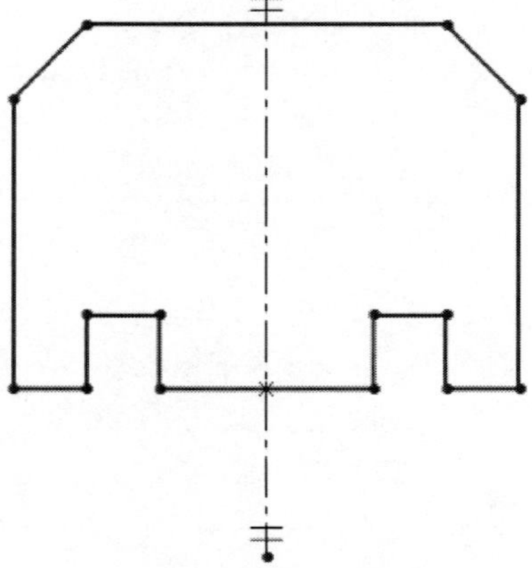

Figure 4-9 *Sketch after completing the outer profile of the sketch*

Next, you will draw the sketch of the inner cavity. To draw the sketch of the inner cavity, you will start with the lower horizontal line.

11. Specify the start point of the line in the mirror line, refer to Figure 4-10 and move the cursor horizontally toward the right. Specify the endpoint when the length of the line displays a value close to 15.

12. Move the line cursor vertically upward. Specify the endpoint when the length of the line displays a value close to 10.

13. Move the line cursor horizontally toward the left. Specify the endpoint when the length of the line displays a value close to 10.

14. Move the line cursor vertically downward. Specify the endpoint when the length of the line displays a value close to 5.

15. Move the line cursor horizontally toward the left. Specify the endpoint when the line cursor snaps to the mirror line.

16. Right-click to display the shortcut menu and then choose the **Select** option from it to exit the **Line** tool. The sketch after completing the inner cavity is shown in Figure 4-10.

17. Choose the **Circle** button from the **Sketch CommandManager** to invoke the **Circle** tool.

18. Move the circle cursor to the point where the inferencing lines originating from the endpoints of the right inclined line intersects.

19. Specify the center of the circle at this point and move the circle cursor toward the left to define the radius of the circle. Press the left mouse button when the radius of the circle displays a value close to 5 and exit the **Circle** tool.

The circle is automatically mirrored on the other side of the mirror line. The sketch after drawing the circle is shown in Figure 4-11.

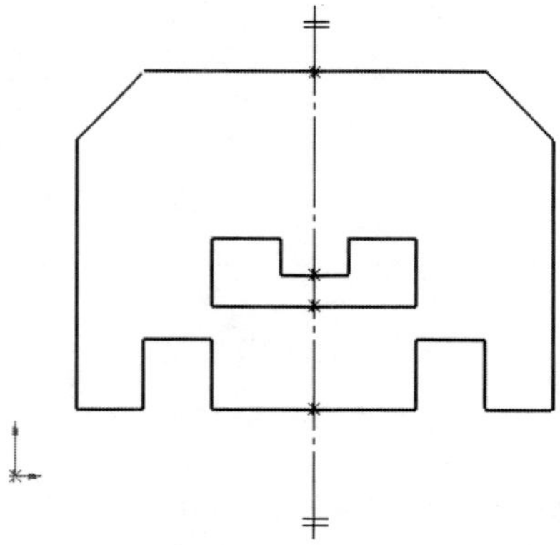

Figure 4-10 *Sketch after drawing the inner cavity*

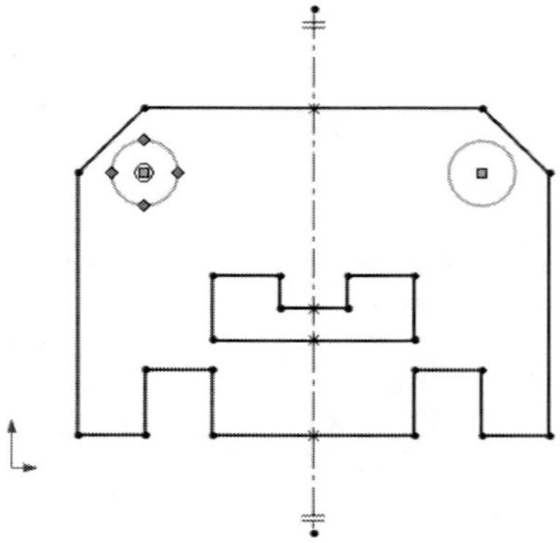

Figure 4-11 *Sketch after drawing the circle*

20. Exit the **Dynamic Mirror** tool by choosing **Tools > Sketch Tools > Dynamic Mirror** from the SOLIDWORKS menus.

Adding Relations to the Sketch

After drawing the sketch, you need to add relations to it by using the **Add Relations PropertyManager**. Relations are applied to a sketch to constrain its degree of freedom, reduce the number of dimensions in the sketch, and capture the design intent of the sketch.

1. As the circles are still selected, press the ESC key to remove the circles created previously from the selection set.

2. Choose the **Add Relation** button from the **Display/Delete Relations** flyout in the **Sketch CommandManager**; the **Add Relations PropertyManager** is displayed. Also, the confirmation corner is displayed at the upper right corner of the drawing area.

3. Select the center point of the circle on the right and then the lower endpoint of the right inclined line. The names of the selected entities are displayed in the **Selected Entities** rollout of the **Add Relations PropertyManager**.

   The relations that can be applied to the two selected entities are displayed in the **Add Relations** rollout of the **Add Relations PropertyManager**, as shown in Figure 4-12. The **Horizontal** option is highlighted, suggesting that the horizontal relation is the most appropriate relation for the selected entities.

4. Choose the **Horizontal** button from the **Add Relations** rollout to apply the **Horizontal** relation to the selected entities.

5. Move the cursor to the drawing area and right-click to display the shortcut menu. Choose the **Clear Selections** option from the shortcut menu to remove the selected entities from the selection set.

6. Select the center point of the circle on the right and then the upper endpoint of the right inclined line; the relations that can be applied to the selected entities are displayed and the **Vertical** button is highlighted in the **Add Relations** rollout.

7. Choose the **Vertical** button from the **Add Relations PropertyManager**. Right-click in the drawing area and choose the **Clear Selections** option.

8. Select the entities, as shown in Figure 4-13. Choose the **Equal** button from the **Add Relations** rollout in the **Add Relations PropertyManager**.

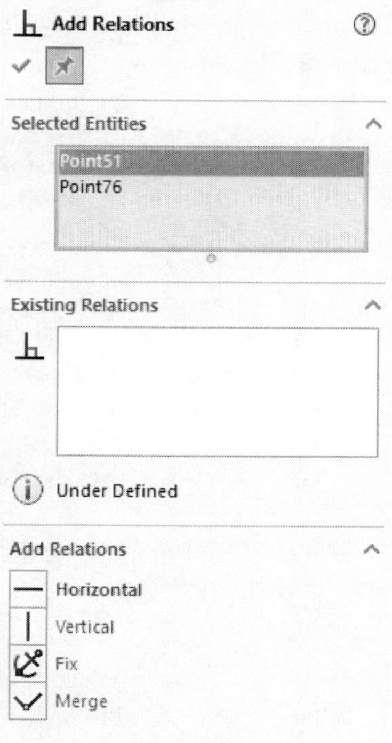

Figure 4-12** The **Add Relations PropertyManager

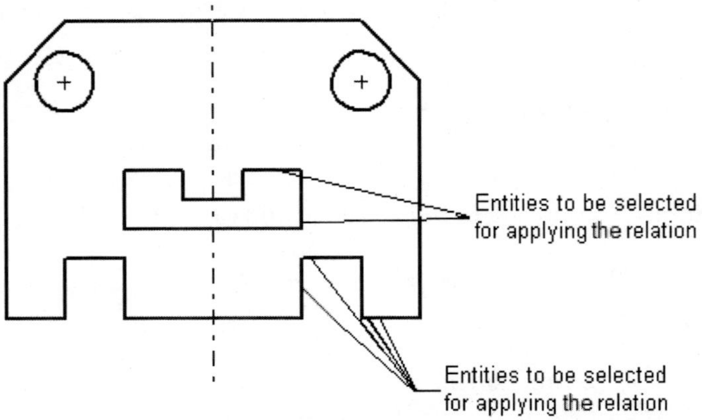

Figure 4-13 *Entities to be selected to apply the* ***Equal*** *relation*

9. Choose the **OK** button from the **Add Relations PropertyManager** or choose **OK** from the confirmation corner to close the PropertyManager. Click anywhere in the drawing area to clear the selected entities.

Applying Dimensions to the Sketch

Next, you will apply dimensions to the sketch and fully define it. As mentioned earlier, the sketched entities are shown in blue indicating that the sketch is underdefined. After required dimensions are applied, the sketched entities will turn black indicating that the sketch is fully defined now.

1. Choose **Options** from the Menu Bar; the **System Options - General** dialog box is displayed. Select the **Input dimension value** check box, if it is cleared, and then choose **OK** from the **System Options - General** dialog box.

   This check box is selected to invoke the **Modify** dialog box. This dialog box is used to enter a new dimension value and modify the sketch as you place the dimension.

2. Choose the **Smart Dimension** tool from the **Smart Dimension** flyout in the **Sketch CommandManager**; the select cursor is replaced by the dimension cursor. You can also right-click in the drawing area and choose the **Smart Dimension** option from the shortcut menu displayed or use the Mouse Gesture to invoke this **Smart Dimension** tool.

3. Move the cursor to the origin and click when an orange circle is displayed.

4. Select the outer left vertical line (Figure 4-14, line 1); a horizontal dimension is attached to the cursor.

5. Place it at a suitable location; the **Modify** dialog box is displayed. Enter **10** in this dialog box and press ENTER.

   Next, you need to apply a vertical dimension between the origin and the left horizontal line.

6. Select the origin and the left most horizontal line (line 2 in Figure 4-14) with the dimension cursor; a vertical dimension is attached to the cursor.

7. Place the vertical dimension at a suitable location; the **Modify** dialog box is displayed. Type **10** in this dialog box and press ENTER.

8. Move the dimension cursor to the lower right horizontal line (Figure 4-14, line 3); the line is highlighted.

9. Select the line; a linear dimension is attached to the cursor.

10. Move the cursor downward and click to place the dimension below the line, refer to Figure 4-15. As you place the dimension, the **Modify** dialog box is displayed.

11. Enter **10** as the value of dimension in this dialog box and press ENTER; the dimension is placed and the length of the line is modified to 10.

12. Select the lower-middle horizontal line (Figure 4-14, line 4) by using the dimension cursor; a dimension is attached to the cursor.

13. Move the cursor downward and click to place the dimension. Enter **30** in the **Modify** dialog box and press ENTER.

14. Select the left vertical line (Figure 4-14, line 1) by using the dimension cursor; a dimension is attached to the cursor.

15. Move the cursor to the left and then click to place the dimension. Enter **40** in the **Modify** dialog box and press ENTER.

16. Select the right inclined line; a dimension is attached to the cursor. Move the cursor vertically upward to apply the horizontal dimension to the selected line. Click to place the dimension at an appropriate place, see Figure 4-15.

17. Enter **10** in the **Modify** dialog box and press ENTER.

18. Again, select the right inclined line; a dimension is attached to the cursor. Move the cursor horizontally toward the right to apply the vertical dimension for the selected line. Click to place the dimension at an appropriate place, see Figure 4-15.

19. Enter **10** in the **Modify** dialog box and press ENTER.

20. Move the cursor to the left circle and when the circle is highlighted, select it; a diametric dimension is attached to the cursor. Next, move the cursor outside the sketch.

21. Place the diametric dimension. Enter **10** in the **Modify** dialog box and press ENTER.

22. Select the lower horizontal line (Figure 4-14, line 5) of the inner cavity sketch; a linear dimension is attached to the cursor. Select the lower right horizontal line (Figure 4-14, line 3) of the outer loop.

    A vertical dimension between the lower horizontal line of the inner cavity sketch and the lower right horizontal line of the outer sketch is attached to the cursor.

23. Move the cursor horizontally toward the right and place the dimension. Enter **15** in the **Modify** dialog box and press ENTER.

24. Select the inner right vertical line (Figure 4-14, line 6) of the inner cavity sketch and place the dimension outside the sketch. Enter **5** in the **Modify** dialog box and press ENTER.

25. Select the upper-middle horizontal line (Figure 4-14, line 11) of the inner cavity sketch and place the dimension above the slot. Enter **10** in the **Modify** dialog box and press ENTER.

    Now, all entities are displayed in black. This indicates that the sketch is fully defined. If the sketch is not fully defined, you may need to apply collinear relation between lines 7 and 8, lines 9 and 10, lines 2, 3, and 4. The fully defined sketch after applying all required relations and dimensions is shown in Figure 4-15.

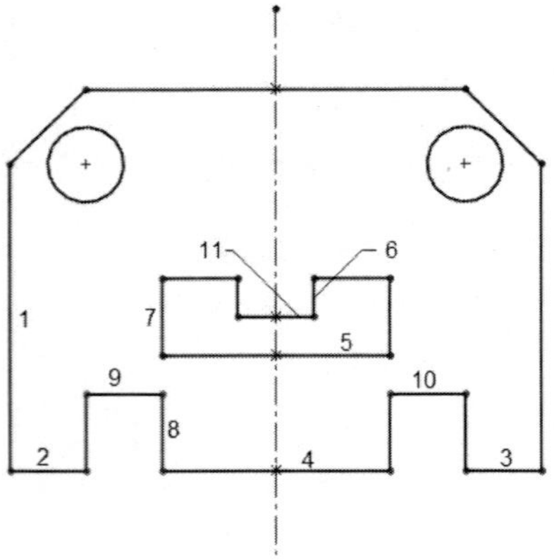

Figure 4-14 The lines to be dimensioned

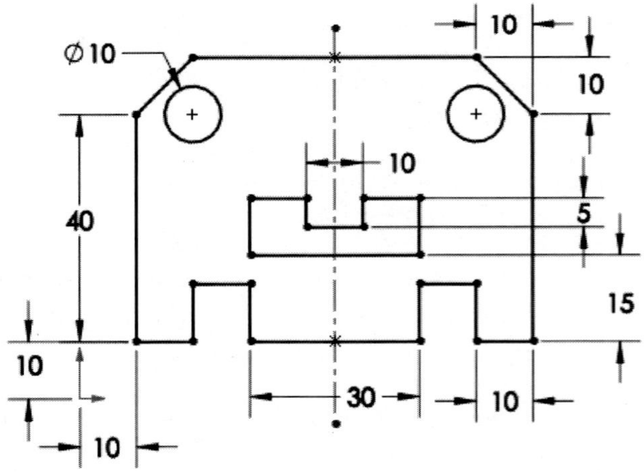

Figure 4-15 Fully defined sketch after applying all the required relations and dimensions

Saving the Sketch

1. Choose the **Save** button from the Menu Bar to invoke the **Save As** dialog box. Browse to the *\Documents\SOLIDWORKS* folder. Choose the **New Folder** button from the **Save As** dialog box. Enter **c04** as the name of the folder and press ENTER.

2. Enter **c04_tut01** as the name of the document in the **File name** edit box and choose the **Save** button. The document will be saved at the location *\Documents\SOLIDWORKS\c04*.

3. Close the document by choosing **File > Close** from the SOLIDWORKS menus.

Tutorial 2

In this tutorial, you will draw the sketch of the revolved model shown in Figure 4-16. The sketch of the feature is shown in Figure 4-17. The solid model is given for your reference only.

(Expected time: 30 min)

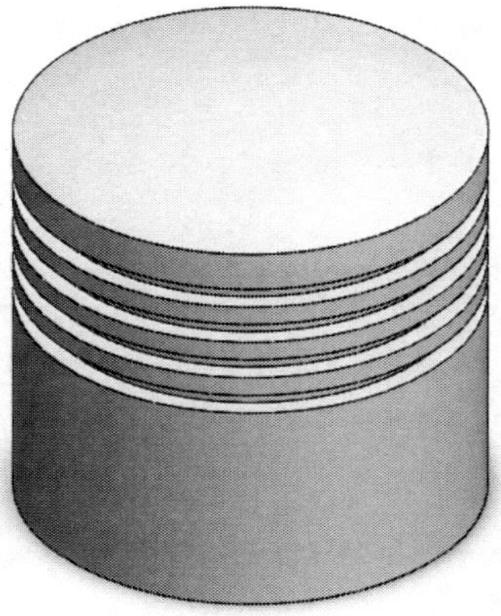

Figure 4-16 Solid model of the piston

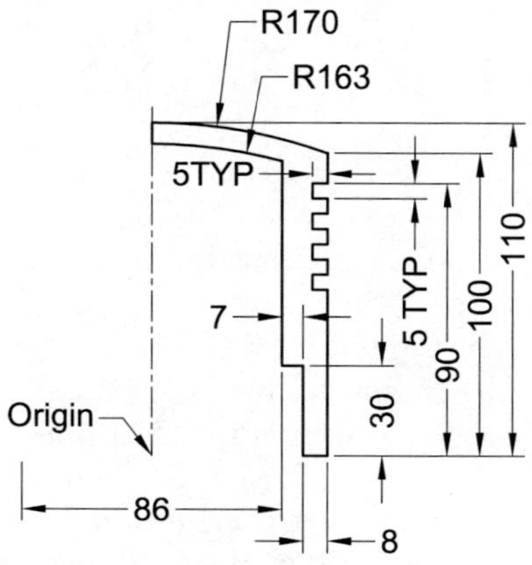

Figure 4-17 The sketch of the base feature

The following steps are required to complete this tutorial:

a. Start a new part document and then invoke the sketching environment.
b. Draw a centerline to add the linear diametric dimensions to the sketch of the piston.
c. Create the sketch by using various sketching tools, refer to Figures 4-18 and 4-19.
d. Offset the required lines by using the **Offset Entities** tool, refer to Figure 4-21.
e. Draw arcs and trim unwanted entities, refer to Figure 4-22.
f. Add relations to the sketch.
g. Add dimensions to the sketch to fully define it, refer to Figure 4-24.

Starting a New Part Document

1. Choose the **New** button from the Menu Bar; the **New SOLIDWORKS Document** dialog box with the **Part** button chosen is invoked.

2. Choose the **OK** button; a new SOLIDWORKS part document is started.

3. Choose the **Sketch** button from the **Sketch CommandManager** and select the **Front Plane** to invoke the sketching environment.

4. Invoke the **System Options - General** dialog box and set units for this tutorial.

Drawing the Sketch

To draw the sketch of the revolved model, you need to draw a centerline around which the sketch of the base feature will be revolved. In this tutorial, you will add the dimensions for the entities while drawing the sketch.

1. Invoke the **System Options - General** dialog box and select the **Sketch** option from the left side of the dialog box. The **System Options - Sketch** dialog box will be invoked. In this dialog box, select the **Enable on screen numeric input on entity creation** and **Create dimension only when the value is entered** check boxes and choose the **OK** button to save the settings.

2. Choose the **Centerline** tool from the **Sketch CommandManager**; the **Insert Line PropertyManager** will be displayed. 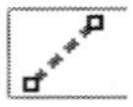

3. Draw a vertical centerline of length 120 starting from the origin. Next, press the ESC key.

   Now, you need to draw the sketch of the piston.

4. Right-click in the drawing area, and then choose the **Line** option from the shortcut menu. Move the cursor to a location whose coordinates are close to 58 mm, 0 mm, and 0 mm. Click at this point to specify the start point of the line.

5. Move the cursor vertically upward and enter **100** mm in the **Dimension** edit box displayed. Right-click and then choose the **End chain** option from the shortcut menu, refer to Line 1 in Figure 4-18.

6. Move the cursor to the lower endpoint of the line drawn earlier. Specify the start point of the line when the endpoint is highlighted. Move the cursor horizontally toward the left and draw a horizontal line of dimension 8 mm, refer to Line 2 in Figure 4-18.

7. Move the line cursor vertically upward and draw a vertical line of dimension 30 mm, refer to Line 3 in Figure 4-18.

8. Move the cursor horizontally toward the left and draw a horizontal line of dimension 7 mm. The sketch after drawing the horizontal line is shown in Figure 4-18.

9. Move the line cursor vertically upward and draw a vertical line of dimension 70 mm.

10. Right-click and then choose the **Select** option from the shortcut menu. Next, choose **3 Point Arc** tool from the **Arc** flyout in the **Sketch CommandManager**. Move the cursor near the upper endpoint of the right vertical line.

11. Specify the first point of the arc when the endpoint is highlighted. Move the cursor horizontally toward left; a reference arc is attached to the cursor. Specify the second point of the arc when the coordinates are close to -58 mm, 100 mm, and 0 mm, refer to Figure 4-19.

12. Move the cursor vertically upward and enter **170** in the **Dimension** edit box, refer to Figure 4-19.

13. Invoke the **Line** tool. Move the line cursor near the outer vertical line on the right and specify the start point of the line such that the start point is on the vertical line and its X and Y coordinates are close to 58, 90.

14. Move the cursor horizontally toward the left and then draw a line of dimension **5** mm. Right-click and then choose the **Select** option to end the creation of the line. The sketch after drawing the horizontal line is shown in Figure 4-19.

15. Choose **Zoom To Fit** from the **View (Heads-Up)** toolbar to fit the sketch into the drawing area.

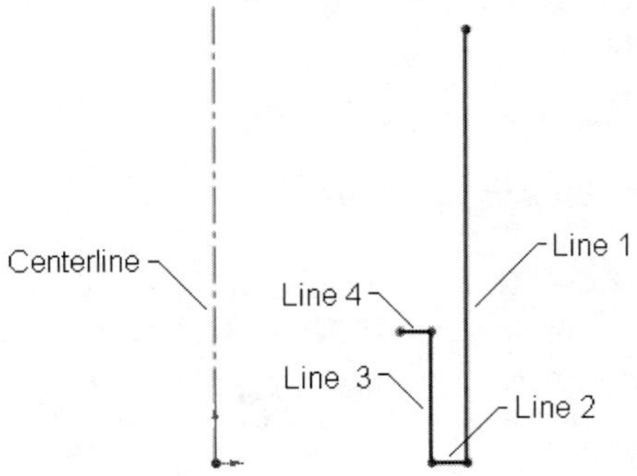

Figure 4-18 Sketch drawn using the ***Line*** *tool*

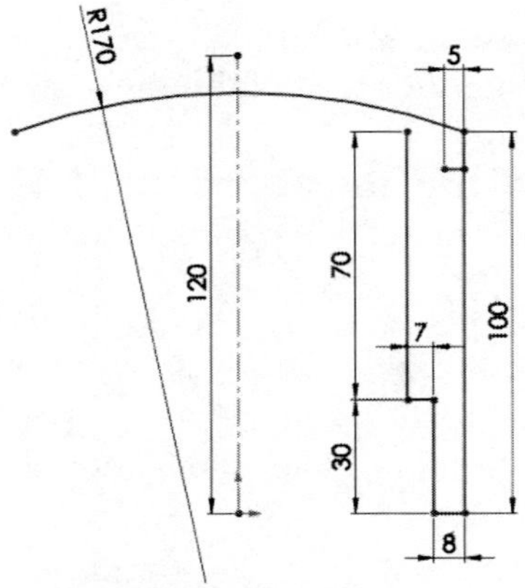

Figure 4-19 Sketch after drawing arc and line

Offsetting Lines

You need to offset the entities created earlier using the **Offset Entities** tool.

1. Select the line of 5 mm length created earlier.

2. Choose the **Offset Entities** button from the **Sketch CommandManager**; the **Offset Entities PropertyManager** is displayed. Also, the confirmation corner is displayed at the upper right corner of the drawing area.

3. Choose the **Keep Visible** button in the **Offset Entities PropertyManager** to pin the PropertyManager.

4. Set the value in the **Offset Distance** spinner to **5**. Next, select the **Reverse** check box to offset the entity in the reverse direction; preview of the entity to be offset is modified in the drawing area.

5. Choose the **OK** button from the **Offset Entities PropertyManager**.

   You will notice that an entity is created at an offset distance of 5 mm from the original entity. Also, a dimension of value 5 mm is applied between the newly created entity and the original entity with value 5. This dimension is the offset distance between the two entities.

6. Select the newly created entity. Move the cursor vertically downward; preview of the entity and the direction of the offset creation are also displayed. Press the left mouse button to offset the selected line.

Repeat this procedure to offset the entities until you have got eight entities including the original entity.

7. Set the value in the **Offset Distance** spinner to **7** and clear the **Select chain** check box if it is selected. Next, select the upper arc; preview of the offset arc is displayed in the drawing area.

8. Choose the **OK** button twice from the **Offset Entities PropertyManager**. The sketch after creating the offset entities is shown in Figure 4-20.

Completing the Remaining Sketch

Next, you need to complete the remaining sketch using the **Line** tool.

1. Invoke the **Line** tool by using the Mouse Gesture. Move the line cursor close to the left endpoint of the original line that was used to create offset lines. When the endpoint is highlighted, specify the start point. Specify the endpoint of the line at the endpoint of the last offset line. The sketch after drawing the vertical line is shown in Figure 4-21.

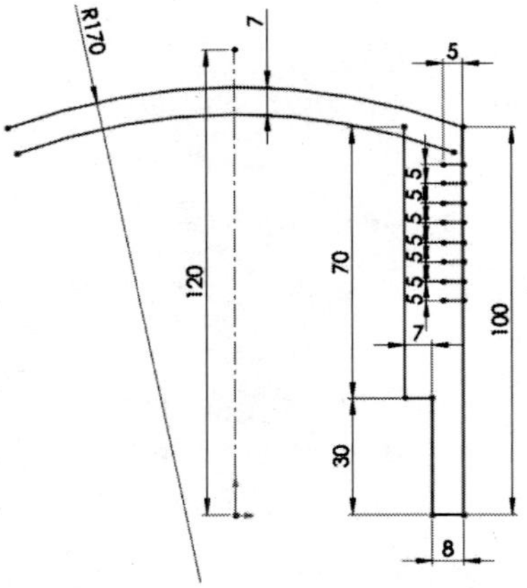

Figure 4-20 Sketch after offsetting the entities

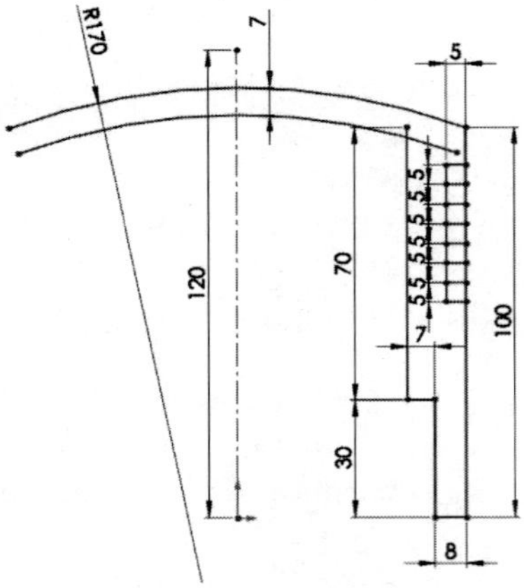

Figure 4-21 Sketch after drawing the vertical line

2. Move the cursor to the intersection point of the upper arc and the centerline. When the cursor snaps to the intersection, draw a vertical line that snaps to the intersection point of the lower arc and the centerline. Next, press the ESC key to exit the tool.

Trimming Unwanted Entities

Next, you need to trim the unwanted entities by using the **Trim Entities** tool.

1. Choose the **Trim Entities** tool from the **Sketch CommandManager**.

2. Choose the **Trim to closest** button from the **Trim PropertyManager** and trim the unwanted entities by using the left mouse button. While trimming the arc or line, if you get a message stating that the trimming will delete the relations, choose **Yes** to trim the entities. The sketch after trimming the unwanted entities is shown in Figure 4-22.

Adding Relations to the Sketch

Next, you will add the required relations to the sketched entities.

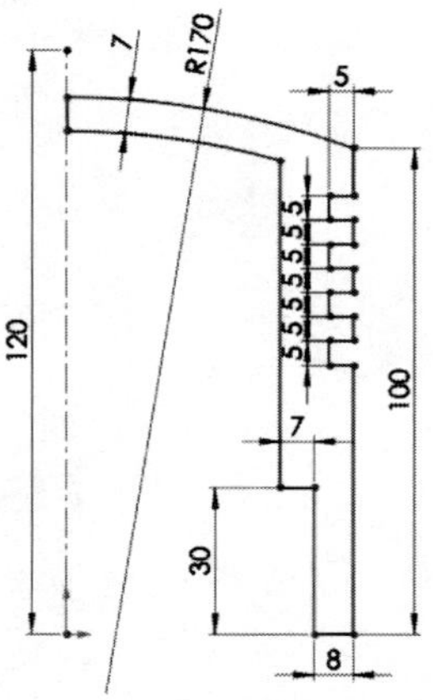

Figure 4-22 Sketch after trimming the unwanted entities

1. Right-click and then choose the **Select** option from the shortcut menu. Press and hold the CTRL key and select one of the endpoints of the lower horizontal line and then select the origin. Now, release the CTRL key after the selection; a pop-up toolbar is displayed. Choose the **Make Horizontal** option from the pop-up toolbar to add the Horizontal relation to the selected entities. Click anywhere in the drawing area to clear the selection set.

2. Press and hold the CTRL key and then select the horizontal lines created by offsetting. Release the CTRL key after selecting the lines; a pop-up toolbar is displayed. Choose the **Make Equal** option from it to apply the Equal relation.

3. Zoom out the drawing using the **Zoom In/Out** tool.

Adding Dimensions to the Sketch

After drawing, editing, and applying relations to the sketch, you need to add dimensions to fully define the sketch. Note that as in this tutorial, some dimensions of the sketch entities are applied automatically while drawing, you only need to apply the dimensions of the sketch that were not applied.

1. Select the dimension with value 7 which is placed between the upper arcs, and press the DELETE key to delete this dimension.

   This dimension needs to be deleted because during the design and manufacturing, the dimension between the tangents should be avoided.

2. Invoke the **Smart Dimension** tool by using the Mouse Gesture.

3. Select the inner vertical line and then the centerline, refer to Figure 4-23. Move the cursor to the other side of the centerline and click to place the dimension below the sketch. Next, enter **86** in the **Modify** dialog box and press ENTER.

4. Add the remaining dimensions to fully define the sketch, as shown in Figure 4-24.

5. Select one of the dimensions created after offsetting the entities. Press and hold the left mouse button and drag the cursor toward the right. Then, left-click to place the dimension at an appropriate place.

6. Arrange all the dimensions using the above procedure. The fully defined sketch is shown in Figure 4-24. In this figure, the display of relations is turned off.

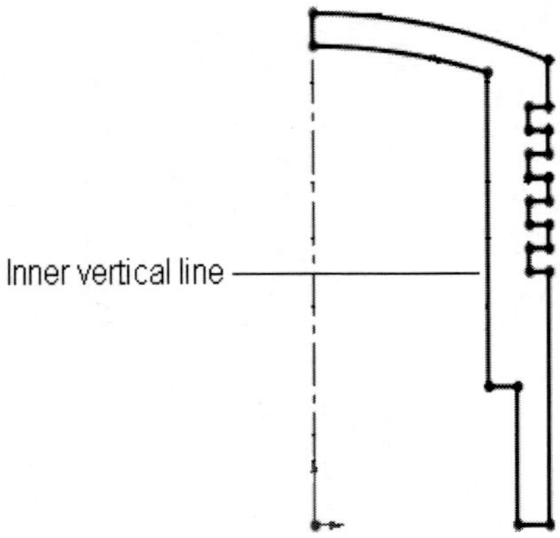

Figure 4-23 Reference to create the dimension

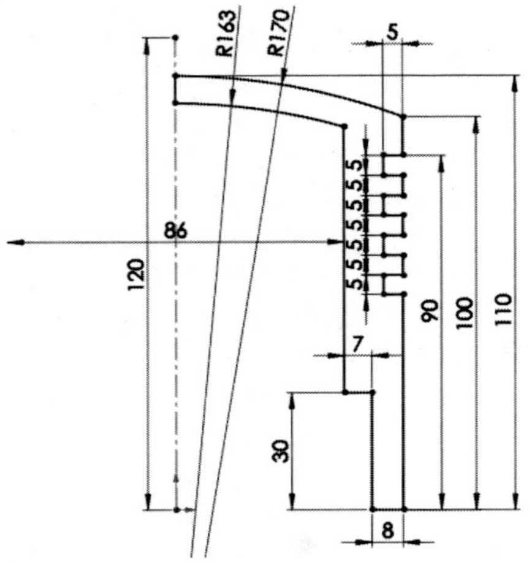

Figure 4-24 Fully defined sketch

Saving the Sketch

1. Choose the **Save** button from the Menu Bar and save the sketch with the name *c04_tut02* at the location *\Documents\SOLIDWORKS\c04*

2. Choose **File > Close** from the SOLIDWORKS menus to close the document.

EXERCISES

Exercise 1

Create the sketch of the model shown in Figure 4-25. Apply the required relations and dimensions to the sketch and fully define it. The sketch is shown in Figure 4-26. The solid model is given for your reference only. **(Expected time: 30 min)**

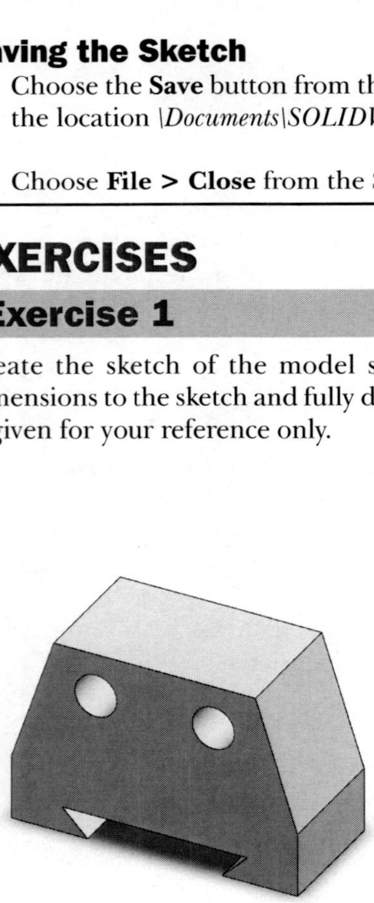

Figure 4-25 Solid model for Exercise 1

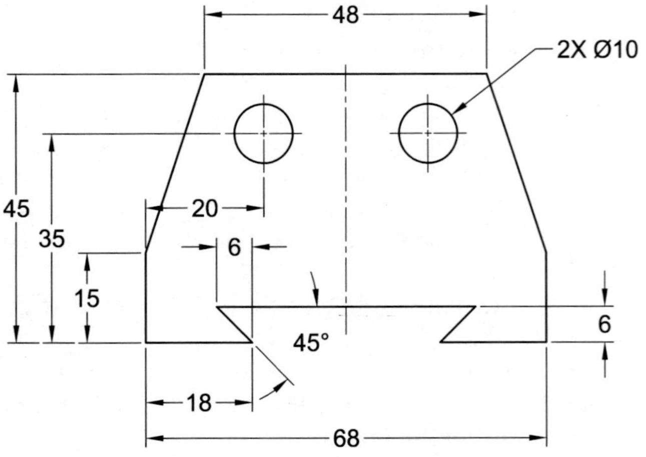

Figure 4-26 Sketch for Exercise 1

Exercise 2

Create the sketch of the model shown in Figure 4-27. Apply the required relations and dimensions to the sketch and fully define it. The sketch is shown in Figure 4-28. The solid model is given for your reference only. (**Expected time: 30 min**)

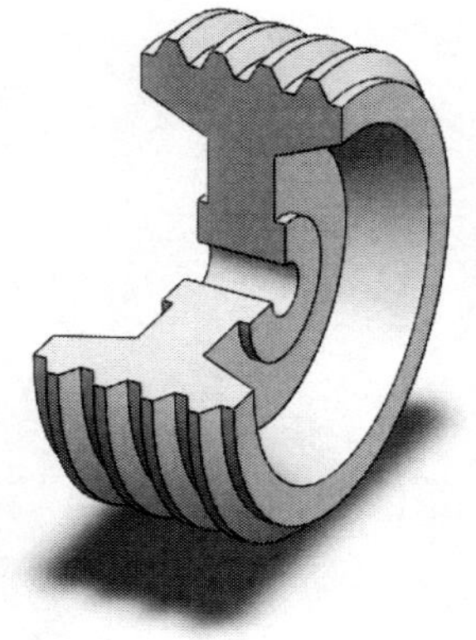

Figure 4-27 *Solid model for Exercise 2*

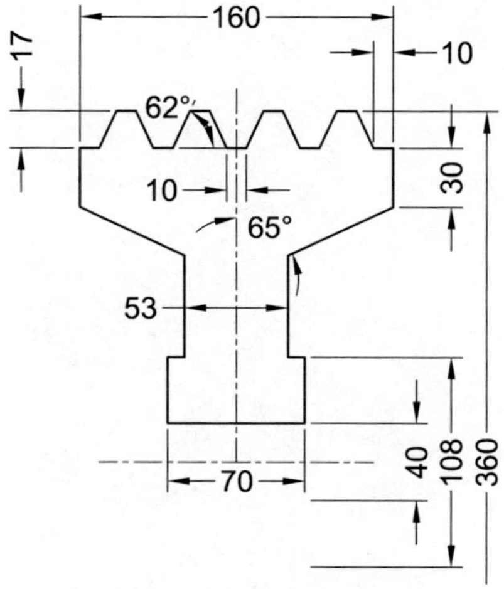

Figure 4-28 *Sketch for Exercise 2*

This page is intentionally left blank

Chapter 5

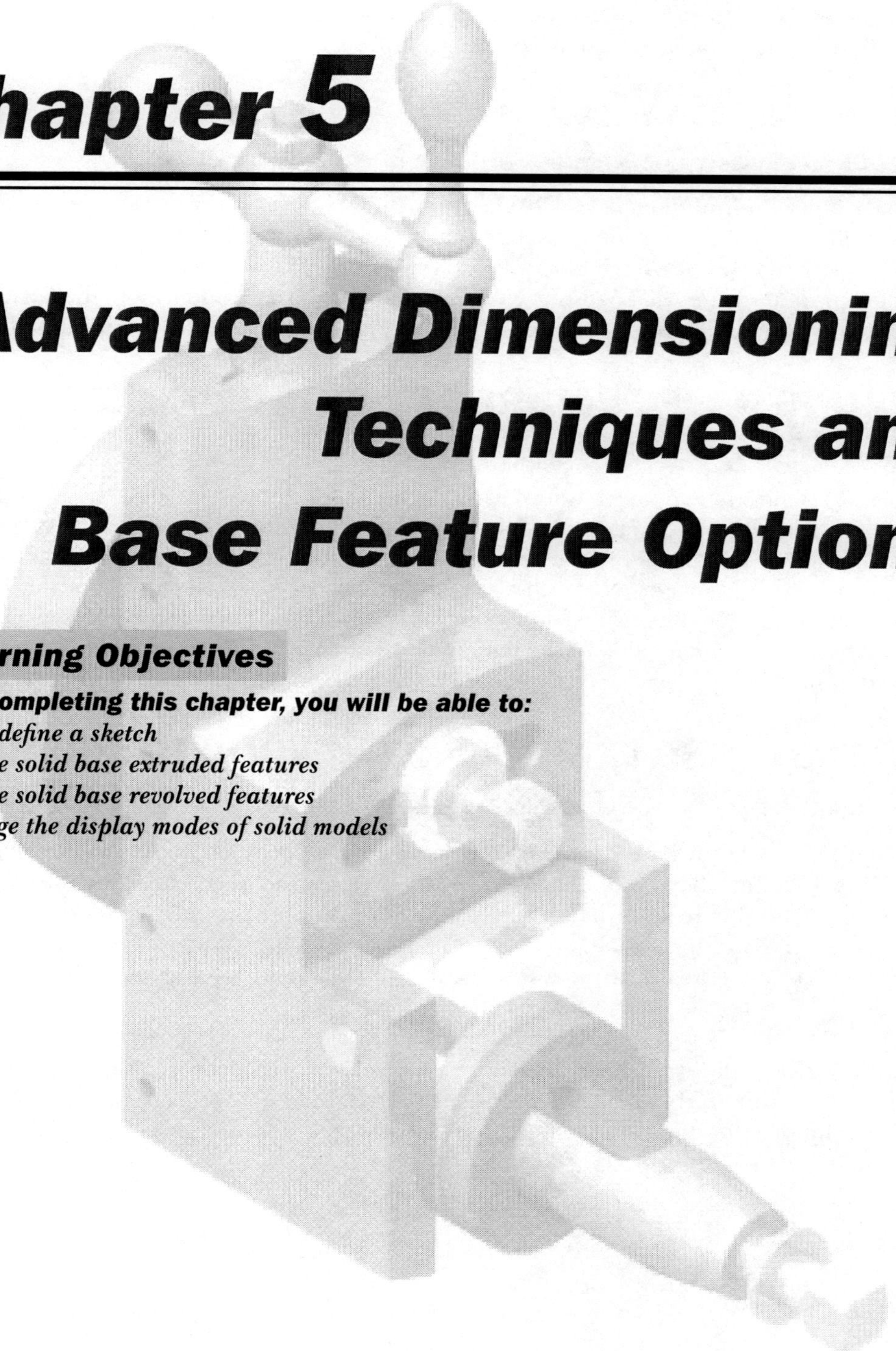

Advanced Dimensioning Techniques and Base Feature Options

Learning Objectives

After completing this chapter, you will be able to:

- *Fully define a sketch*
- *Create solid base extruded features*
- *Create solid base revolved features*
- *Change the display modes of solid models*

ADVANCED DIMENSIONING TECHNIQUES

In this chapter, you will learn about the advanced dimensioning techniques that are used to dimension the sketches. In SOLIDWORKS, you can apply all possible relations and dimensions to a sketch by using a single tool to make it fully defined.

Fully Defining the Sketches

SOLIDWORKS menus:	Tools > Dimensions > Fully Define Sketch

The **Fully Define Sketch** tool is used to apply relations and dimensions to a sketch automatically. To fully define a sketch, draw the sketch using the standard sketching tools. Choose **Fully Define Sketch** tool from the **Dimensions/Relations** toolbar; the **Fully Define Sketch PropertyManager** will be displayed. You can also right-click and choose the **Fully Define Sketch** option from the shortcut menu to display this PropertyManager after selecting the sketch.

CREATING BASE FEATURES BY EXTRUDING SKETCHES

CommandManager:	Features > Extruded Boss/Base

The sketches that you have drawn can be converted into base features by extruding the sketch using the **Extruded Boss/Base** tool from the **Features CommandManager**. After drawing the sketch, choose the **Features** tab from the **CommandManager** to display the **Features CommandManager**. Next, choose the **Extruded Boss/Base** tool from the **Features CommandManager**; the sketching environment will be closed and the part modeling environment will be invoked. Also, the preview of the feature that is created using the default options will be displayed in the trimetric view. The trimetric view gives a better display of the solid feature.

CREATING BASE FEATURES BY REVOLVING SKETCHES

CommandManager:	Features > Revolved Boss/Base

You can create the base feature by revolving a sketch about an axis with the help of the **Revolved Boss/Base** tool. This tool is available in the **Features CommandManager**. Choose this tool to revolve the sketch about a revolution axis. The revolution axis could be an axis, an entity of the sketch, or an edge of another feature. Note that whether you use a centerline or an edge to revolve the sketch, the sketch should be drawn on one side of the centerline or the edge.

In SOLIDWORKS, the right-hand thumb rule is followed for determining the direction of revolution. This rule states that if the thumb of your right hand points in the direction of the axis of revolution, the direction of the curled fingers will determine the default direction of revolution, refer to Figure 5-1.

Figure 5-1 *The right-hand thumb rule*

TUTORIALS

Tutorial 1

In this tutorial, you will create the model shown in Figure 5-2. Its dimensions are shown in Figure 5-3. The extrusion depth of the model is 20 mm. After creating the model, add a color of your choice to it and apply a decal as well. **(Expected time: 45 min)**

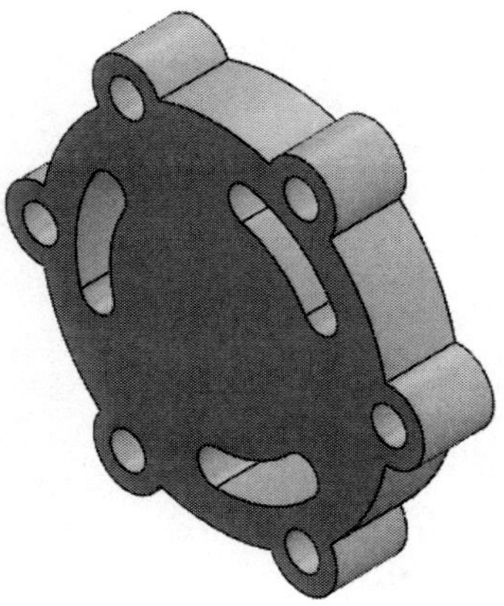

Figure 5-2 *Model for Tutorial 1*

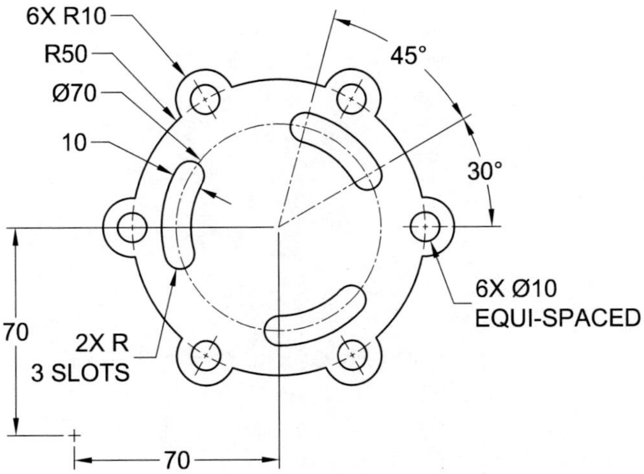

Figure 5-3 *Dimensions of the model for Tutorial 1*

The following steps are required to complete this tutorial:

a. Start a new SOLIDWORKS part document and then invoke the sketching environment.
b. Create the outer loop and then create the sketch of three inner cavities. Finally, draw six circles inside the outer loop for the holes, refer to Figures 5-4 through 5-7.
c. Invoke the **Extruded Boss/Base** tool and extrude the sketch through a distance of 20 mm, refer to Figure 5-8.

d. Rotate the view by using the **Rotate View** tool.
e. Change the current view to isometric view and then save the document.

Starting a New Part Document

1. Choose the **New** button from the Menu Bar and start a new part document using the **New SOLIDWORKS Document** dialog box.

2. Choose the **Sketch** button from the **Sketch CommandManager** and then select the **Front Plane**; the sketching environment is invoked.

Drawing the Outer Loop

This is the same sketch that was created in Tutorial 2 of Chapter 3. In this tutorial, you will create only the outer loop using the steps that were discussed in Chapter 3. It is recommended that you add relations and dimensions to it to make it fully defined.

1. Follow the steps that were discussed in Tutorial 2 of Chapter 3 to create the outer loop.

2. Add dimensions to it to fully define the sketch, as shown in Figure 5-4.

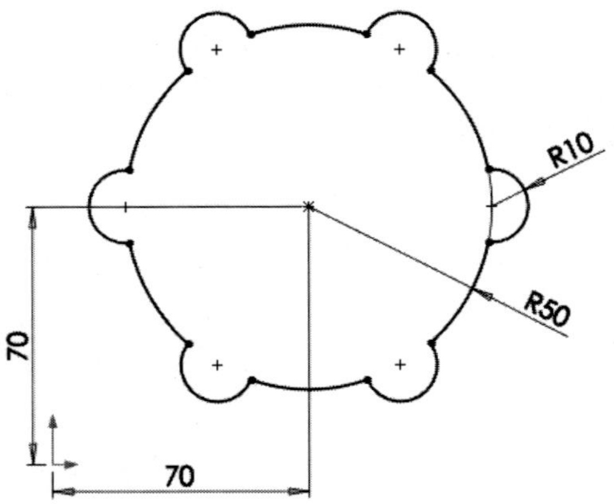

Figure 5-4 *Sketch after creating the outer loop*

Drawing the Sketch of Inner Cavities

Now, you need to draw the sketch of inner cavities. Draw the sketch of one of the cavities and then add required relations and dimensions to it. Next, you need to create a circular pattern of this cavity. The number of instances in the circular pattern is 3.

1. Choose the **Centerpoint Arc Slot** button from the **Sketch CommandManager**. Next, draw a slot with its center at the center point of the larger arc of diameter 100 mm. Make sure the start point and endpoint of the slot arc are in the first quadrant.

2. Complete the slot and then add dimensions to it, as shown in Figure 5-5. The slot turns black indicating that it is fully defined.

Next, you need to create a circular pattern of the inner cavity.

3. Select the sketch of the inner cavity and then invoke the **Circular Pattern PropertyManager**.

4. Drag the center of the circular pattern to the center of the circle of diameter 100 mm.

5. Set the value **3** in the **Number of Instances** spinner of the **Parameters** rollout and then choose the **OK** button to create the circular pattern, as shown in Figure 5-6.

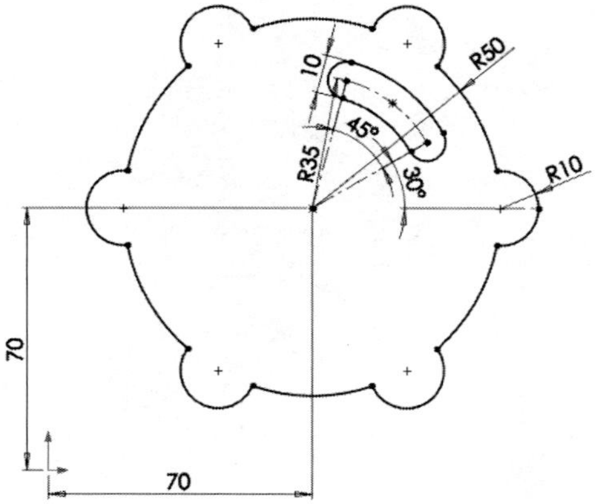

Figure 5-5 *Sketch after the first slot is drawn*

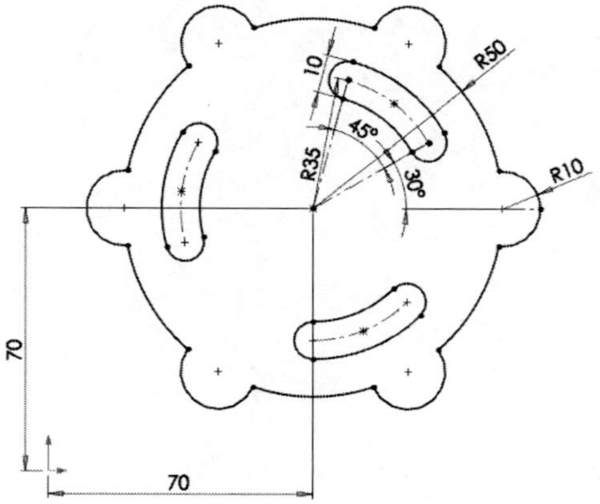

Figure 5-6 *Sketch after creating the circular pattern of the inner cavity*

6. Draw a circle of 10 mm diameter which is concentric to the circle of 20 mm diameter and then create a circular pattern of the circle. This completes the sketch of the model. The final sketch of the inner cavity is shown in Figure 5-7.

Note

If the status of the sketch drawn shows ***Under Defined*** *in the Status Bar and the circles created by using the circular pattern appears in blue color, you need to drag any of the circles to check its degree of freedom. The moment you drag a circle, the status of the sketch updates to* ***Fully Defined*** *and the color of the entities turns to black.*

Extruding the Sketch

The next step after creating the sketch is to extrude it. You can extrude the sketch using the **Extruded Boss/Base** tool.

1. Choose the **Features** tab from the CommandManager to display the **Features CommandManager** and then choose the **Extruded Boss/Base** tool.

The current view is changed to the trimetric view and the **Boss-Extrude PropertyManager** is displayed. Also, the preview of the model created using the default values is displayed in the drawing area.

2. Set **20** in the **Depth** spinner and then choose the **OK** button to extrude the sketch.

3. Choose **View Settings > Shadows In Shaded Mode** from the **View (Heads-Up)** toolbar to display the model with shadow.

4. Press SPACEBAR and then click on the **Isometric** option in the **Orientation** dialog box to change the current view to the isometric view. The complete model for Tutorial 1 is shown in Figure 5-8.

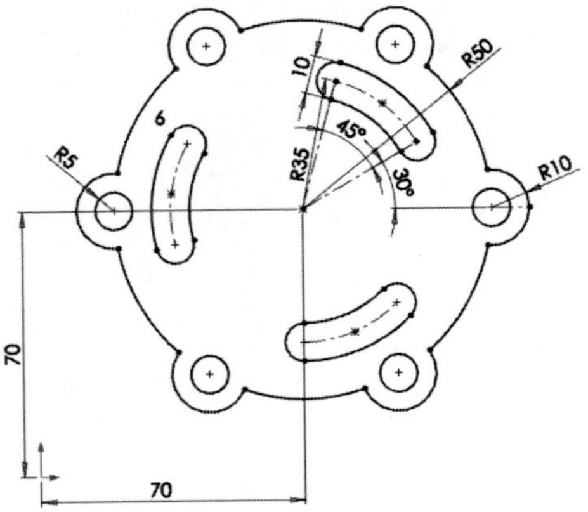

Figure 5-7 *Sketch after drawing the inner cavity*

Figure 5-8 *Complete model for Tutorial 1*

Adding Color to the Front Face

1. Choose the **DisplayManager** tab to view the DisplayManager.

2. Choose the **Open Appearance Library** button; the **Appearances, Scenes, and Decals** task pane is displayed.

3. Select the **Color** thumbnail in the lower portion of the task pane, and drag and drop it on the front face of the model and do not move the cursor; a pop-up toolbar with six options is displayed.

4. Select the **Face (Boss-Extrude-1)** option; a color node is added to the **Appearances** section.

   If the default color set in your system is dull, you may not able to view the color applied. To change the color, you need to edit it.

5. Double-click on the color node in the **DisplayManager**; the front face of the model will be highlighted and the **color PropertyManager** is displayed.

6. Select a color of your choice and choose the **OK** button; the new color is applied.

Applying Decal to the Back Face

1. Choose the **View Decals** button from the **DisplayManager**; the **Decals** area is displayed.

2. Choose the **Open Decal Library** button; the **Appearances, Scenes, and Decals** task pane with default decals is displayed.

   Now, you need to place the back face of the model parallel to the screen. Before you start rotating the view of the model, it is recommended that you turn off the display of the shadow.

3. Choose **View Settings > Shadows In Shaded Mode** from the **View (Heads-Up)** toolbar to turn off the display of the shadow.

4. Rotate the model such that the back face of the model is visible. Then, set the view orientation to normal; the back face is set parallel to the screen.

5. Select the **Design with SOLIDWORKS** logo in the lower portion of the task pane. Now, drag and drop it on the back face of the model; the **Decals PropertyManager** is displayed.

6. Scale the decal and place it at the center of the face, as shown in Figure 5-9, and choose the **OK** button.

7. Change the current view to the isometric view using the **Orientation** dialog box.

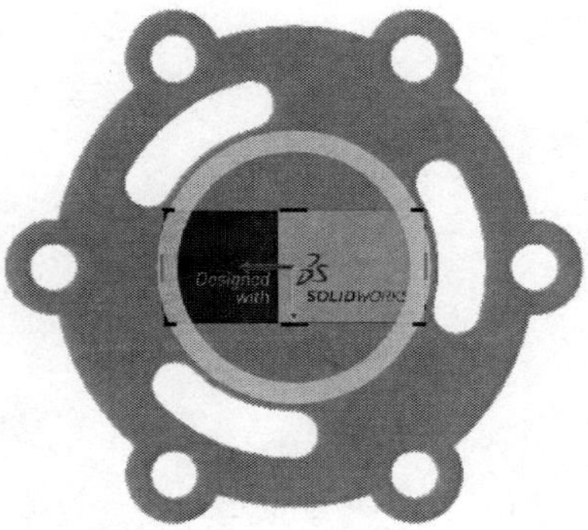

Figure 5-9 *Decal placed at the center of the face*

Note

1. To remove the color applied, right-click on the color node in the ***DisplayManager*** *and choose the* ***Remove Appearance*** *option from the menu displayed.*

2. On dragging and dropping the ***color*** *thumbnail from the* ***Appearances, Scenes, and Decals*** *task pane to the face of the model, a pop-up toolbar will be displayed with six options. Click on these options one by one to check their effect.*

3. To edit the decal applied, expand the ***Decals*** *node in the* ***DisplayManager****. Next, right-click on the* ***logos*** *subnode and choose the* ***Edit Decal*** *option from the shortcut menu displayed. On doing so, the* ***Decals PropertyManager*** *will be displayed. You can set the properties of decal by choosing the* ***Mapping*** *and* ***Illumination*** *tabs. To change the image, choose the* ***Image*** *tab and specify the location of the new image in the* ***Decal Preview*** *rollout.*

Saving the Model

1. Choose the **Save** button from the Menu Bar and save the model with the name *c05_tut01* at the location given next.

   \Documents\SOLIDWORKS\c05

2. Choose **File > Close** from the SOLIDWORKS menus to close the document.

Tutorial 2

In this tutorial, you will open the sketch drawn in Exercise 2 of Chapter 4. You will then create a thin feature by revolving the sketch through an angle of 270 degree, as shown in Figure 5-10. You will offset the sketch outward while creating the thin feature. After creating the model, you will turn on the option to display shadows and also apply the Copper material to the model. Additionally, you will determine the mass properties. **(Expected time: 30 min)**

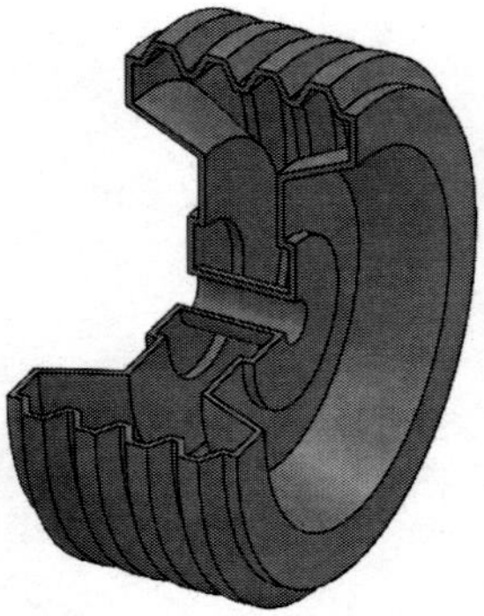

Figure 5-10 Revolved model for Tutorial 2

The following steps are required to complete this tutorial:

a. Open Exercise 2 of Chapter 4, refer to Figure 5-11.
b. Save it in the folder of the current chapter.
c. Invoke the **Revolved Boss/Base** tool and revolve the sketch through an angle of 270 degree, refer to Figure 5-13.
d. Change the current view to the isometric view and then display the model in the shadow, refer to Figure 5-14.
e. Assign copper material to the model, refer to Figure 5-15, and then check the mass properties.

Opening Exercise 3 of Chapter 4

As the required document is saved in the *\Documents\SOLIDWORKS\c04* folder, you need to select this folder and then open the *c04_exr02.sldprt* document.

1. Choose the **Open** button from the Menu Bar to display the **Open** dialog box. The *c05* folder is the current folder in this dialog box.

2. Browse and select the *c04* folder. All the documents that were created in Chapter 4 are displayed in this folder.

3. Select the *c04exr2.sldprt* document and then choose the **Open** button; the document is opened in the sketching environment.

Saving the Document in the c05 Folder

As mentioned earlier, it is recommended that you save the document with a new name in the folder of the current chapter so that the original document is not affected.

1. Choose **File > Save As** from the SOLIDWORKS menus to display the **Save As** dialog box. As the *c04* was the folder selected last, it is the current folder.

2. Browse to the folder c05 and then double-click on it to make it the current folder.

3. Enter **c05_tut02** as the name of the document in the **File name** edit box. Choose the **Save** button to save the document. The sketch is displayed in the drawing area, as shown in Figure 5-11.

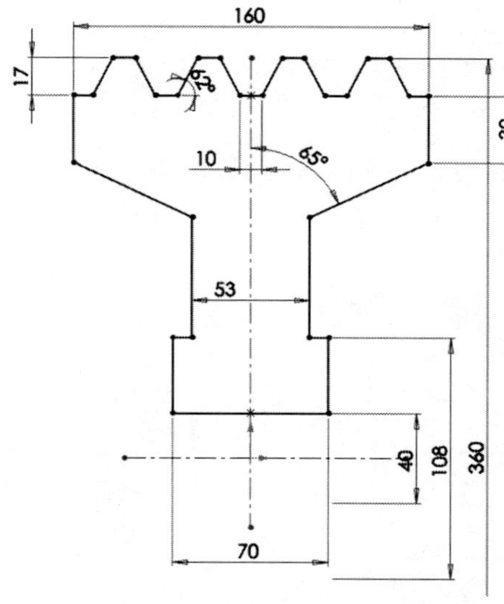

Figure 5-11 Sketch for the revolved model

Revolving the Sketch

The sketch consists of two centerlines. The vertical centerline will be used to mirror the sketched entities and the horizontal centerline will be used to apply linear diameter dimensions. You need to revolve the sketch about the horizontal centerline.

1. Choose the **Features** tab from the **CommandManager** to display the **Features CommandManager**. Now, choose the **Revolved Boss/Base** button; the sketch is automatically oriented in the trimetric view and the **Revolve PropertyManager** is displayed. As the sketch has two centerlines, SOLIDWORKS cannot determine which one of them has to be used as the axis of revolution. As a result, you are prompted to select the axis of revolution.

2. Select the horizontal centerline which is used to create linear diameter dimensions as the axis of revolution. On doing so, a preview of the complete revolved feature in temporary shaded graphics is displayed in the drawing area. As the preview of the model is not displayed properly in the current view, you need to zoom the drawing.

3. Choose the **Zoom to Fit** button from the **View (Heads-Up)** toolbar or press the F key on the keyboard.

4. In the **Direction1** rollout of the **Revolve PropertyManager**, select the **Blind** option from the **Revolve Type** drop-down list if it is not selected by default. Next, set the value of the **Direction 1 Angle** spinner to **270** and click anywhere on the screen; preview of the revolved model is also modified accordingly.

   Note that if the horizontal centerline is drawn from left to right, then the direction of revolution has to be reversed to get the required model. To get the desired direction, you need to follow the right-hand thumb rule. You can reverse the direction of revolution using the **Reverse Direction** button available on the left of the **Revolve Type** drop-down list.

5. Select the **Thin Feature** check box to expand the **Thin Feature** rollout, as shown in Figure 5-12. Set **5** in the **Direction 1 Thickness** spinner. You will notice that preview of the thin feature is shown outside the original sketch.

6. Choose the **OK** button; the revolved feature is created, as shown in Figure 5-13.

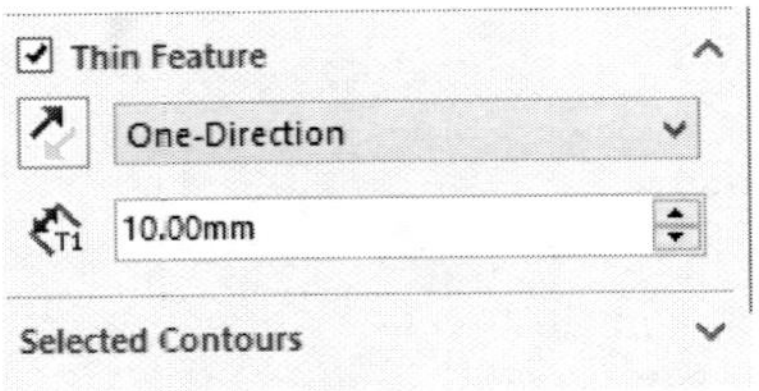

***Figure 5-12** The **Thin Feature** rollout*

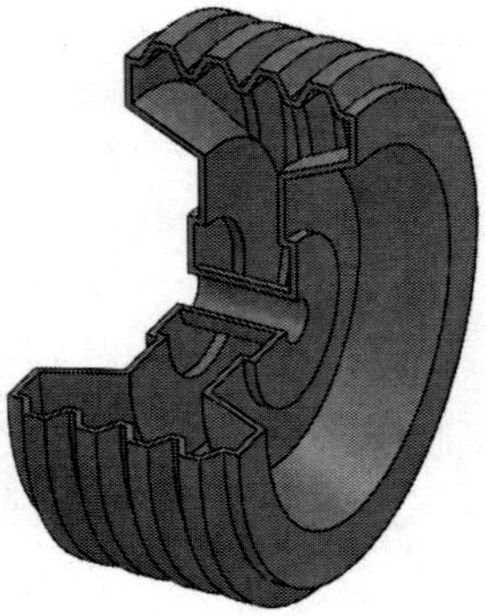

***Figure 5-13** Model created by revolving the sketch*

7. Choose the **View Orientation** button from the **View (Heads-Up)** toolbar; a flyout is displayed. Choose the **Isometric** option from the flyout to orient the model to the isometric view.

Rotating the View

Next, you need to rotate the view so that you can view the model from all directions. As mentioned earlier, the view can be rotated using the **Rotate View** tool.

1. Press the middle mouse button and move the cursor; the arrow cursor is replaced by the rotate view cursor.

2. Hold the middle mouse button and drag the cursor in the drawing area to rotate the view.

3. Press SPACEBAR to invoke the **Orientation** dialog box. In this dialog box, click on the **Isometric** option; the model orients to the isometric view.

Displaying the Shadow

As mentioned in the tutorial description, you need to display the shadow of the model if not displayed by default. You can turn on the display of the shadow using the **View (Heads-Up)** toolbar.

1. Choose **View Settings > Shadows In Shaded Mode** from the **View (Heads-Up)** toolbar to display the model with shadow, as shown in Figure 5-14.

Assigning Materials to the Model

As mentioned earlier, you can invoke this PropertyManager by using the **Material** node in the **FeatureManager Design Tree**. You can also assign a material to a model by choosing **Edit > Appearance > Material** from the SOLIDWORKS menus.

1. Right-click on the **Material <not specified>** node in the **FeatureManager Design Tree** and choose the **Edit Material** option; the **Material** dialog box is displayed.

2. Click on the  sign located on the left of the **Copper Alloys** option from the list of materials available in the left area of the dialog box; the tree view expands and materials in this family are displayed.

3. Select the **Copper** option and choose the **Apply** button from the **Material** dialog box and then choose the **Close** button to exit. The model after assigning the material is shown in Figure 5-15.

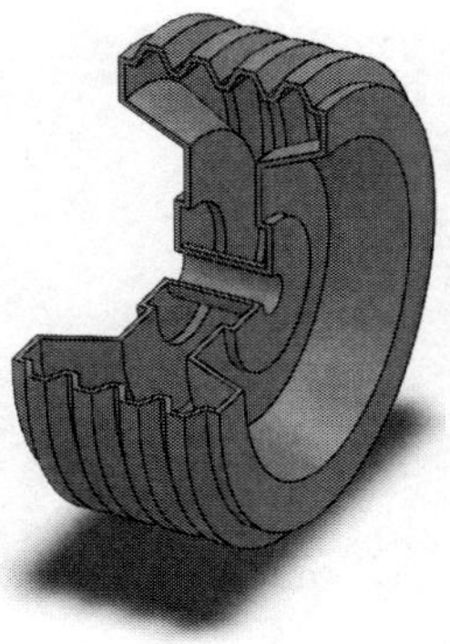

***Figure 5-14** Model with the display of shadow turned on*

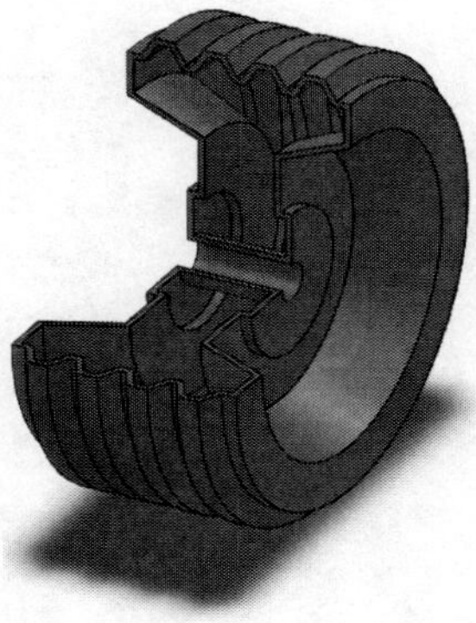

***Figure 5-15** Model after assigning the copper material*

Determining the Mass Properties of the Model

As discussed earlier, to determine the mass properties of the current model, you need to invoke the **Mass Properties** dialog box.

1. Choose the **Mass Properties** tool from the **Evaluate CommandManager**; the **Mass Properties** dialog box with the mass properties of the current model is displayed, as shown in Figure 5-16.

Saving the Model

As the name of the document is specified in the beginning, you need to choose the **Save** button to save the document.

1. Choose the **Save** button from the Menu Bar to save the model. If the **SOLIDWORKS** warning box is displayed, choose **Yes** from it to rebuild the model before saving. The model is saved at the following location *Documents\SOLIDWORKS\c05*.

2. Choose **File > Close** from the SOLIDWORKS menus to close the document.

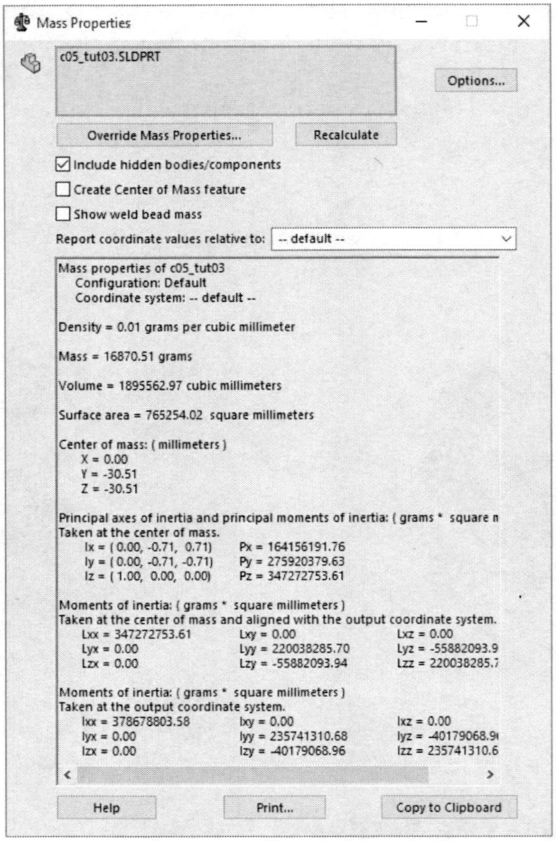

Figure 5-16 *The* ***Mass Properties*** *dialog box*

EXERCISES

Exercise 1

Create the model shown in Figure 5-17. The sketch of the model is shown in Figure 5-18. Create the sketch and dimension it using the fully defined sketch option. The extrusion depth of the model is 15 mm. After creating the model, rotate the view. **(Expected time: 30 min)**

Figure 5-17 *Model for Exercise 1*

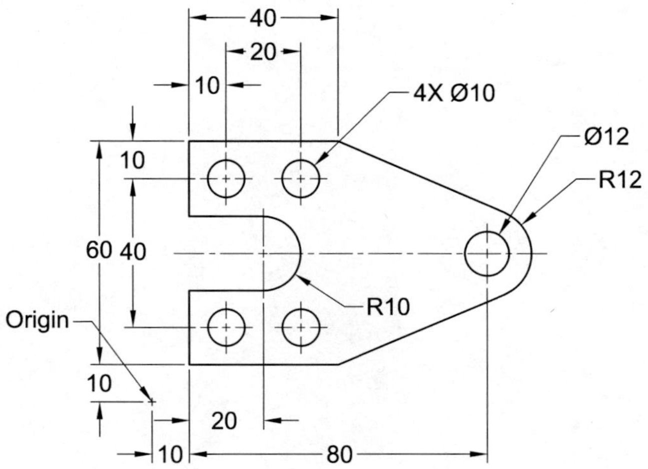

Figure 5-18 *Sketch of the model for Exercise 1*

Exercise 2

Create the model shown in Figure 5-19. The sketch of the model is shown in Figure 5-20. Create the sketch and dimension it using the fully defined sketch option. The Direction 1 angle of the model is 270 degrees. **(Expected time: 30 min)**

Figure 5-19 *Model for Exercise 2*

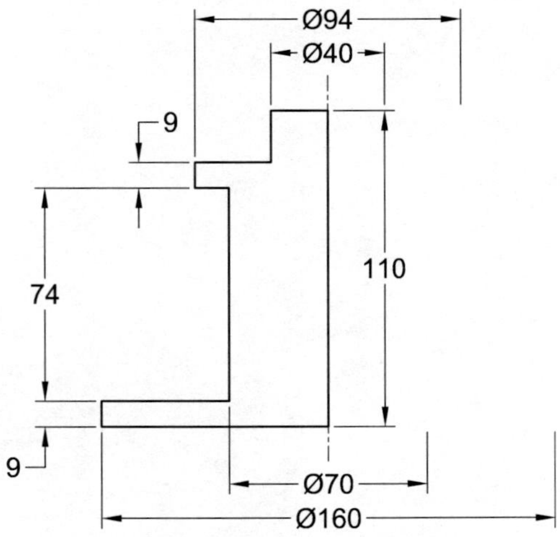

Figure 5-20 *Sketch of the model for Exercise 2*

This page is intentionally left blank

Chapter 6

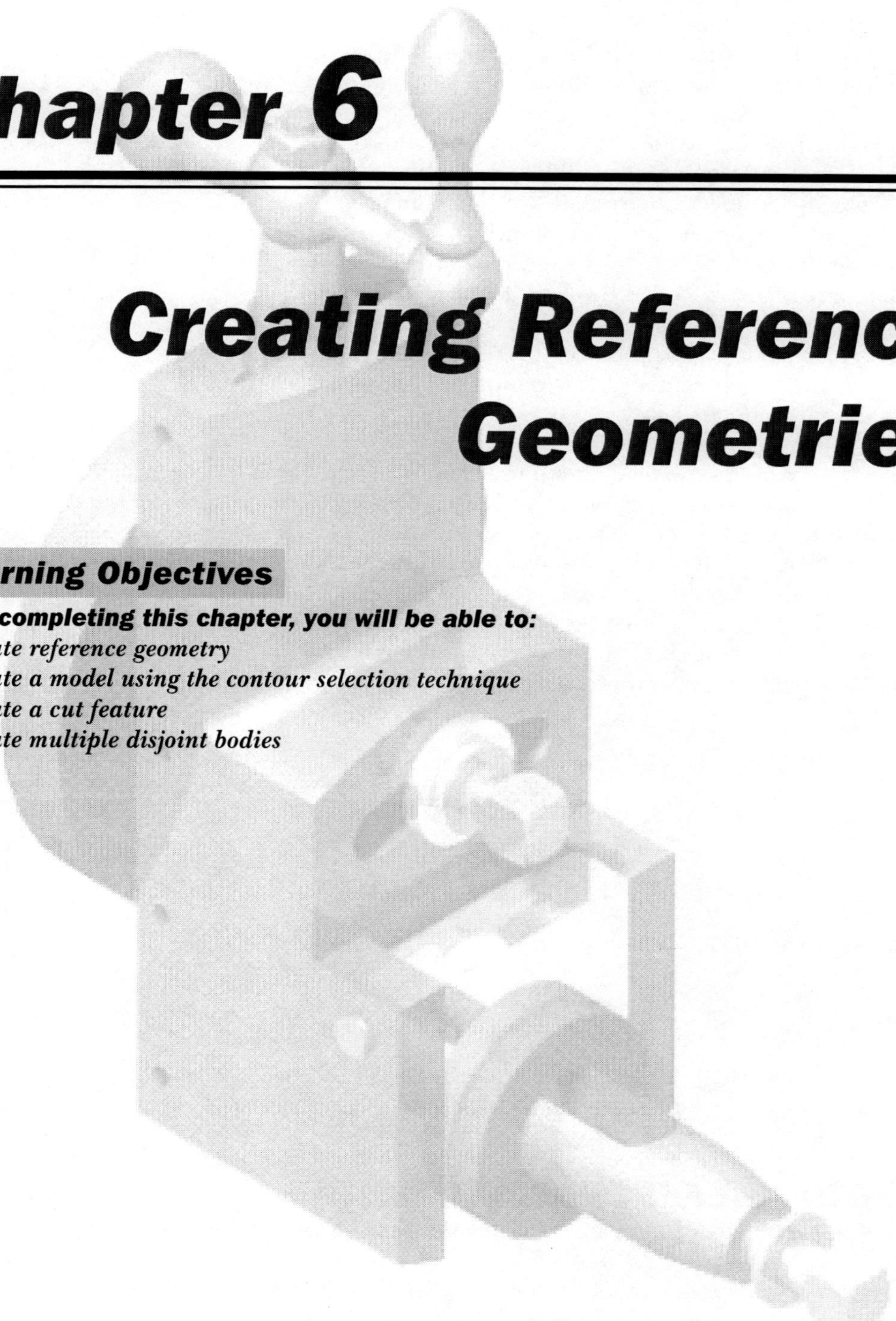

Creating Reference Geometries

Learning Objectives

After completing this chapter, you will be able to:

- *Create reference geometry*
- *Create a model using the contour selection technique*
- *Create a cut feature*
- *Create multiple disjoint bodies*

IMPORTANCE OF SKETCHING PLANES

In the earlier chapters, you created basic models by extruding or revolving the sketches. All those models were created on a single sketching plane, the **Front Plane**. But most mechanical designs consist of multiple sketched features, referenced geometries, and placed features. These features are integrated together to complete a model. Most of these features lie on different planes. When you start a new SOLIDWORKS document and try to invoke a sketching plane, you are prompted to select the plane on which you want to draw the sketch. On the basis of design requirements, you can select any plane to create the base feature. To create additional sketched features, you need to select an existing plane or a planar surface, or you need to create a plane that will be used as a sketching plane. For example, consider the model shown in Figure 6-1.

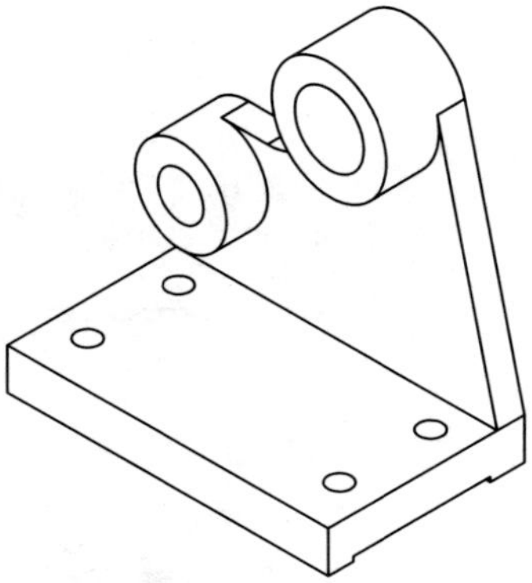

Figure 6-1 *A multifeatured model*

The base feature of this model is shown in Figure 6-2. The sketch for the base feature is drawn on the **Top Plane**. After creating the base feature, you need to create the other sketched features, applied features, and referenced features, see Figure 6-3. The boss features and cut features are the sketched features that require sketching planes on which you draw the sketches of the features.

It is evident from Figure 6-3 that the features added to the base feature are not created on the same plane on which the sketch for the base feature is created. Therefore, to draw the sketches of other sketched features, you need to define other sketching planes.

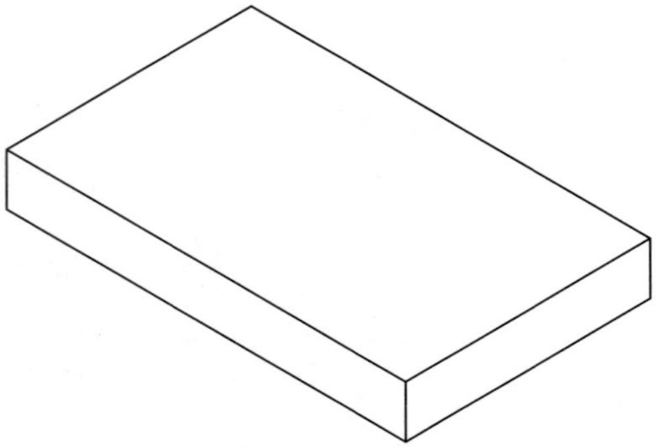

Figure 6-2 *Base feature of the model*

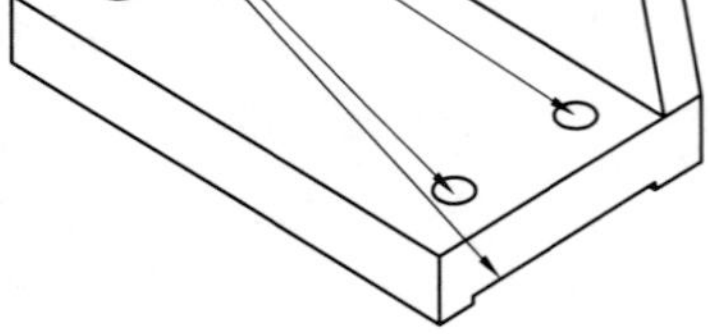

Figure 6-3 *Model after adding other features*

CREATING REFERENCE GEOMETRY

The reference geometry features are those that are available only to assist you in creating models. The reference geometries in SOLIDWORKS include planes, axes, points, and coordinate systems. These reference geometries act as reference for drawing the sketches for the sketched features, defining the sketch plane, and assembling the components. They also act as references for various applied features and sketched features. These features have no mass or volume. You must have a good understanding of these geometries because they are widely used in creating complex models. The tools to create the reference geometries are grouped together in the **Reference Geometry** flyout in the **Features CommandManager**, refer to Figure 6-4. The reference planes and the procedure of creating new planes is discussed next.

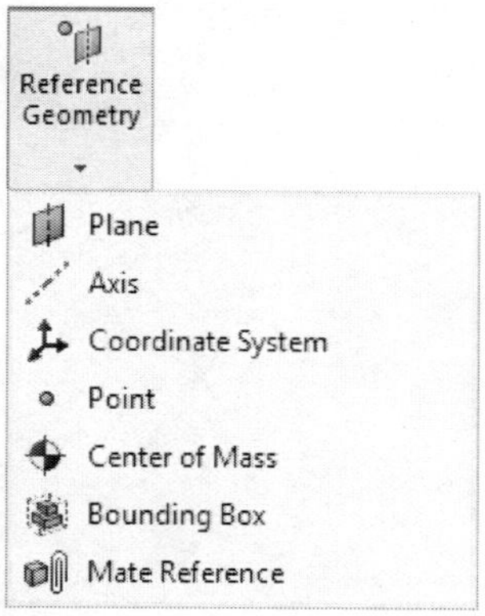

Figure 6-4 *Tools in the* ***Reference Geometry*** *flyout*

CREATING CUT FEATURES

Cut extrude is a material removal process. You can define a cut feature by extruding a sketch, revolving a sketch, sweeping a section along a path, lofting sections, or by using a surface. You will learn more about sweep, loft, and surface in the later chapters. The cut feature can be created only if a base feature exists. The extruded and revolved cut features are discussed next.

Creating Extruded Cuts

CommandManager:	Features > Extruded Cut
SOLIDWORKS menu:	Insert > Cut > Extrude

To create an extruded cut feature, create a sketch for the cut feature and then choose the **Extruded Cut** button from the **Features CommandManager**; the **Cut-Extrude PropertyManager** will be displayed, as shown in Figure 6-5.

Creating Revolved Cuts

CommandManager:	Features > Revolved Cut
SOLIDWORKS menu:	Insert > Cut > Revolve

Revolved cuts are used to remove the material by revolving a sketch around a selected axis. Similar to the revolved boss/base feature, you can define the revolution axis using a centerline or using an edge in the sketch. When you invoke the **Revolved Cut** tool, the **Cut-Revolve PropertyManager** will be displayed, as shown in Figure 6-6.

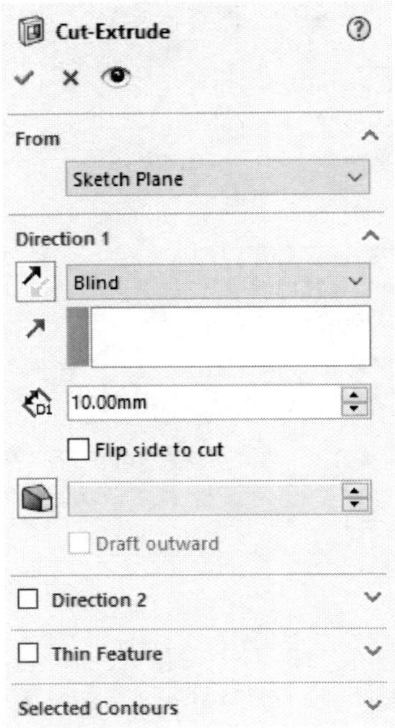

Figure 6-5** The **Cut-Extrude PropertyManager

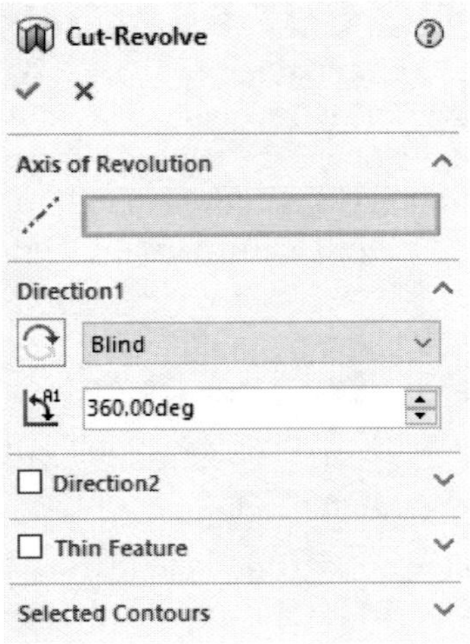

Figure 6-6** The **Cut-Revolve PropertyManager

TUTORIALS

Tutorial 1

In this tutorial, you will create the model shown in Figure 6-7. The dimensions of the model are also shown in the same figure. Create the model by extruding the contours of the sketch.

(Expected time: 30 min)

It is clear from the above figure that the given model is a multifeatured model. It consists of various extruded features. In conventional methods, you need to create a separate sketch for each sketched feature and then convert it into a feature. But in this tutorial, you will draw the sketch of the front view of the model and extrude it by selecting different contours in it.

The following steps are required to complete this tutorial:

a. Create the sketch on the default plane and apply the required relations and dimensions to it, refer to Figure 6-8.
b. Invoke the **Extrude Boss/Base** tool and extrude the selected contour, refer to Figures 6-9 and 6-10.
c. Select the second set of contours and extrude them to the required distance, refer to Figures 6-11 and 6-12.
d. Select the third set of contours and extrude them to the required distance, refer to Figures 6-13 and 6-14.
e. Save and close the document.

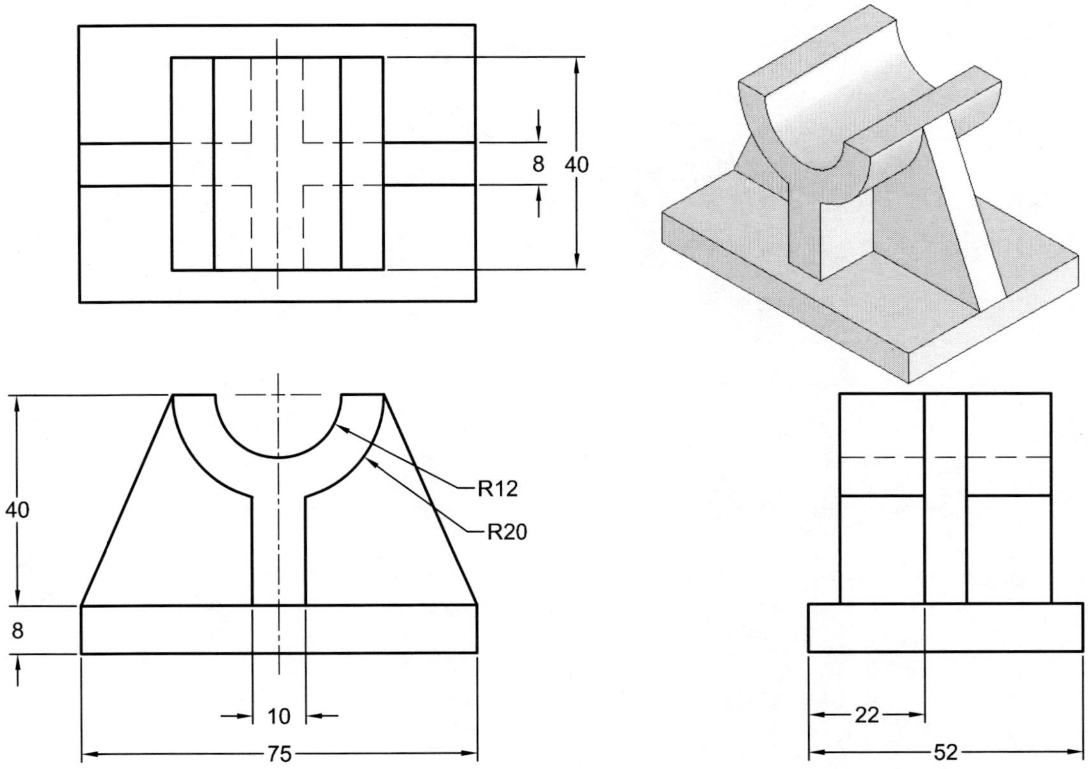

Figure 6-7 Dimensions and views for Tutorial 1

Creating the Sketch of the Model

1. Start a new SOLIDWORKS part document using the **New SOLIDWORKS Document** dialog box.

2. Draw the sketch of the front view of the model on the Front Plane. Apply the required relations and dimensions to the sketch, as shown in Figure 6-8. Make sure that you do not exit the sketching environment.

Selecting and Extruding the Contours of the Sketch

In this tutorial, you need to use the contour selection method to create the model. Therefore, you first need to select one of the contours from the given sketch and then extrude it. For a better view, you can orient the sketch to Isometric view.

1. Choose **View Orientation > Isometric** from the **View (Heads-Up)** toolbar; the sketch is displayed in the isometric view.

2. Right-click in the drawing area to invoke the shortcut menu. Expand the shortcut menu, if required. Choose the **Contour Select Tool** option; the select cursor is replaced by the contour selection cursor and the selection confirmation corner is displayed.

3. Move the cursor to the lower rectangle of the sketch; the rectangle is highlighted. This indicates that this rectangle is a closed profile.

4. Click on the highlighted rectangular area; the lower rectangular area is selected as a contour, as shown in Figure 6-9.

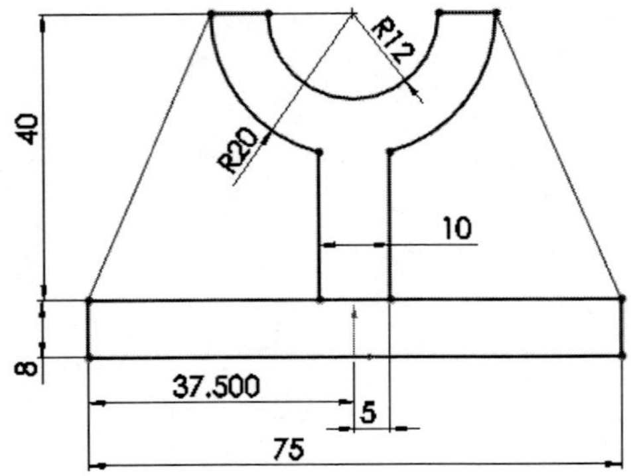

Figure 6-8 *Fully defined sketch for creating the model*

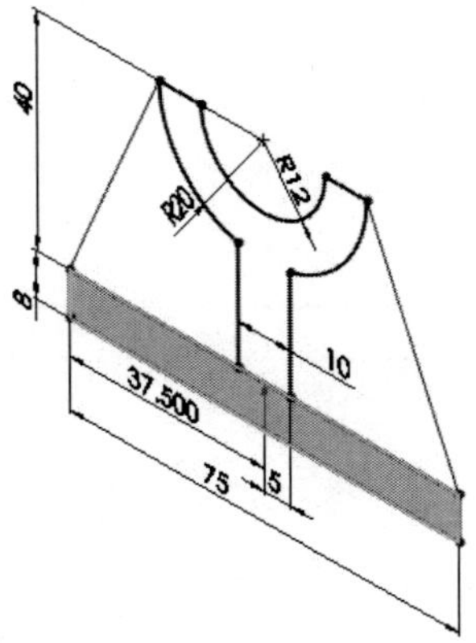

Figure 6-9 *Lower rectangle selected as a contour*

5. Choose the **Extruded Boss/Base** tool from the **Features CommandManager** or from the pop-up toolbar; the **Boss-Extrude PropertyManager** is invoked and preview of the base feature is displayed in the drawing area in temporary graphics.

   The name of the selected contour is displayed in the selection box of the **Selected Contours** rollout.

6. Right-click in the drawing area and choose the **Mid Plane** option from the shortcut menu; preview of the feature is modified.

7. Set **52** in the **Depth** spinner and choose the **OK** button from the **Boss-Extrude PropertyManager**; the selected contour is extruded, as shown in Figure 6-10.

8. Again, right-click in the drawing area and choose the **Contour Select Tool** option from the shortcut menu; the select cursor is replaced by the contour selection cursor.

9. Select an entity of the sketch using the contour selection cursor to invoke the selection mode of the sketch.

10. Select the middle contour of the sketch using the left mouse button; the selected region is highlighted, as shown in Figure 6-11.

11. Invoke the **Extruded Boss/Base** tool. Right-click in the drawing area and choose the **Mid Plane** option from the shortcut menu.

12. Set the value of the **Depth** spinner to **40** and choose the **OK** button from the **Boss-Extrude PropertyManager**. The feature created by extruding the middle contour is shown in Figure 6-12.

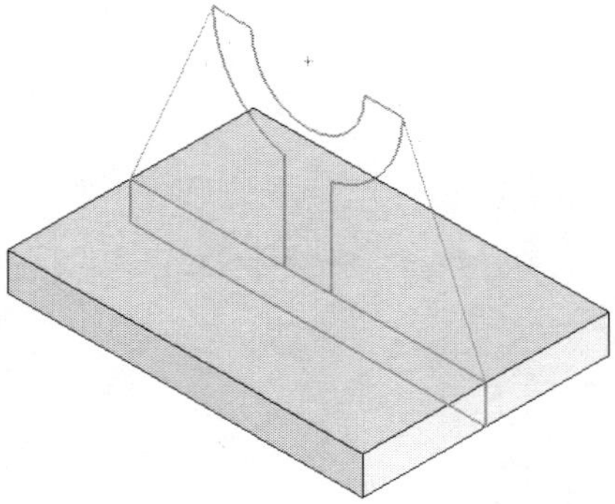

Figure 6-10 Base feature of the model

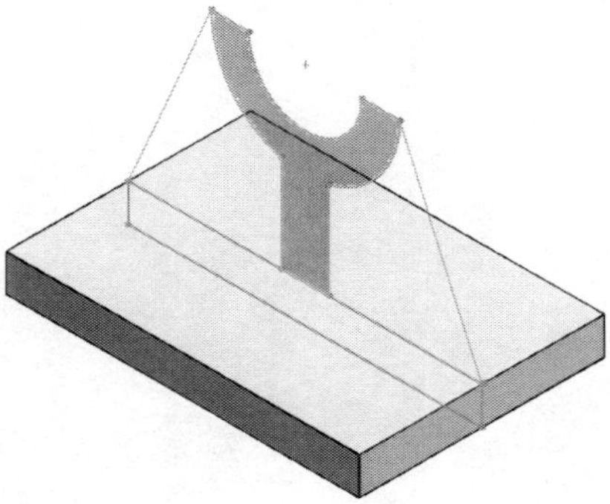

Figure 6-11 Middle contour selected using the ***Contour Select Tool***

13. Choose the **Contour Select Tool** option again and then select a sketched entity. Next, select the contour on the right. Press and hold the CTRL key and then select the contour on the left side, see Figure 6-13.

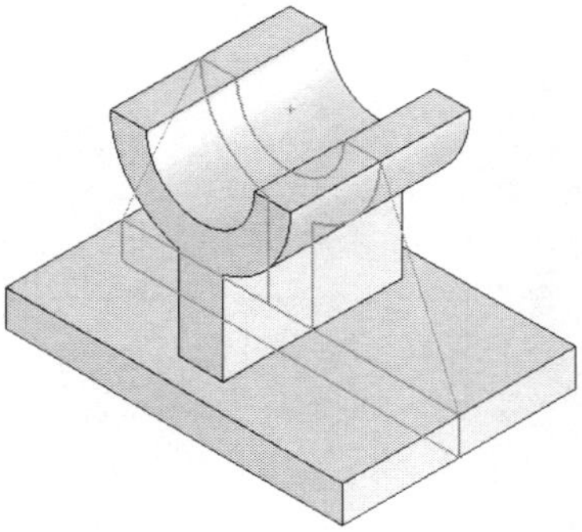

Figure 6-12 Second feature created by extruding the middle contour

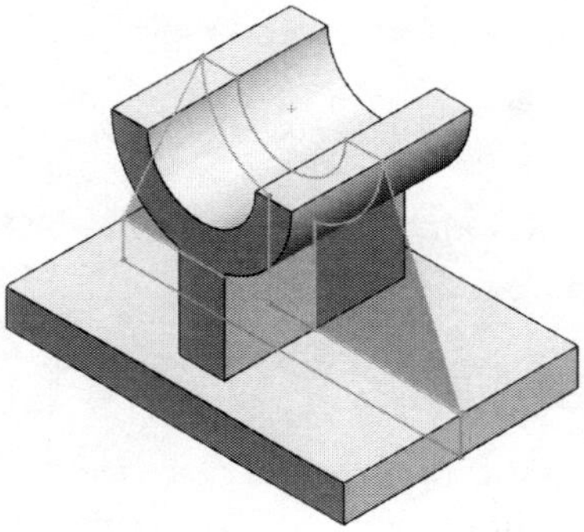

Figure 6-13 The right and the left contours selected

14. Invoke the **Extruded Boss/Base** tool. Right-click and choose the **Mid Plane** option from the shortcut menu.

15. Set the value of the **Depth** spinner to **8** and choose the **OK** button from the **Boss-Extrude PropertyManager**.

    The model is completed, but the sketch is still displayed in the model. Therefore, you need to hide the sketch.

16. Move the cursor to any of the sketched entities and when the entity is highlighted, select it; a pop-up toolbar is displayed. Choose the **Hide** button from the pop-up toolbar.

    The isometric view of the final model after turning off the display of the sketch is shown in Figure 6-14. The **FeatureManager Design Tree** displaying various parts of the model is shown in Figure 6-15.

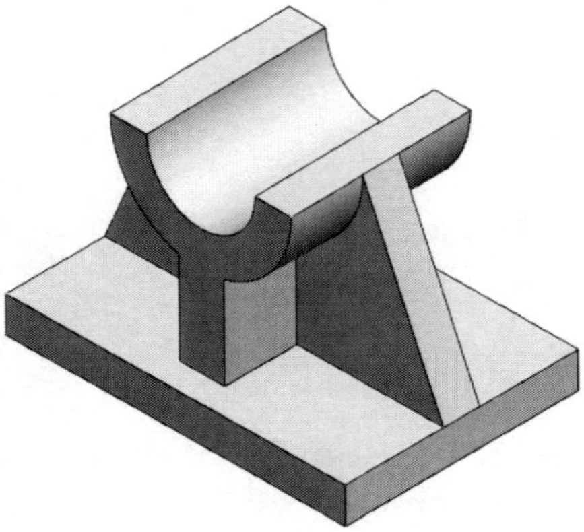

Figure 6-14 Final solid model

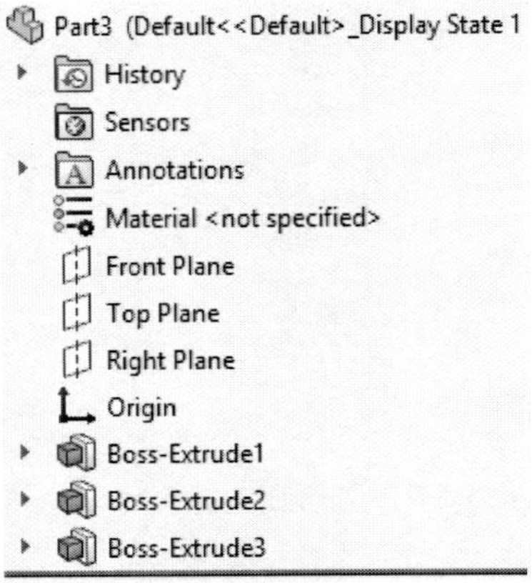

Figure 6-15** The **FeatureManager Design Tree

Saving the Model

1. Choose the **Save** button from the Menu Bar and save the model with the name *c06_tut01* at the location given below:

   \Documents\SOLIDWORKS\c06

2. Choose **File > Close** from the SOLIDWORKS menus to close the document.

Tutorial 2

In this tutorial, you will create the model shown in Figure 6-16. You will use a combination of the conventional modeling method and the contour selection method to create this model. The views and dimensions of the model are given in the same figure. **(Expected time: 30 min)**

The following steps are required to complete this tutorial:

a. Draw the sketch of the front view of the model, refer to Figure 6-17.
b. Extrude the selected contours, refer to Figures 6-18 through 6-20.
c. Add the recess feature to the model by drawing the sketch on the right planar face, refer to Figures 6-21 and 6-22.
d. Create four holes using the cut feature on the top face of the base feature, refer to Figures 6-23 and 6-24.
e. Save and close the document.

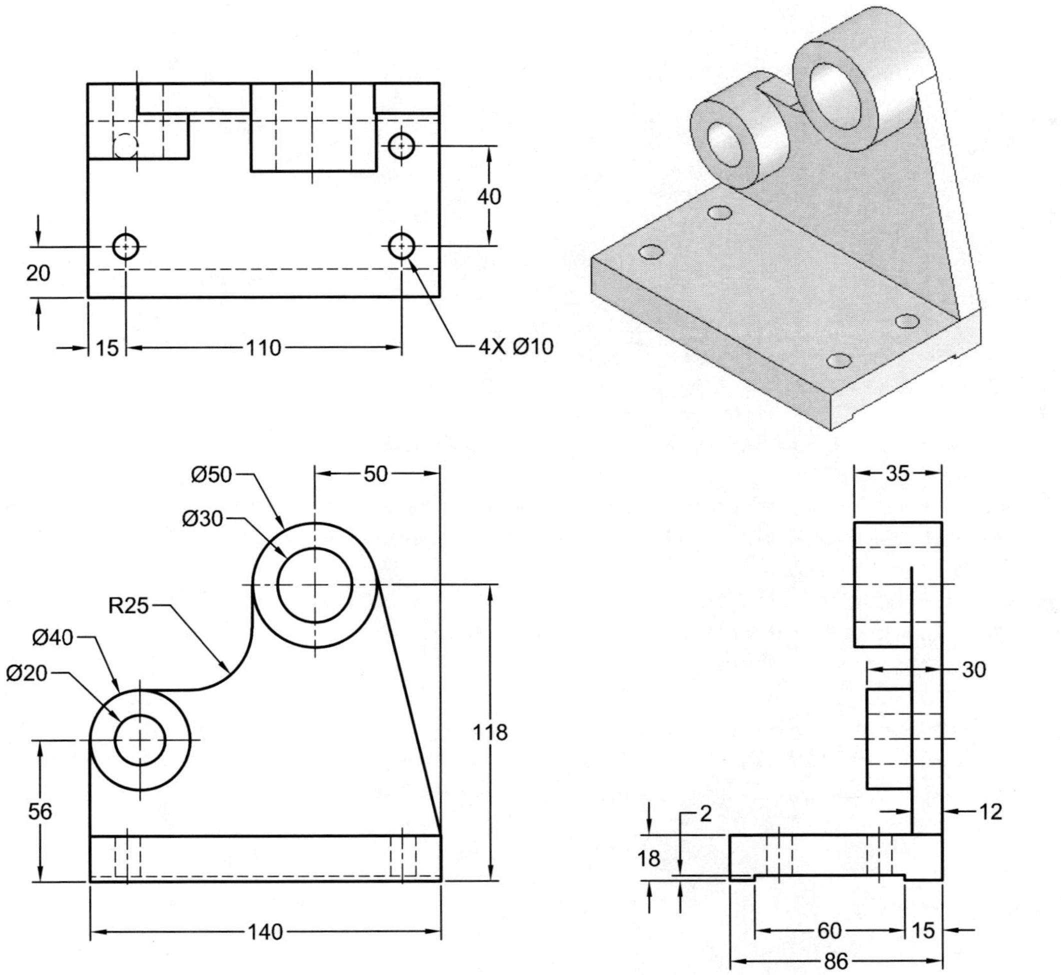

Figure 6-16 Solid model for Tutorial 2

Drawing the Sketch for the Contour Selection Modeling

1. Start a new SOLIDWORKS part document. Draw the sketch of the front view of the model on the Front Plane using the sketching tools.

2. Apply the required relations and dimensions to fully define the sketch, refer to Figure 6-17.

   You need to orient the view to isometric because it helps you in selecting the contours.

3. Press the SPACEBAR key and change the current view to the isometric view.

4. Right-click in the drawing area and choose the **Contour Select Tool** option from the shortcut menu; the select cursor is replaced by the contour selection cursor.

5. Now, by using the contour selection cursor, select the area enclosed by the lower rectangle, as shown in Figure 6-18.

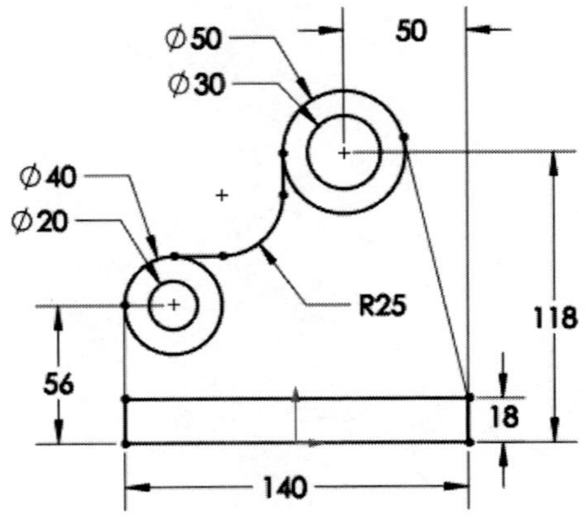

Figure 6-17 The fully defined sketch

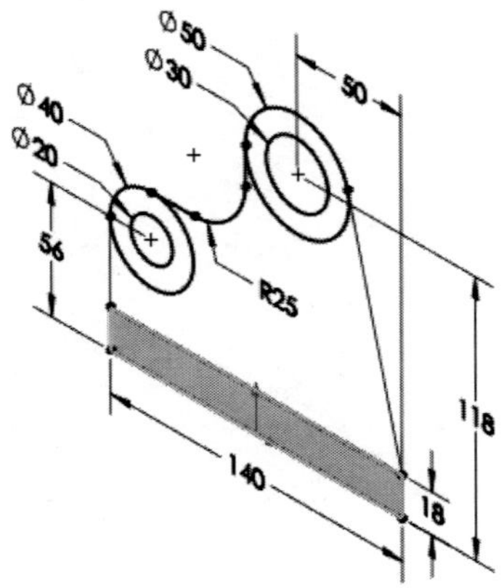

Figure 6-18 Lower rectangle selected as a contour

6. Choose the **Extruded Boss/Base** tool from the **Features CommandManager** or from the pop-up toolbar; the **Boss-Extrude PropertyManager** is displayed.

7. Set the value in the **Depth** spinner to **86** and then choose the **OK** button from the **Boss-Extrude PropertyManager**. The base feature created after extruding the selected contour is shown in Figure 6-19.

8. Use the **Contour Select Tool** and **Extruded Boss/Base** tools to create other features and then hide the sketch. For the depth of the extruded features, refer to Figure 6-16. The model created after extruding all contours and hiding the sketch is shown in Figure 6-20.

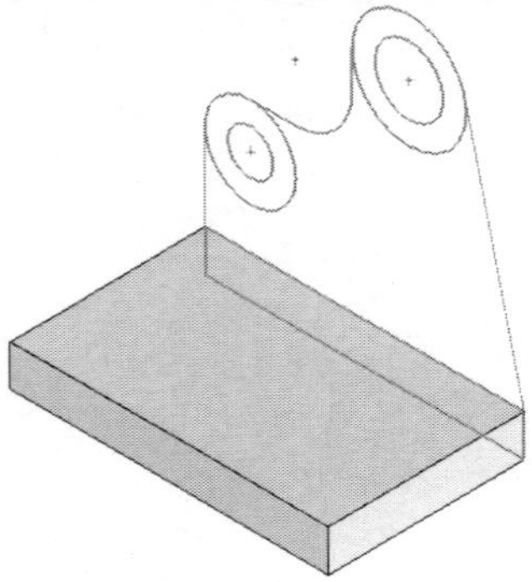

Figure 6-19 Base feature created after extruding the selected contour

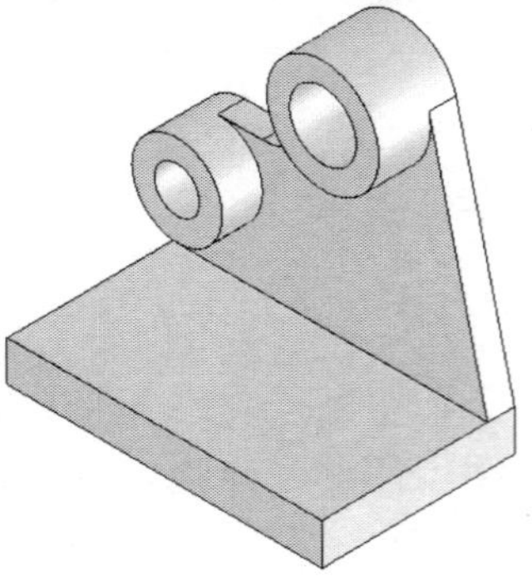

Figure 6-20 Model created after extruding all contours

Creating the Recess on the Base of the Model

After creating the extruded features of the model, you need to create the recess provided at the base of the model. The recess is created as a cut extrude feature. This cut extrude feature is created by drawing a sketch on the right planar face of the model.

1. Select the right planar face of the base feature as the sketching plane; the selected face is highlighted and a pop-up toolbar is displayed.

2. Choose the **Sketch** option from the pop-up toolbar to invoke the sketching environment.

Now, you need to orient the view such that the selected face is normal to your eye view.

3. Choose **View Orientation > Normal To** from the **View (Heads-Up)** toolbar to orient the selected plane normal to the view.

4. Draw the sketch for the recess using the sketching tools and apply the required relations and dimensions to it. The fully defined sketch for the cut feature is shown in Figure 6-21.

5. Choose the **Extruded Cut** tool from the **Features CommandManager** to invoke the **Cut-Extrude PropertyManager**. The preview of the cut feature is displayed in the drawing area in temporary graphics.

6. Right-click in the drawing area and choose the **Through All** option from the shortcut menu.

7. Choose **OK** from the **Cut-Extrude PropertyManager** to complete the feature creation. The isometric view of the model after creating the cut feature is shown in Figure 6-22.

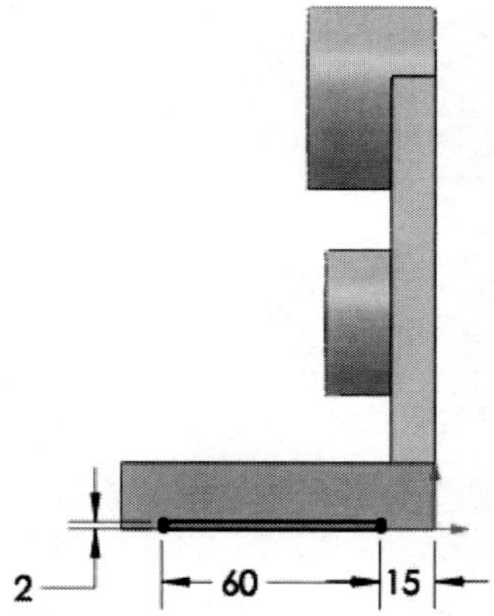

Figure 6-21 *Sketch for the cut feature*

Figure 6-22 *Cut feature added to the model*

Creating Holes

Next, you need to create holes at the base of the model. These holes will be created as the extruded cut features. You need to draw the sketch of the hole feature on the top planar face of the base feature of the model. To draw the sketch of holes, you first need to draw a circle and then create the pattern of remaining circles.

1. Select the top planar face of the base feature; a pop-up toolbar is displayed. Choose the **Sketch** tool from the pop-up toolbar.

2. Orient the current view normal to the viewing direction. Draw a circle of 10 mm diameter and pattern it using the **Linear Sketch Pattern** tool. You may need to apply horizontal relations between the center points of the top circles to fully define the sketch. The fully defined sketch is shown in Figure 6-23.

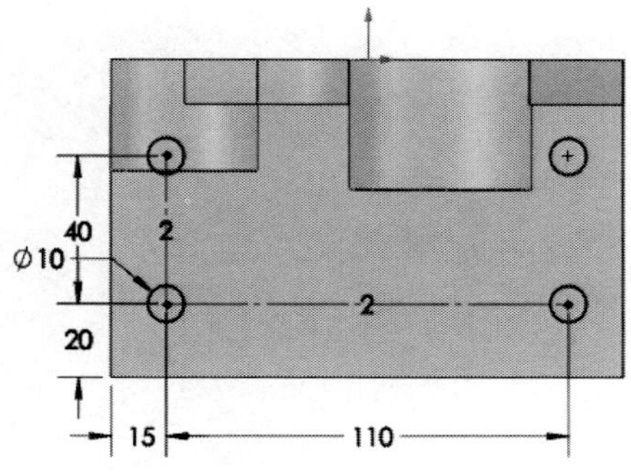

Figure 6-23 *Holes sketched for the cut feature*

3. Change the current view to the isometric view and then choose the **Extruded Cut** button from the **Features CommandManager** to invoke the **Cut-Extrude PropertyManager**.

4. Right-click in the drawing area and choose the **Through All** option from the shortcut menu. Choose the **OK** button from the **Cut-Extrude PropertyManager**. The isometric view of the final model after hiding the sketch is shown in Figure 6-24. The **FeatureManager Design Tree** displaying various features of the model is shown in Figure 6-25.

Figure 6-24 *Final solid model*

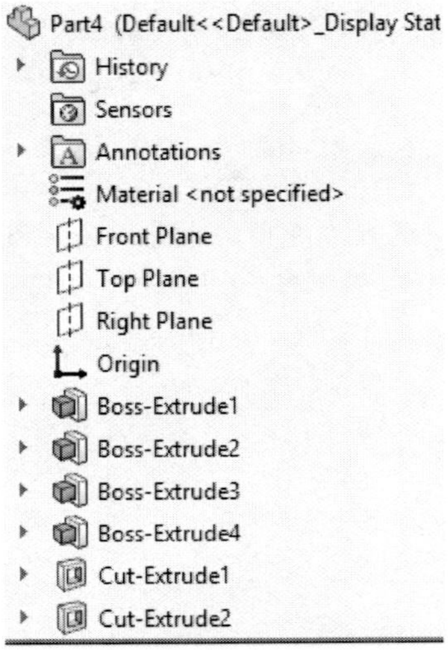

Figure 6-25 *The* ***FeatureManager Design Tree***

Saving the Model

1. Choose the **Save** button from the Menu Bar and save the model with the name *c06_tut02* at the location given below.

   \Documents\SOLIDWORKS\c06

2. Choose **File > Close** from the SOLIDWORKS menus to close the file.

EXERCISES

Exercise 1

Create the solid model shown in Figure 6-26. The dimensions of the model are given in Figure 6-27. **(Expected time: 30 min)**

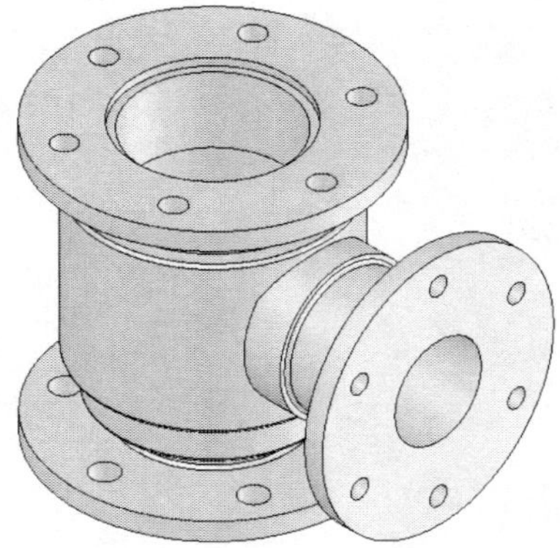

Figure 6-26 *Model for Exercise 1*

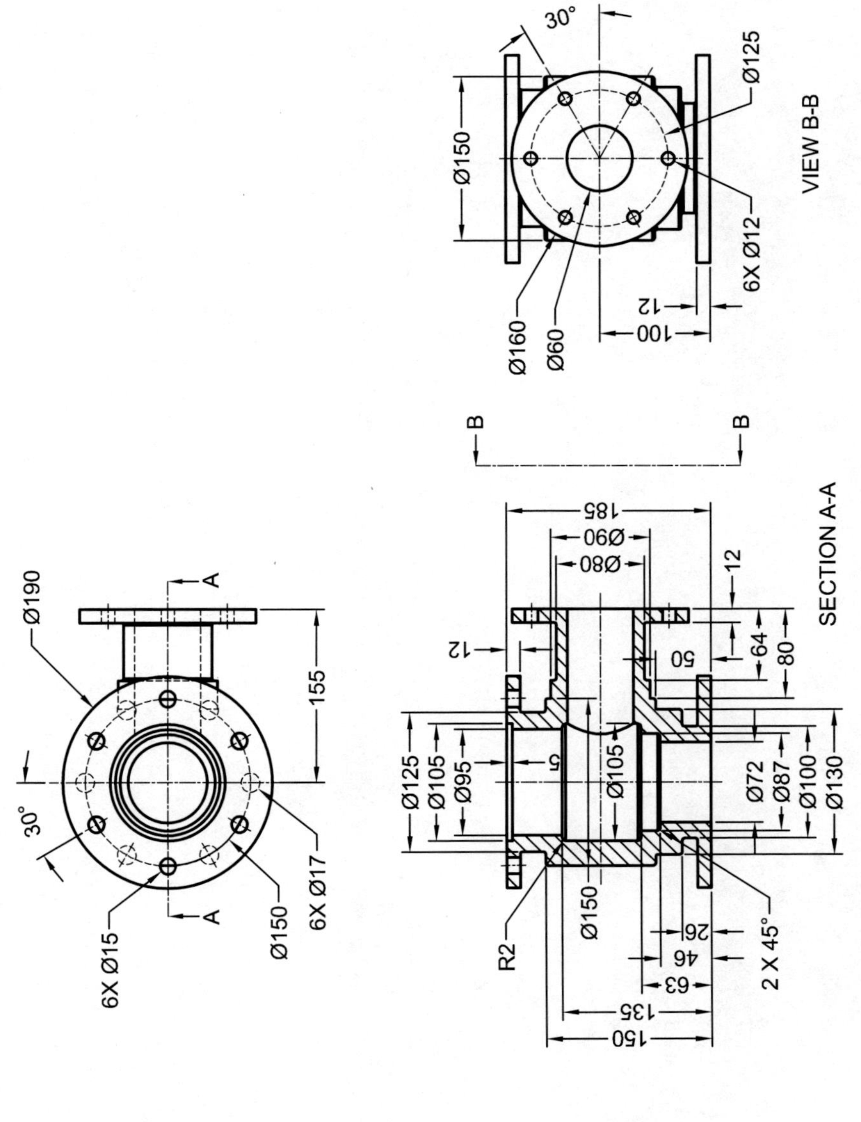

Figure 6-27 *Views and dimensions of the model for Exercise 1*

Exercise 2

Create the model shown in Figure 6-28. The dimensions of the model are given in Figure 6-29. **(Expected time: 30 min)**

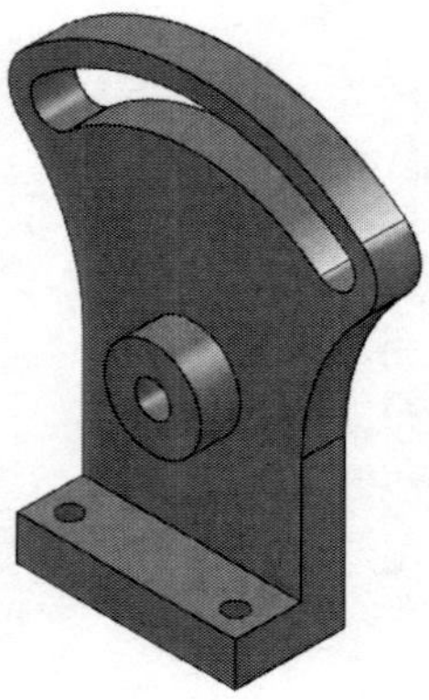

Figure 6-28 Model for Exercise 2

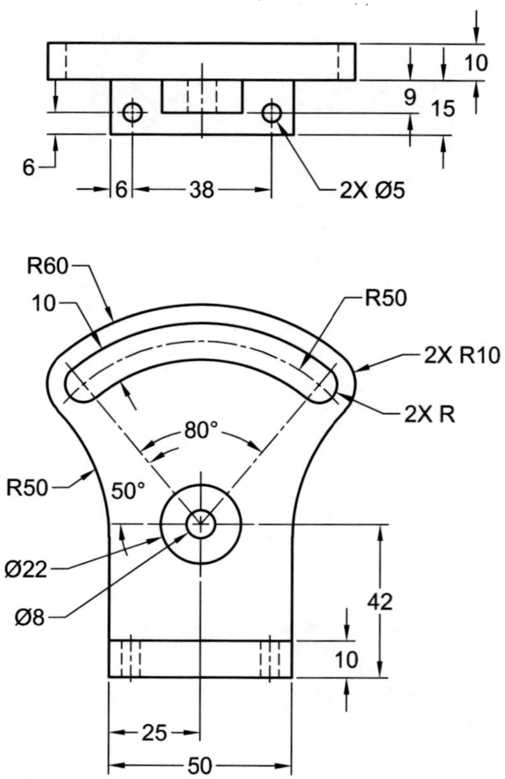

Figure 6-29 Dimensions of the model for Exercise 2

Chapter 7

Advanced Modeling Tools-I

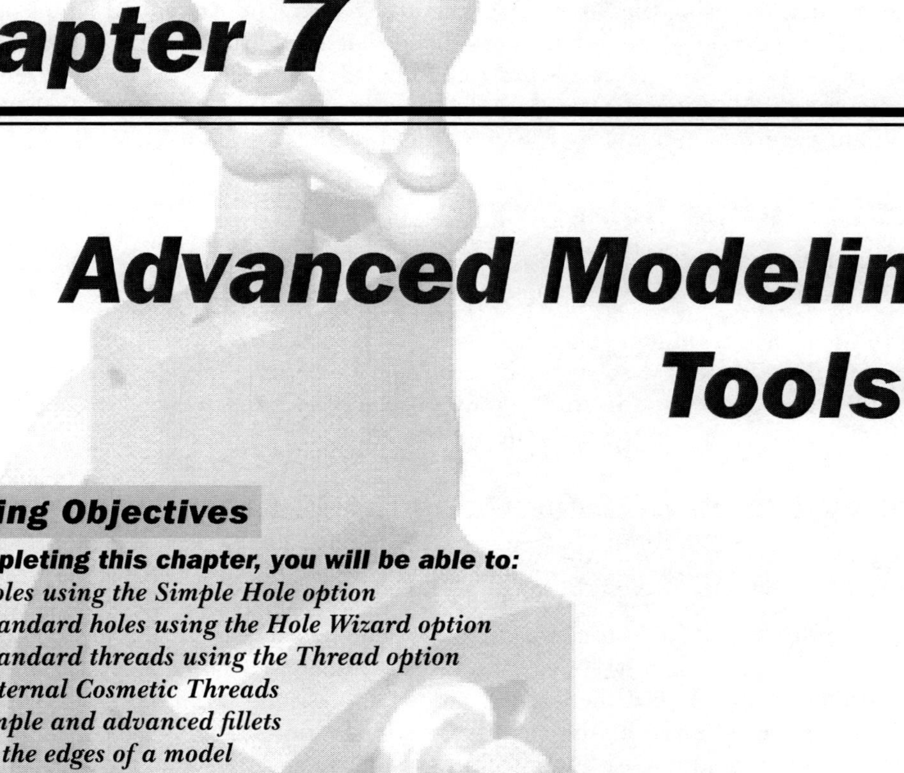

Learning Objectives

After completing this chapter, you will be able to:

- *Create holes using the Simple Hole option*
- *Create standard holes using the Hole Wizard option*
- *Create standard threads using the Thread option*
- *Apply External Cosmetic Threads*
- *Apply simple and advanced fillets*
- *Chamfer the edges of a model*
- *Create the shell feature*
- *Create the wrap feature*

ADVANCED MODELING TOOLS

This chapter discusses various advanced modeling tools available in SOLIDWORKS that assist you in creating a better and accurate design by capturing the design intent in a model. In the previous chapters, you have learned to create holes using the **Extruded Cut** tool. In this chapter, you will learn to create holes using the **Simple Hole**, **Hole Wizard**, and **Advanced Holes** options. The hole wizard is used to create standard holes that are defined based on the industrial standards, screw types, and sizes. The **Hole Wizard** tool of SOLIDWORKS is one of the largest standard industrial virtual hole generation tools available in any CAD package. You will also learn about some other advanced modeling tools such as the thread, fillet, chamfer, shell, and wrap.

Creating Simple Holes

CommandManager: Features > Simple Hole *(Customize to add)*
SOLIDWORKS menus: Insert > Features > Simple Hole

In the previous chapter, you have learned to create holes by extruding a circle using the **Extruded Cut** tool. Now, you will learn how to create a hole feature using the **Simple Hole** tool. If you use this tool, you do not need to draw the sketch of a hole. The holes created using this option act as applied features.

Creating Standard Holes Using the Hole Wizard

CommandManager: Features > Hole Wizard flyout > Hole Wizard
SOLIDWORKS menus: Insert > Features > Hole Wizard

The **Hole Wizard** tool is used to add standard holes such as the counterbore, countersink, drilled, tapped, legacy holes, etc. You can also add a user-defined counterbore drill holes, simple holes, simple drilled holes, tapered holes, and so on. You can control all the parameters of the holes including the termination options. You can also modify the holes according to your requirement after placing them. Therefore, you can place the standard parametric holes using this tool. You can select a face or a plane to place the hole even before invoking this tool.

Creating Advanced Holes

CommandManager: Features > Hole Wizard flyout > Advanced Hole
SOLIDWORKS menus: Insert > Features > Advanced Hole

The **Advanced Hole** tool is used to add combination of standard holes such as counterbore, countersink, drilled, straight, and tapered tap from near and side faces of the model.

Creating Threads

CommandManager: Features > Hole Wizard flyout > Thread
SOLIDWORKS menus: Insert > Features > Thread

The **Thread** tool is used to create helical threads on cylindrical faces. You can also store custom thread profile as a library feature and then use it later as per your requirement. This tool enables you to modify the parameters of the threads even after their creation. Thus, you can create standard parametric threads with this tool.

Adding External Cosmetic Threads

CommandManager: Features > Cosmetic Thread (*Customize to Add*)
SOLIDWORKS menus: Insert > Annotations > Cosmetic Thread

In SOLIDWORKS, you can also add external cosmetic threads to cylindrical surfaces to give them a realistic look like threads. To do so, you need to invoke the **Cosmetic Thread** tool. This tool is not available by default in the **Features CommandManager**. Therefore, you need to add this tool from the list of the **Annotation** commands given in the **Customize** dialog box.

Creating Fillets

CommandManager: Features > Fillet flyout > Fillet
SOLIDWORKS menus: Insert > Features > Fillet/Round

In SOLIDWORKS, you can add fillets to a model as a feature using the **Manual** or **FilletXpert** option. The **Manual** option is used to fillet an internal or external face or the edge of a model. You can preselect the face, edge, or feature to which the fillet has to be added. You can also select the entity to be filleted after invoking the **Fillet** tool.

Creating Fillets Using the FilletXpert

CommandManager: Features > Fillet flyout > Fillet
SOLIDWORKS menus: Insert > Features > Fillet/Round

The **FilletXpert** tool allows you to create single or multiple fillets, change the existing fillet, and create fillets at corners.

Creating Chamfers

CommandManager: Features > Fillet flyout > Chamfer
SOLIDWORKS menus: Insert > Features > Chamfer

Chamfering is a process in which the sharp edges are beveled in order to reduce the area of stress concentration. This process also eliminates the undesirable sharp edges and corners. In SOLIDWORKS, a chamfer is created using the **Chamfer** tool.

Creating Shell Features

CommandManager: Features > Shell
SOLIDWORKS menus: Insert > Features > Shell

Shelling is a process in which the material is scooped out from a model. The resulting model will be a hollow model with walls of a specified thickness and a cavity inside. The selected face or the faces of the model are also removed in this operation. If you do not select a face to be removed, a closed hollow model will be created. You can also specify multiple thicknesses to the walls.

Creating Wrap Features

CommandManager: Features > Wrap
SOLIDWORKS menus: Insert > Features > Wrap

The **Wrap** tool is used to emboss, deboss, or scribe a closed multiloop sketch on a selected planar, curved, or splined face. There are two methods for creating wrap features: Analytical and Spline Surface. For planar or curved surfaces, use Analytical wrap and for spline surfaces, use Spline Surface wrap. For Analytical wrap, the planar or curved face must be tangent to the plane on which the selected sketch is created. Using the Spline Surface method, you cannot create wrap around bodies. You can create a wrap feature on multiple faces. You can also create this type of geometry by extruding the sketch from a selected surface using the **Extruded Boss/Base** or **Extruded Cut** tool. There are two main differences between the emboss and the deboss features created using the **Wrap** tool and those created using the **Extruded Boss/Base** or **Extruded Cut** tool. The first difference is in the method of projection. The projection of the geometry using the **Wrap** tool follows the rule of true length, which means the actual length of the geometry remains the same after projecting it on the surface. The projection of the geometry created using the **From** option of the **Extruded Boss/Base** or **Extruded Cut** tool follows the rule of true projection. Therefore, the original size of the geometry is distorted. The second difference is the direction of the side faces of the geometry. The side faces of the geometry created using the **Wrap** tool are always normal to the reference surface, while those created using the **From** option of the **Extruded Boss/Base** or the **Extruded Cut** tool are normal to the sketching plane or parallel to the direction vector.

TUTORIALS

Tutorial 1

In this tutorial, you will create the model of the Plummer Block Casting shown in Figure 7-1. The dimensions of the model are shown in Figure 7-2. **(Expected time: 30 min)**

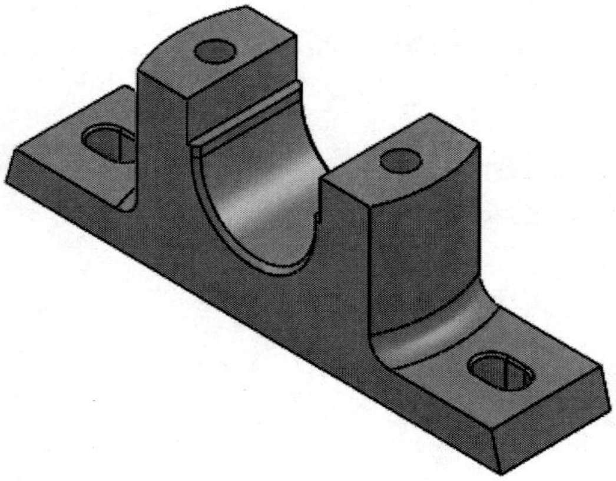

Figure 7-1 *Solid model of the Plummer Block Casting*

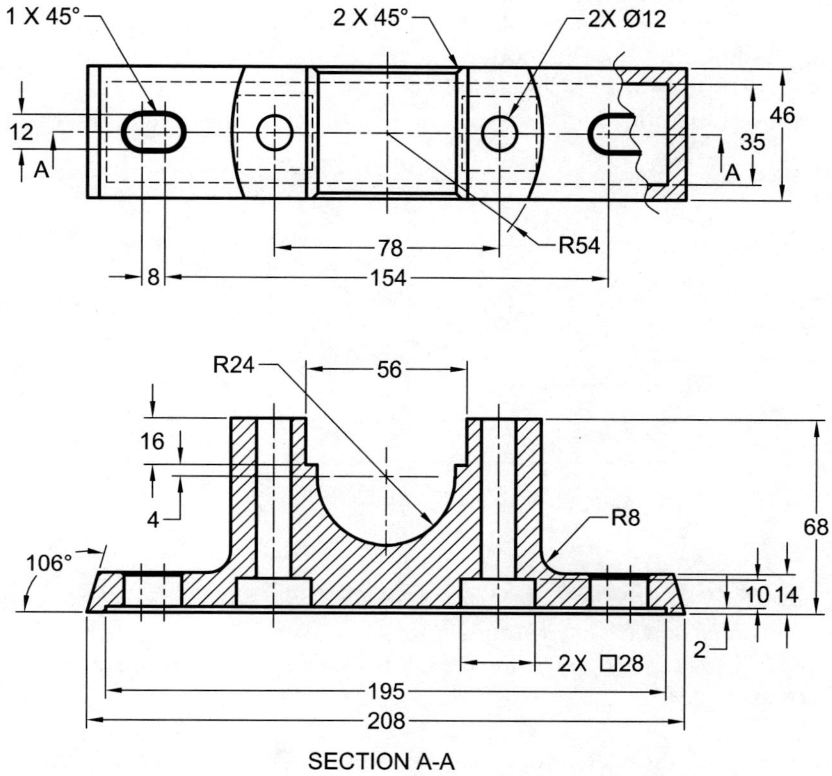

Figure 7-2 Dimensions of the Plummer Block Casting

The following steps are required to complete this tutorial:

a. Create the base feature of the model on the Front Plane, refer to Figures 7-3 and 7-4.
b. Create the second feature, which is a cut feature on the top planar face of the base feature, refer to Figures 7-5 through 7-6.
c. Create the rectangular recess as a cut feature at the bottom of the base.
d. Create the square cuts that will act as the recess for the head of the square head bolts, refer to Figure 7-8.
e. Create the hole feature using the **Hole PropertyManager** and modify the placement of the hole feature, refer to Figure 7-9.
f. Create the cut feature on the second top planar face of the base feature, refer to Figure 7-10.
g. Add the fillet feature to the model, refer to Figures 7-11 and 7-12.
h. Add chamfers to the model, refer to Figures 7-13 and 7-14.
i. Save the model.

Creating the Base Feature

1. Start a new SOLIDWORKS part document using the **New SOLIDWORKS Document** dialog box.

2. Invoke the **Extruded Boss/Base** tool and select the **Front Plane** as the sketching plane.

3. Draw the sketch of the front view of the model. Use the **Dynamic Mirror Entities** tool to capture the design intent of the model.

4. Add the required relations and dimensions to fully define the sketch, refer to Figure 7-3.

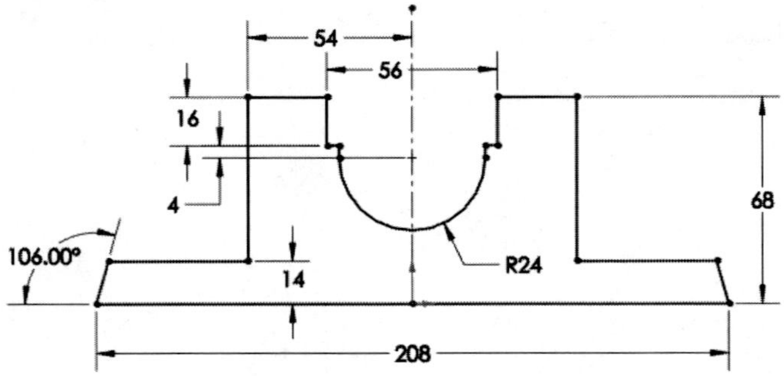

Figure 7-3 *Fully defined sketch of the base feature*

Note

After adding the dimensions to the sketch, select the radial dimension and add tolerance to them using the ***Dimension PropertyManager****, as discussed in the earlier chapters. Also, change the radial dimension of the arc into the diameter dimension by using the* ***Leaders*** *tab of the* ***Dimension PropertyManager****.*

You need to extrude the sketch to a distance of 46 mm using the **Mid Plane** option so that the parts to be assembled have the default planes at the center of the model.

5. Exit the sketching environment to display the **Boss-Extrude PropertyManager**.

6. Right-click in the drawing area and choose the **Mid Plane** option from the shortcut menu to extrude the sketch symmetrically on both the sides of the sketching plane.

7. Set **46** in the **Depth** spinner and choose the **OK** button from the **Boss-Extrude PropertyManager**.

8. Change the view orientation to isometric. The resulting base feature after extruding the sketch to a given depth is shown in Figure 7-4.

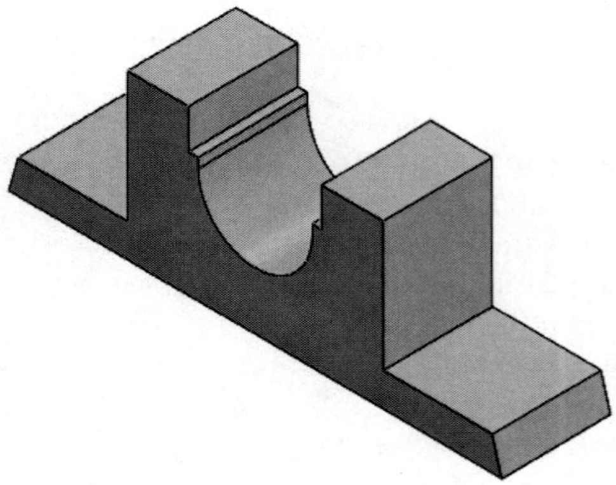

Figure 7-4 *Isometric view of the base feature of the model*

Tip

Sometimes you need to set tolerance for the dimensions of a feature. For example, consider a case where the depth of the extruded feature has a tolerance applied to it. In such cases, double-click on the extruded feature in the ***FeatureManager Design Tree*** *or in the drawing area. On doing so, all dimensions applied to the model will be displayed. Double-click on the dimension that reflects the depth of the extruded feature and apply the tolerance using the* ***Dimension PropertyManager****.*

Creating the Second Feature

The second feature of the model is a cut-extrude feature. The sketch of the cut-extrude feature will be drawn on the top planar face of the base feature. This sketch will be extruded up to the specified plane to create the resulting cut feature.

1. Invoke the **Extruded Cut** tool and select the top planar face of the base feature as the sketching plane; the sketching plane orients parallel to screen.

2. Draw the sketch for the cut feature using the sketching tools. Apply the required relations to the sketch.

   The sketch of the cut-extrude feature is shown in Figure 7-5.

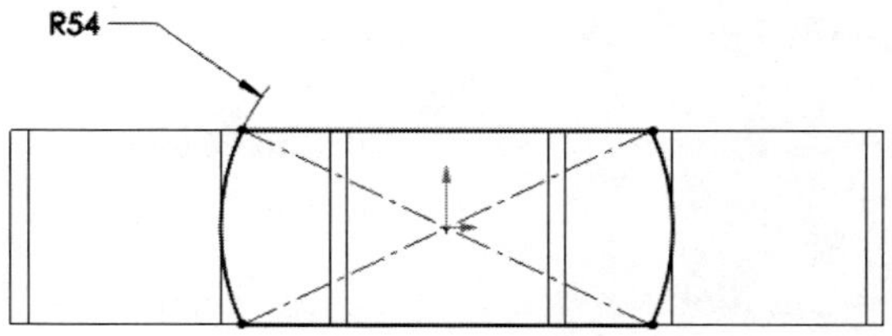

Figure 7-5 *Sketch of the second feature*

3. Exit the sketching environment to display the **Cut-Extrude PropertyManager**.

   The preview of the cut feature is displayed in the drawing area with the default values of the **Blind** option. You need to extrude the cut feature up to the selected surface. Therefore, orient the model in the isometric view so that the feature termination surface can be selected easily in this view.

4. Change the view orientation to isometric.

   You will observe that the preview of the cut feature is inside the model. You need to remove the outer part of the sketch profile by changing the side of the cut feature.

5. Select the **Flip side to cut** check box; the preview of the cut feature is modified dynamically.

6. Right-click in the drawing area and choose **Up To Surface** from the shortcut menu; you are prompted to select a face or a surface to complete the specification of direction 1.

7. Select the surface for feature termination, refer to Figure 7-6. Right-click and choose **OK** from the shortcut menu to exit the tool.

   The model after creating the cut feature is shown in Figure 7-7.

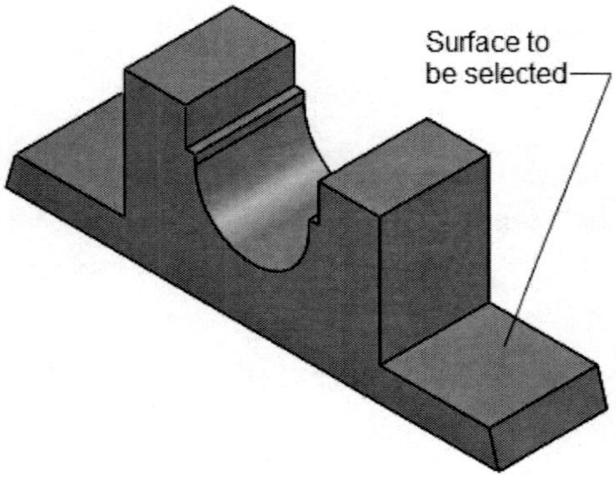

Figure 7-6 *Surface to be selected for the cut feature*

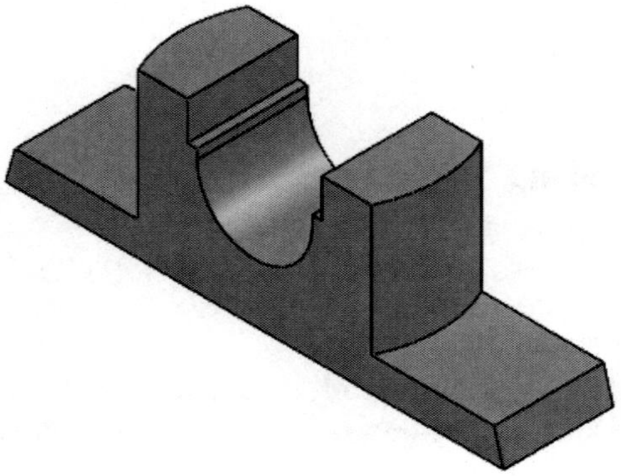

Figure 7-7 *Model after creating the cut feature*

Creating the Rectangular Recess

The third feature is the rectangular recess. This feature will be created by using a rectangular cut feature created on the bottom face of the model.

1. Orient the model using the **Rotate View** tool and select the bottom face of the base feature as the sketching plane.
2. Choose the **Extruded Cut** tool; the sketching environment is invoked.
3. Use standard sketching tools to draw a rectangle of dimension 195 mm x 35 mm as the sketch for the rectangular recess. Apply the required relations and dimensions to the sketch.
4. Exit the sketching environment.
5. Set the value to **2** in the **Depth** spinner and choose **OK** from the **Cut-Extrude PropertyManager**.

Creating the Recess for the Head of the Square-headed Bolt

It is evident from Figure 7-2 that the bolt to be inserted in the part is a square-headed bolt. Therefore, you need to create the recess for the head of this bolt. A square of 28 mm length will be used to create this recess.

1. Rotate the model and select the upper face of the recess created in the previous section as the sketching plane. Invoke the **Extruded Cut** tool; the model orients normal to the sketching plane.

2. Draw a sketch using the sketching tools. The sketch includes two squares of 28 mm length with the distance between their centers as 78 mm, refer to Figure 7-2. Apply the required relations and dimensions to fully define the sketch.

3. Exit the sketching environment and extrude the sketch to a depth of 10 mm. Exit the tool. The rotated model after adding this cut feature is shown in Figure 7-8.

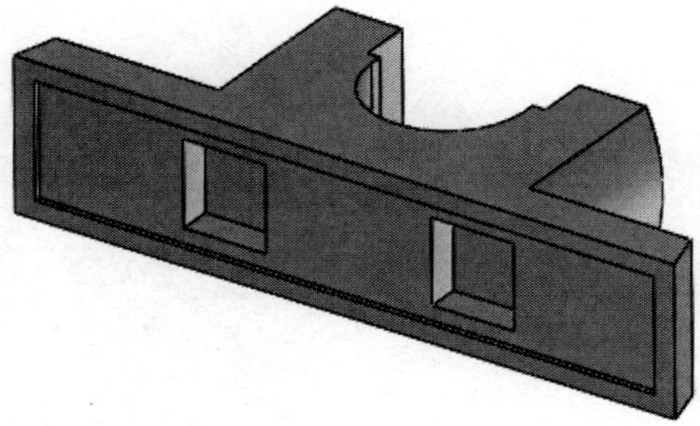

Figure 7-8 *Model after creating recess for the square-headed bolts*

Creating the Hole Features

After creating the recess for the head of the square-headed bolt, you need to create a hole. You can create a hole using the **Hole Wizard** tool. However, for this tutorial, you will create the hole using the **Simple Hole** tool.

1. Select the top planar surface of the base feature as the placement plane for the hole feature.

2. Choose **Insert > Features > Simple Hole** from the SOLIDWORKS menus to invoke the **Hole PropertyManager**; the preview of the hole feature with the default settings is displayed in the drawing area.

3. Right-click in the drawing area and choose the **Through All** option from the shortcut menu to specify the feature termination.

4. Set **12** in the **Hole Diameter** spinner.

5. Choose the **OK** button from the **Hole PropertyManager**.

6. Select **Hole1** from the **FeatureManager Design Tree**; a pop-up toolbar is displayed.

7. Choose **Edit Sketch** from the pop-up toolbar; the sketching environment is displayed. Apply relations and dimensions (refer to Figure 7-2) to locate the hole feature.

8. Press CTRL+B to rebuild the model.

9. Use the same procedure as discussed above to create the second hole feature on the model. The model after creating both hole features is displayed in Figure 7-9.

10. Create the cut feature on the second top planar face of the base feature. Refer to Figure 7-6 for the surface to be selected and Figure 7-2 for the dimensions of the sketch.

    The model after creating the cut features is shown in Figure 7-10.

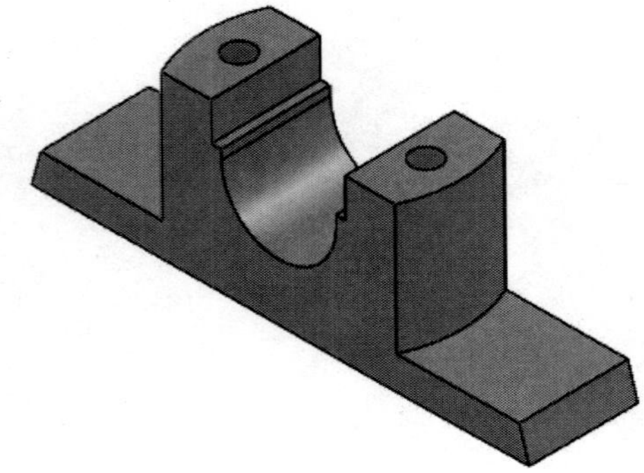

Figure 7-9 Model after creating the hole features

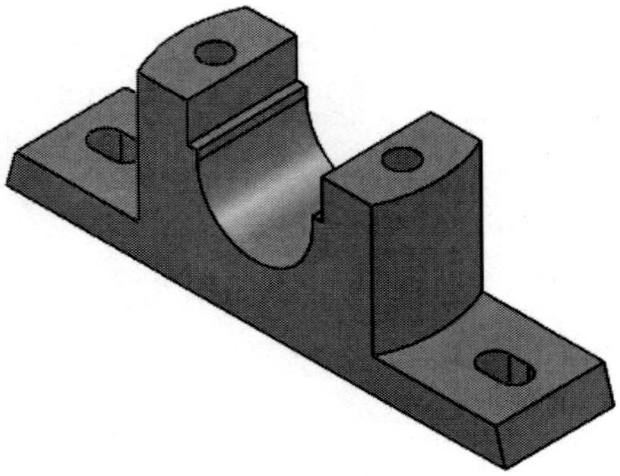

Figure 7-10 Model after creating the cut features

Tip
In this tutorial, you will draw the sketch of the slot on one side, mirror it, and then create the cut feature. You can also create a cut feature on one side and then mirror it on the other side. Mirroring a feature will be discussed in the later chapters.

Adding a Fillet to the Model

Now, you need to add the fillet feature to the model.

1. Choose the **Fillet** button from the **Fillet** flyout in the **Features CommandManager** to invoke the **Fillet PropertyManager**. But if the **FilletXpert PropertyManager** is displayed then choose the **Manual** button.

On invoking the **Fillet PropertyManager**, you are prompted to select the edges, faces, features, or loops to be filleted. It is evident from the model that you need to select only the edges to apply the fillet feature.

2. Choose the **Constant Size Fillet** button from the **Fillet PropertyManager** and then select the edges to be filleted, as shown in Figure 7-11.

As soon as you select the edges, the preview of the fillet feature with default values appears in the drawing area. Also, a radius callout is displayed along the selected edge.

Note
If the preview of the fillet feature is not displayed in the drawing area, select the ***Full Preview*** *radio button in the* ***Items To Fillet*** *rollout of the* ***Fillet PropertyManager****.*

Now, you need to modify the default radius value of the fillet feature.

3. Set the value to **8** in the **Radius** spinner and choose the **OK** button from the **Fillet PropertyManager**.

The model after adding the fillet is shown in Figure 7-12.

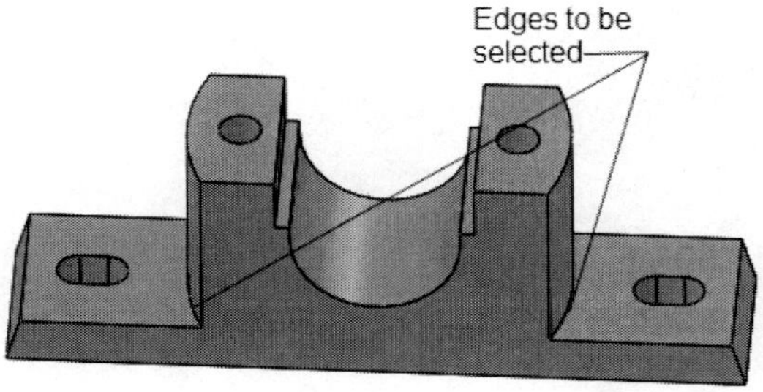

Figure 7-11 Edges to be selected for the fillet feature

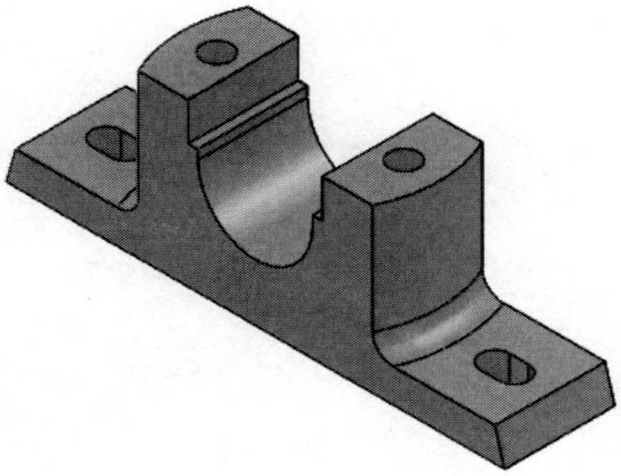

Figure 7-12 Model after adding the fillet

Adding Chamfers to the Model

The next feature that you need to add to this model is a chamfer feature.

1. Choose the **Chamfer** button from the **Fillet** flyout in the **Features CommandManager** to invoke the **Chamfer PropertyManager**.

2. Select the edges of the cut features, as shown in Figure 7-13. On doing so, the edges that are tangent to the selected edges are selected automatically because by default the selection mode of the chamfer feature uses the tangent propagation.

   As soon as you select the edges, the preview of the chamfer feature is displayed in the drawing area with default values. The angle and distance callouts are also displayed. Next, you need to set the chamfer parameters.

   You need to create chamfer by using 1 mm distance and 45 degrees angle. The **Angle Distance** button is selected by default in the **Chamfer Type** rollout. The value of the angle in the **Angle** spinner is set to 45 degrees by default. Therefore, you do not need to modify this value. You need to set only the value of the distance in the **Distance** spinner.

3. Set the value to **1** in the **Distance** spinner and choose the **OK** button from the **Chamfer PropertyManager**. Refer to Figure 7-2 for the parameters of the chamfer feature.

4. Similarly, create the chamfer of dimension 2 X 45° on the base feature, refer to Figure 7-2. The final solid model is shown in Figure 7-14.

Saving the Model

Next, you need to save the document.

1. Choose the **Save** button from the Menu Bar and save the document with the name *c07_tut01* at the location given next.

   \Documents\SOLIDWORKS\c07

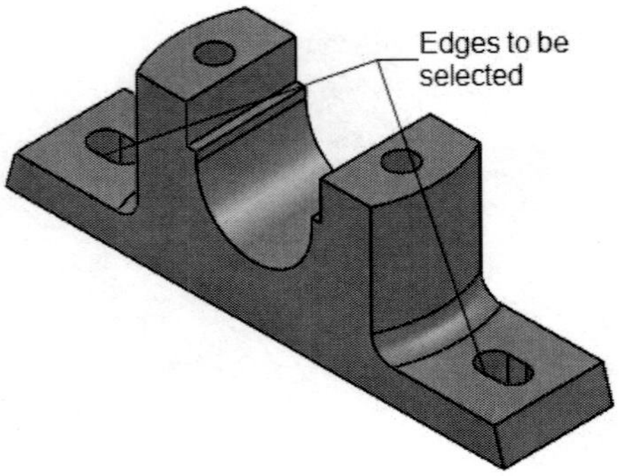

Figure 7-13 Edges to be selected

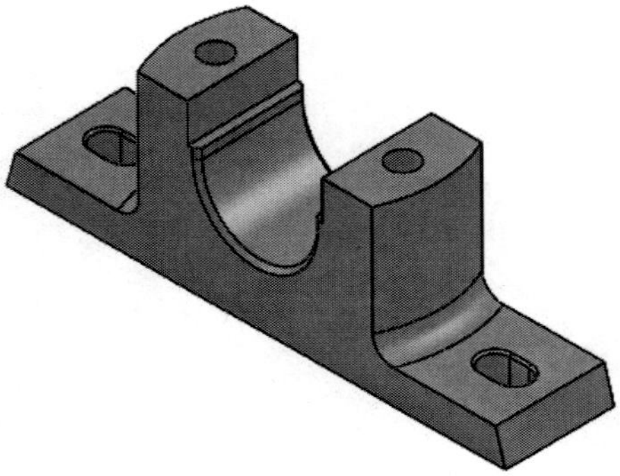

Figure 7-14 Final solid model

2. Choose **File > Close** from the SOLIDWORKS menus to close the file.

Tutorial 2

In this tutorial, you will create the model shown in Figure 7-15. For a better understanding, the section view of the model is shown in Figure 7-16. The dimensions of the model are shown in Figure 7-17. **(Expected time: 30 min)**

The following steps are required to complete this tutorial:

a. Create the base feature of the model by extruding a rectangle of 100 mm x 70 mm to a distance of 20 mm.
b. Add a fillet to the base feature, refer to Figures 7-18 through 7-21.
c. Create the shell feature to create a thin-walled part and remove some of the faces, refer to Figures 7-22 and 7-23.
d. Create a reference plane at an offset distance from the top planar face of the base feature and extrude the sketch created on the new plane, refer to Figure 7-24.
e. Use the **Hole Wizard** tool to add the countersink hole to the model, refer to Figure 7-25.
f. Add a fillet to the extruded feature, refer to Figures 7-26 and 7-27.
g. Create the lip of the component by extruding the sketch, refer to Figure 7-28.
h. Save the model.

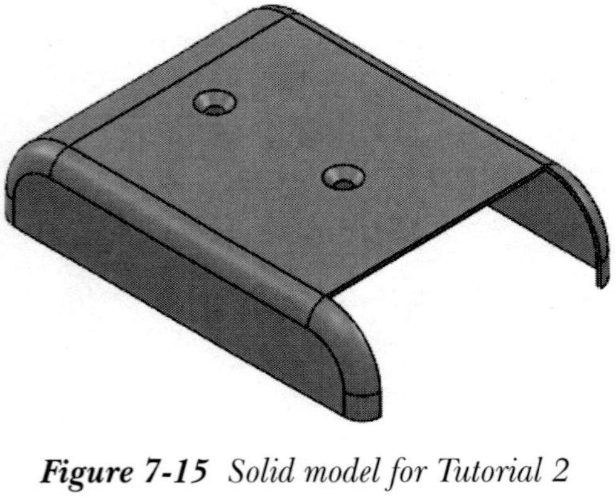

Figure 7-15 Solid model for Tutorial 2

Figure 7-16 Section view of the model

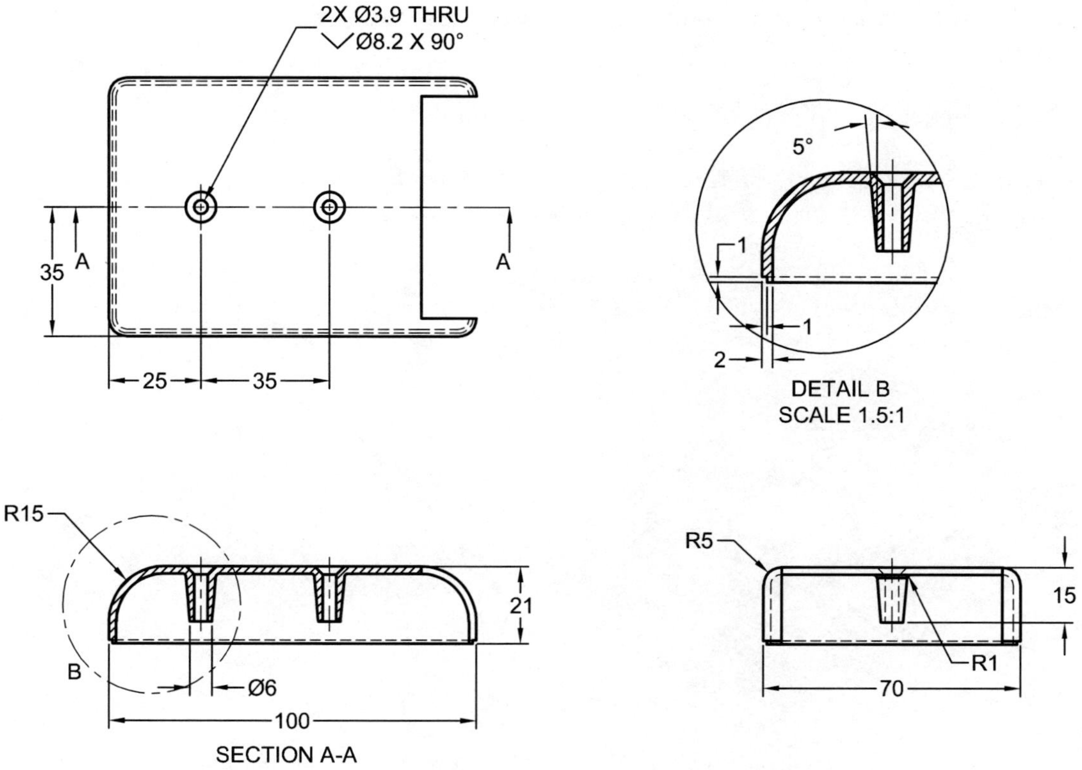

Figure 7-17 *Top view, front section view, and right side view with dimensions*

Creating the Base Feature

The base feature of the model will be created by extruding a rectangle of 100 mm x 70 mm to a distance of 20 mm. It is evident from the model that the sketch of the base feature is created on the Top Plane. Therefore, you need to select the Top Plane as the sketching plane.

1. Start a new SOLIDWORKS part document and invoke the **Extruded Boss/Base** tool. Select the **Top Plane** from the **FeatureManager Design Tree** or from the drawing area.

2. Orient the sketch plane normal to the viewing direction, if it is not oriented by default.

3. By using the **Rectangle** tool, draw a rectangle of size 100 mm x 70 mm. Add the other required dimensions to fully define the sketch.

4. Exit the sketching environment and extrude the rectangle to a depth of 21 mm.

Creating the Fillet Features

After creating the base feature, you need to add fillets to the model. In this model, you need to add three fillet features. Two fillet features will be added at this stage of the design process and the remaining one will be added at the later stage of the design process.

1. Choose the **Fillet** button from the **Features CommandManager** to invoke the **Fillet PropertyManager**.

2. Select the edges of the model, as shown in Figure 7-18.

   As soon as you select the edges of the model, the preview of the fillet with default values and the radius callout is displayed in the drawing area.

3. Set the value in the **Radius** spinner to **15** and choose the **OK** button from the **Fillet PropertyManager**. Figure 7-19 shows the model after adding the first fillet feature.

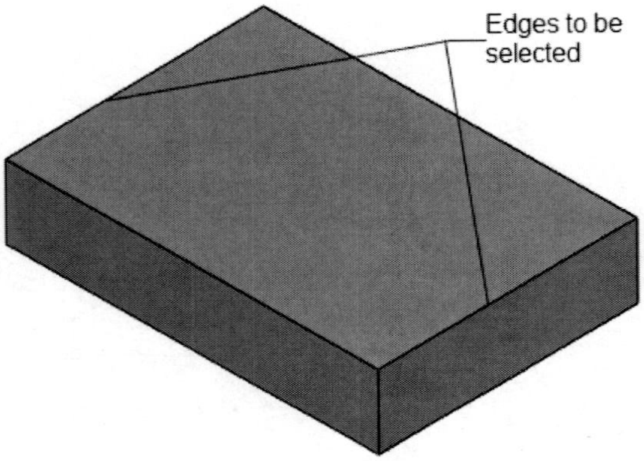

Figure 7-18 Edges to be selected

Figure 7-19 Fillet added to the base feature

Now, you need to add the second fillet feature to the model.

4. Again, invoke the **Fillet PropertyManager** and set the value to **5** in the **Radius** spinner.

5. Select the edges of the model, as shown in Figure 7-20. Right-click and then choose **OK** from the shortcut menu to complete the feature creation. The model after adding the second fillet feature is shown in Figure 7-21.

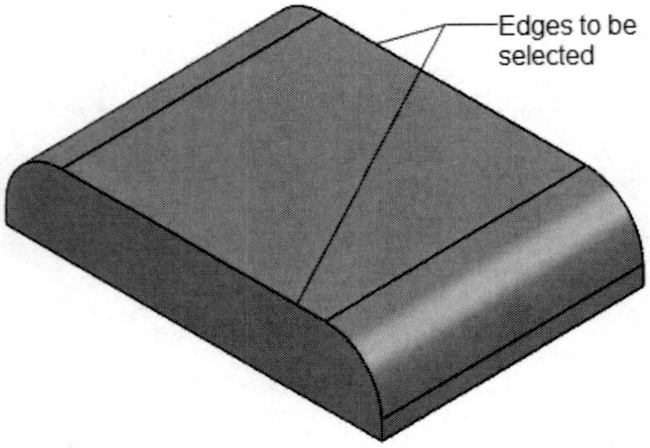

Figure 7-20 Edges to be selected

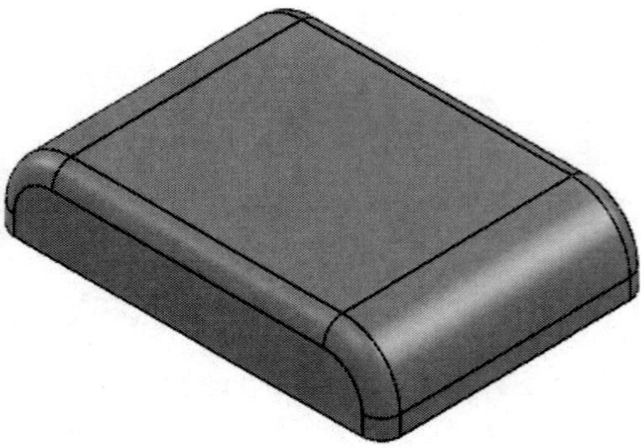

Figure 7-21 Second fillet added to the model

Note

You can also create two fillets in a single step by selecting the ***Multiple radius*** *check box in the* ***Fillet PropertyManager*** *and specifying the appropriate radius.*

Creating the Shell Feature

It is evident from Figures 7-15 and 7-16 that you need to create a thin-walled structure. This will be created using the **Shell** tool. As discussed earlier, the **Shell** tool is used to scoop out the material from the model, leaving behind a thin-walled hollow part.

1. Choose the **Shell** button from the **Features CommandManager**; the **Shell PropertyManager** is displayed and you are prompted to select the faces to be removed. 

2. Rotate the model and select the faces to be removed, as shown in Figure 7-22; the names of the selected faces are displayed in the **Faces to Remove** selection box.

3. Set the value to **2** in the **Thickness** spinner and choose the **OK** button from the **Shell PropertyManager**.

   The model after creating the shell feature is shown in Figure 7-23.

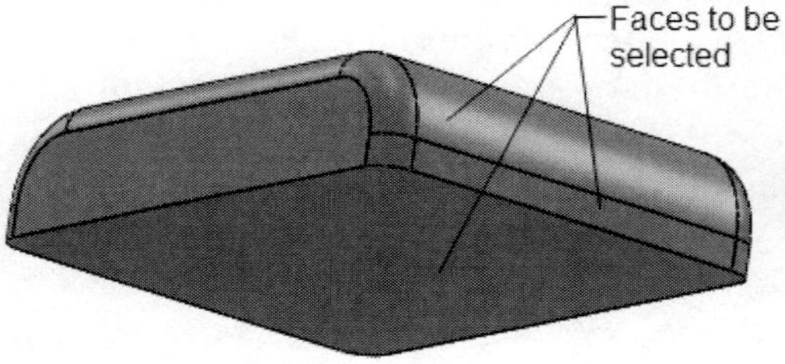

Figure 7-22 *Faces selected to be removed*

Figure 7-23 *Model after creating the shell feature*

Creating the Extruded Feature

The next feature that you need to create is an extruded feature. Before creating this feature, you need to create a reference plane at an offset distance from the top planar face of the base feature.

1. Invoke the **Plane PropertyManager**, select the top planar face of the base feature as the first reference, and create a plane at an offset distance of 15 mm. You need to select the **Flip offset** check box from the **Plane PropertyManager** to reverse the direction of plane creation. Choose the **OK** button from the **Plane PropertyManager** to close it.

2. Invoke the **Extruded Boss/Base** tool, select the newly created plane as the sketching plane, and create the sketch using the standard sketching tools. The sketch consists of two circles of 6 mm diameter. For other dimensions, refer to Figure 7-16.

3. Exit the sketching environment and extrude the sketch using the **Up To Next** option and activate the **Draft on/off** button. Next, add an outward draft of 5-degree in the **Draft Angle**

edit box. Choose the **OK** button from the **Boss-Extrude PropertyManager** to close it. Also, hide the reference plane.

Figure 7-24 shows the rotated model after creating the extruded feature with draft and hiding the reference plane.

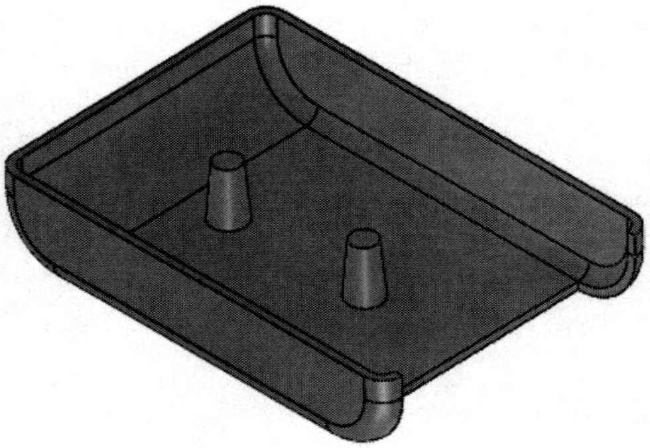

Figure 7-24 *Model after creating the extruded feature*

Adding the Countersink Hole using the Hole Wizard Tool

The next feature that you need to create is a countersink hole, refer to Figure 7-16. In SOLIDWORKS, you are provided with one of the largest standard hole-generating tools known as **Hole Wizard**. You can use the **Hole Wizard** tool to add standard holes to the model so that the holes can accommodate standard fasteners.

1. Choose the **Hole Wizard** button in **Hole Wizard** flyout from the **Features CommandManager**; the **Hole Specification PropertyManager** is displayed.

2. Choose the **Countersink** button from the **Hole Type** rollout. Now, set the parameters to define the standard hole.

3. Select the **ANSI Metric** option from the **Standard** drop-down list.

4. Select the **Flat Head Screw - ANSI B18.6.7M** option from the **Type** drop-down list.

5. Select the **M3.5** option from the **Size** drop-down list and the **Normal** option from the **Fit** drop-down list in the **Hole Specifications** rollout.

6. Select the **Through All** option from the **End Condition** rollout.

7. Choose the **Positions** tab from the **Hole Specification PropertyManager**; you are prompted to select the face to place the hole.

8. Move the point cursor on the top planar face and select the planar face; the select cursor is replaced by the point cursor. Also, you are prompted to use the dimensions and other sketching tools to position the hole.

9. Specify a point for the placement of the left countersink hole. Similarly, specify one more point for the placement of the right countersink hole.

   If you place the points anywhere in the top planar face, you need to add the required relations and dimensions to define the location of these points. Before doing that, you need to change the model display from **Shaded With Edges** to **Hidden Lines Visible** for a better visibility.

10. Right-click in the drawing area and choose the **Select** option.

11. Choose the **Hidden Lines Visible** button from the **Display Style** flyout in the **View (Heads-Up)** toolbar to display the model with the hidden lines visible.

12. Select the left placement point and right-click to choose the **Add Relation** option from the **Sketch Tools** cascade menu to display the **Add Relations PropertyManager**.

13. Select the upper left hidden circle and choose the **Concentric** button from the **Add Relations** rollout.

14. Right-click in the drawing area and choose **Clear Selections** from the shortcut menu. Select the right placement point and the upper right hidden circle. Choose the **Concentric** button from the **Add Relations** rollout. Now, choose **OK** from the confirmation corner.

15. Choose the **OK** button from the **Hole Position PropertyManager**. Next, choose the **Shaded With Edges** button from the **View** toolbar.

   The isometric view of the model after adding the hole feature is shown in Figure 7-25.

Adding a Fillet to the Model

Now, you need to add the fillet to the edges of the extruded feature with the draft that was created earlier.

1. Rotate the model and choose the **Fillet** button; the **Fillet PropertyManager** is displayed.

2. Choose the **FilletXpert** button; the **FilletXpert PropertyManager** is displayed with the **Add** tab chosen by default.

3. Select one of the edges of the draft, as shown in Figure 7-26, and do not move the mouse; a pop-up toolbar is displayed. Remember that if you move the cursor away from the edge after selecting it, the pop-up toolbar will disappear.

Figure 7-25 Model after adding the hole feature using the ***Hole Wizard*** *tool*

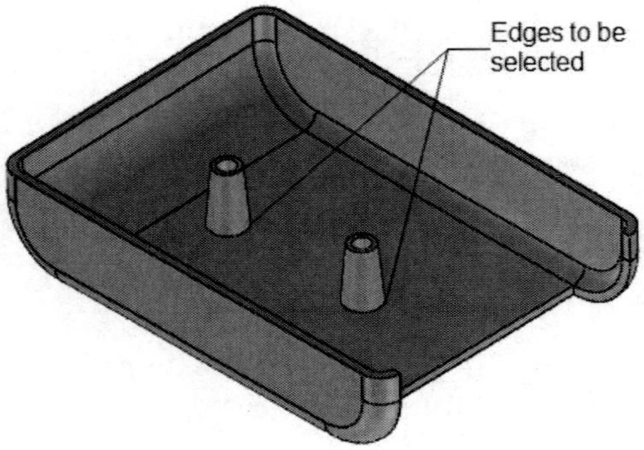

Figure 7-26 Edges to be selected

4. Move the cursor on the **All internal loops of right face, 1 Edge** button in the pop-up toolbar; the edge of the other draft that has to be filleted is highlighted, refer to Figure 7-26.

5. Choose the **All internal loops of right face, 1 Edge** button to select the edge; the name of the edge is displayed in the selection box in the **Items To Fillet** rollout.

6. Set the value of the **Radius** spinner to **1** and choose the **OK** button to end the feature creation. The model after adding the fillet is shown in Figure 7-27.

Adding a Lip to the Model

The last feature that you need to add to the model is a lip. It is created by extruding the open sketch.

1. Invoke the **Extruded Boss/Base** tool and select the bottom face of the base feature as the sketching plane.

2. Right-click on any one of the inner edges of the model on the current sketching plane using the **Select** tool to display a shortcut menu. Now, choose the **Select Tangency** option from the shortcut menu.

   If you need to create a sketch similar to that of an existing entity, it is recommended to convert the existing entity to a sketch by invoking the **Convert Entities** tool.

3. Choose the **Convert Entities** button from the **Sketch CommandManager**; the selected edges are converted into the sketched entities.

4. Exit the sketching environment; the **Boss-Extrude PropertyManager** is displayed and the **Thin Feature** rollout is invoked automatically because you are extruding an open sketch.

5. In the **Direction 1** rollout, set the value to **1** in the **Depth** as well as the **Thickness** spinners of the **Thin Feature** rollout.

6. Choose the **OK** button from the **Boss-Extrude PropertyManager**. The rotated view of the final model is shown in Figure 7-28.

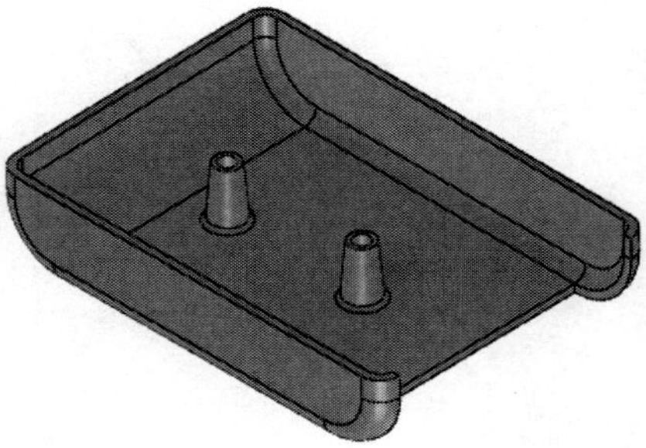

Figure 7-27 *Model after adding the fillet*

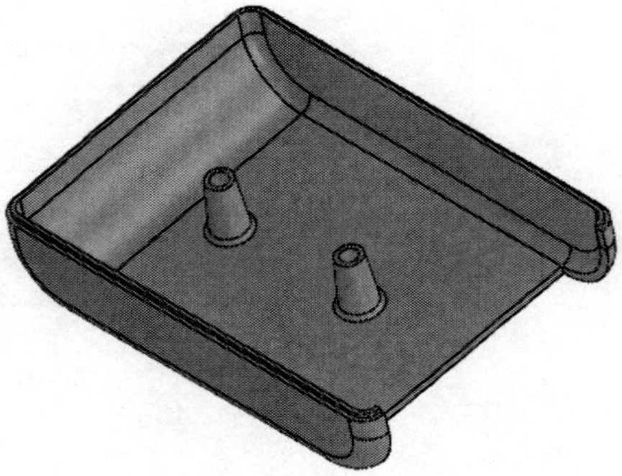

Figure 7-28 *Rotated model displaying maximum features*

Saving the Model

1. Save the part document with the name *c07_tut02* at the following location:

   \Documents\SOLIDWORKS\c07

2. Choose **File > Close** from the SOLIDWORKS menus to close the document.

EXERCISES

Exercise 1

Create the model shown in Figure 7-29. The views and dimensions of the model are shown in Figure 7-30. **(Expected time: 30 min)**

Figure 7-29 *Solid model for Exercise 1*

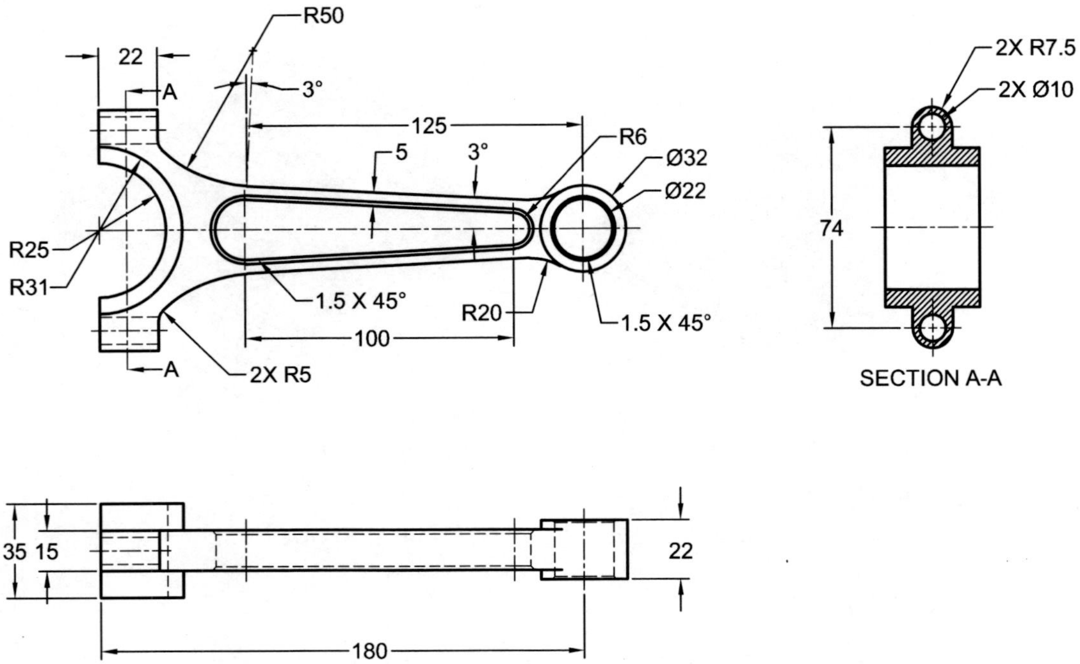

Figure 7-30 *Views and dimensions of the model for Exercise 1*

Exercise 2

Create the model shown in Figure 7-31. The views and dimensions of the model are shown in Figures 7-32. **(Expected time: 30 min)**

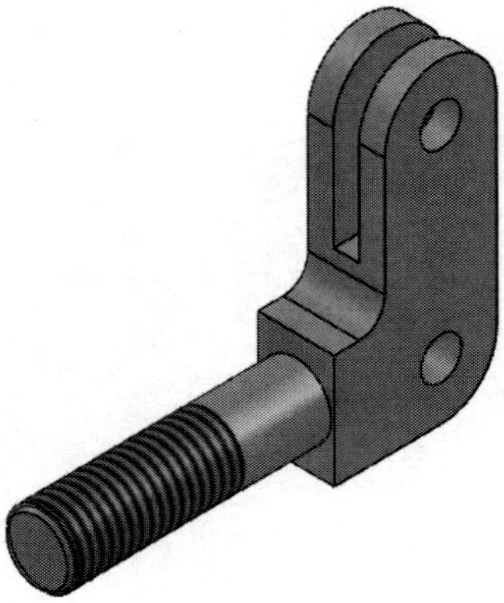

Figure 7-31 *Solid model for Exercise 2*

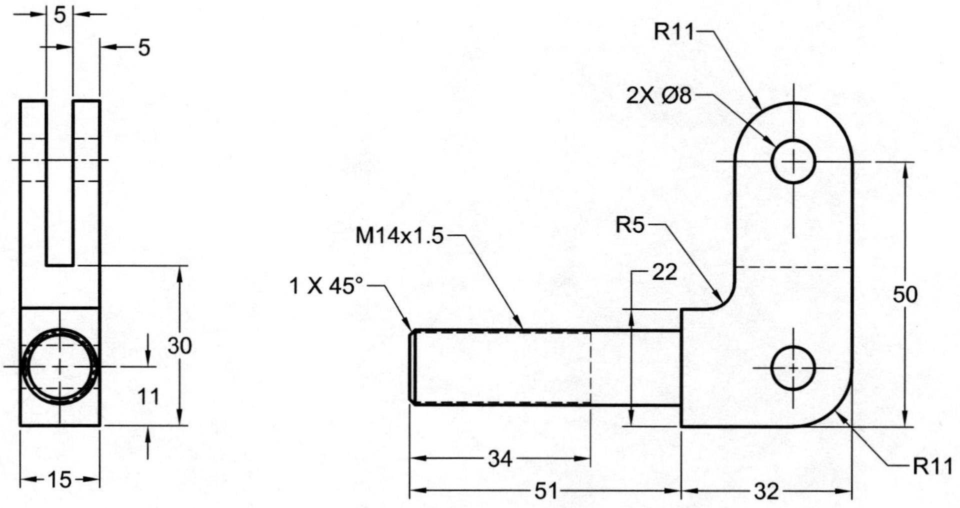

Figure 7-32 Dimensions of the model

This page is intentionally left blank

Chapter 8

Advanced Modeling Tools-II

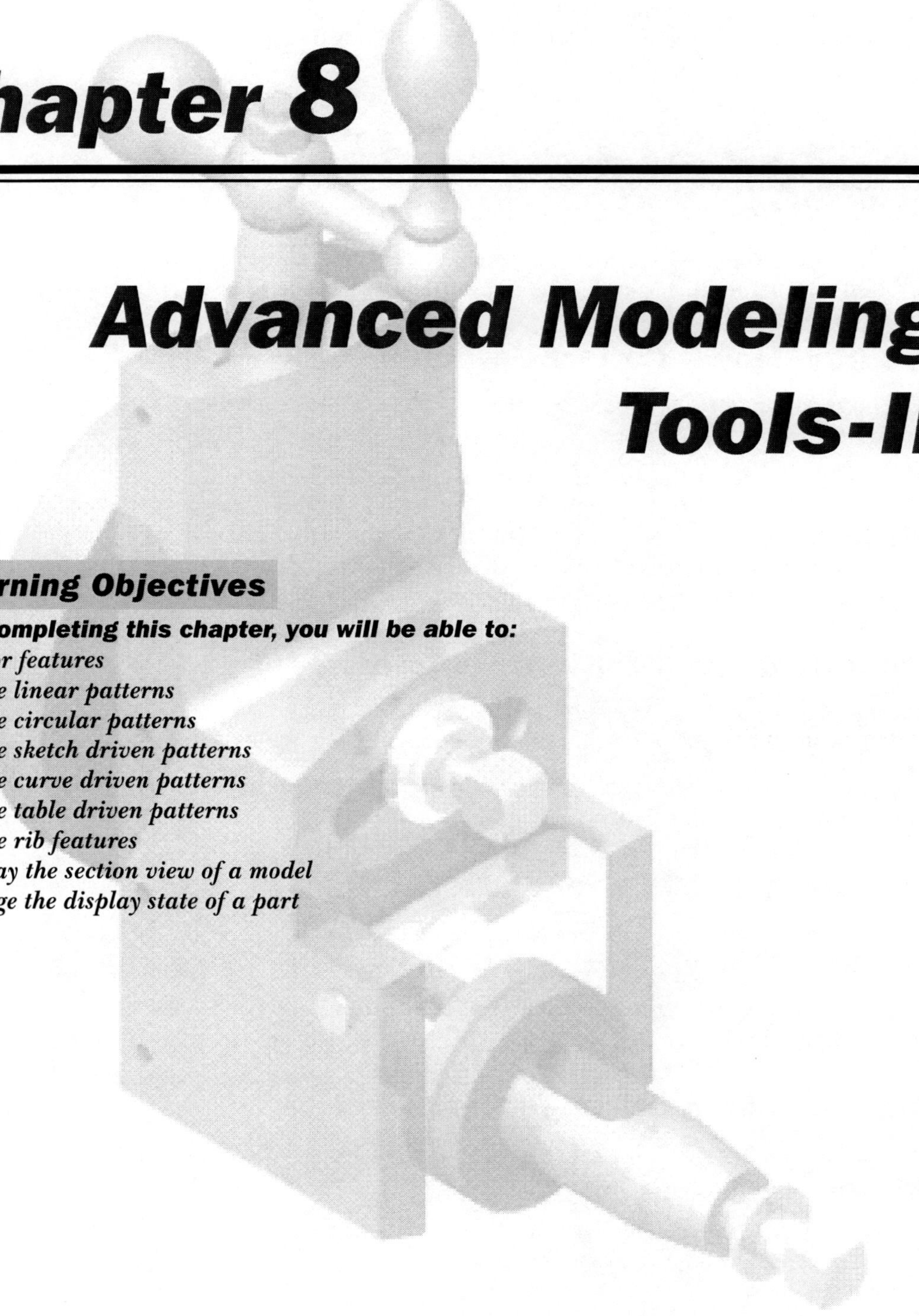

Learning Objectives

After completing this chapter, you will be able to:

- *Mirror features*
- *Create linear patterns*
- *Create circular patterns*
- *Create sketch driven patterns*
- *Create curve driven patterns*
- *Create table driven patterns*
- *Create rib features*
- *Display the section view of a model*
- *Change the display state of a part*

ADVANCED MODELING TOOLS

Some of the advanced modeling tools were discussed in Chapter 7, Advanced Modeling Tools-I. In this chapter, you will learn about some more advanced modeling tools that can be used to capture the design intent of a model. The rest of the advanced modeling tools will be discussed in the later chapters.

Creating Mirror Features

CommandManager:	Features > Linear Pattern flyout > Mirror
SOLIDWORKS menus:	Insert > Pattern/Mirror > Mirror

The **Mirror** tool is used to copy or mirror a selected feature, face, or body about a specified mirror plane which can be a reference plane or a planar face.

Creating Linear Pattern Features

CommandManager:	Features > Linear Pattern flyout > Linear Pattern
SOLIDWORKS Menus:	Insert > Pattern/Mirror > Linear Pattern

As discussed in the previous chapters, you can arrange multiple instances of sketched entities in a particular pattern. Similarly, you can also arrange the features, faces, and bodies in a particular pattern. In SOLIDWORKS, you are provided with various types of patterns such as linear patterns, circular patterns, sketch driven patterns, curve driven patterns, table driven patterns, fill patterns, and variable patterns.

Creating Circular Pattern Features

CommandManager:	Features > Linear Pattern flyout > Circular Pattern
SOLIDWORKS menus:	Insert > Pattern/Mirror > Circular Pattern

In this section, you will learn to create the circular pattern of a feature, face, or body by using the **Circular Pattern** tool.

Creating Sketch Driven Patterns

CommandManager:	Features > Linear Pattern flyout > Sketch Driven Pattern
SOLIDWORKS menus:	Insert > Pattern/Mirror > Sketch Driven Pattern

A sketch driven pattern is created when features, faces, or bodies are to be arranged in a nonuniform manner, which is neither rectangular nor circular.

Creating Curve Driven Patterns

CommandManager:	Features > Linear Pattern flyout > Curve Driven Pattern
SOLIDWORKS menus:	Insert > Pattern/Mirror > Curve Driven Pattern

The **Curve Driven Pattern** tool is used to pattern features, faces, or bodies along a selected reference curve. A reference curve can be a sketched entity, an edge, an open profile, or a closed loop.

Creating Table Driven Patterns

CommandManager: Features > Linear Pattern flyout >Table Driven Pattern
SOLIDWORKS menus: Insert > Pattern/Mirror > Table Driven Pattern

A table driven pattern is created by specifying the X and Y coordinates of the pattern feature with reference to a coordinate system. The instances of the selected features, faces, or bodies are created at the points specified using the X and Y coordinates.

Creating Fill Patterns

CommandManager: Features > Linear Pattern flyout > Fill Pattern
SOLIDWORKS menus: Insert > Pattern/Mirror > Fill Pattern

The **Fill Pattern** tool is used to fill a defined area with the pattern of features, faces, bodies, or predefined shapes. The area to be filled with the pattern of features or holes can be a sketched entity, a face, or a coplanar face.

Creating Variable Patterns

CommandManager: Features > Linear Pattern flyout > Variable Pattern
SOLIDWORKS menus: Insert > Pattern/Mirror > Variable Pattern

The **Variable Pattern** tool is used to pattern features by varying their dimensions. Using this tool, you can create variable patterns for Extrude/Cut Extrude, Revolve/Cut Revolve, Sweep/Cut Sweep, Loft/Cut Loft, Fillet, Chamfer, Dome, and Draft features.

Creating Rib Features

CommandManager: Features > Rib
SOLIDWORKS menus: Insert > Features > Rib

Ribs are defined as the thin-walled structures that are used to increase the strength of the entire structure of a component so that it does not fail under an increased load. In SOLIDWORKS, the ribs are created using an open sketch as well as a closed sketch.

Displaying the Section View of a Model

SOLIDWORKS menus: View > Display > Section View
Toolbar: View (Heads-Up) > Section View

The **Section View** tool is used to display the section view of a model by cutting it using a plane or a face. You can also save the section view with a name to generate the section view directly on the drawing sheet in the drawing mode.

CHANGING THE DISPLAY STATES

In SOLIDWORKS, you can control the transparency of individual features and display a model in different display states in the part environment also. The name of the default display state is displayed in parenthesis next to the name of the model in the **FeatureManager Design Tree**. To display a model in different display states, first you need to create different display states and then change the display of features for each display state created. To create different display states, choose the **ConfigurationManager** tab available in the Manager Pane; the **ConfigurationManager** along with the default display state **<Default>_Display State 1** will be displayed. Right-click on the default display state in the **Display States** area and choose the **Add Display State** option from the shortcut menu; a new display state, **Display State-2**, will be added. Also, you will notice that the default display state, **<Default>_Display State 1**, is greyed out. You can rename the newly created display state by using the F2 key. Similarly, you can add multiple display states to a model and rename them.

TUTORIALS

Tutorial 1

In this tutorial, you will create the model shown in Figure 8-1. The dimensions of the model are shown in same figure. **(Expected time: 30 min)**

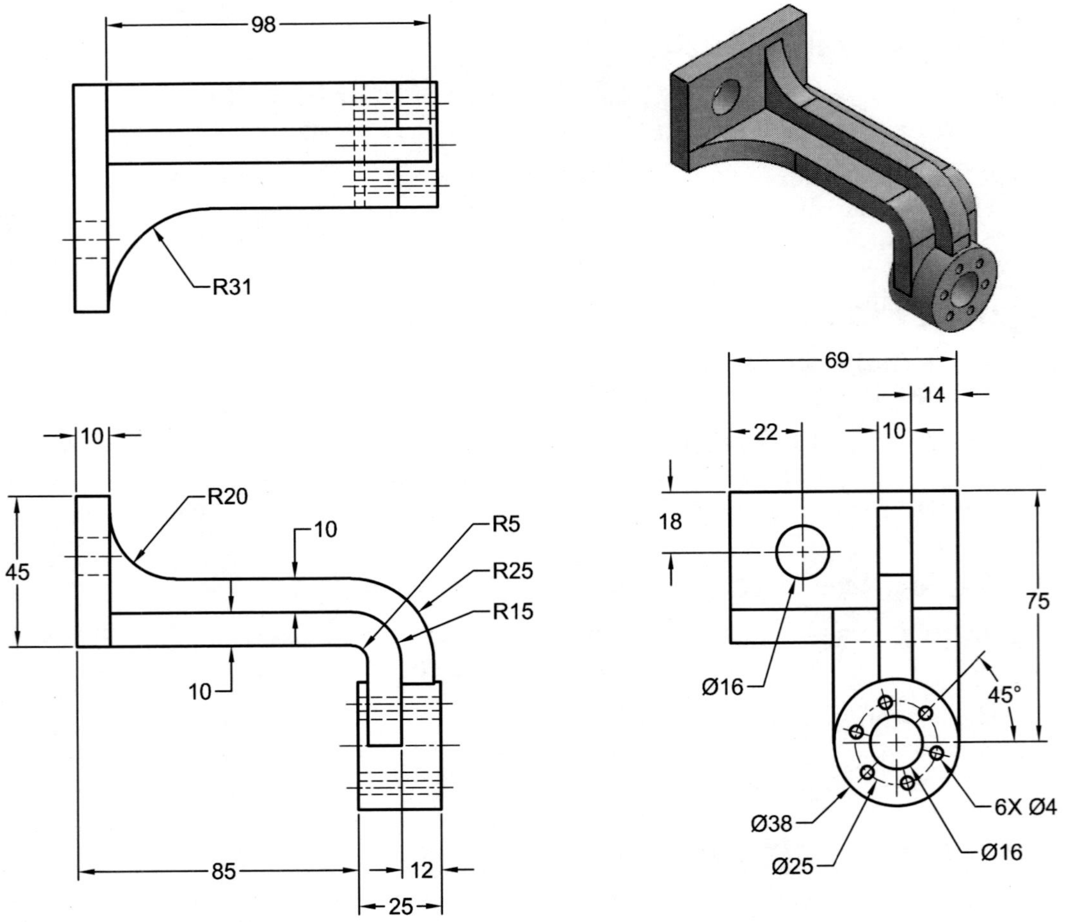

Figure 8-1 *Views and dimensions of the model for Tutorial 1*

The following steps are required to complete this tutorial:

a. Create the base feature of the model by extruding a rectangle of 69 mm x 45 mm, created on the Right Plane to a depth of 10 mm, refer to Figure 8-2.
b. Create the second feature by extruding the sketch created on the back face of the base feature, refer to Figures 8-3 and 8-4.
c. Create the third feature by extruding a circle created on the second feature, refer to Figure 8-5.
d. Create a hole feature placed concentric to the circular feature.
e. Create a hole on the specified Pitch circle, and pattern the hole feature using the circular pattern option, refer to Figure 8-6.
f. Create a hole feature on the base feature, refer to Figure 8-6.
g. Create a fillet feature to add the required fillets, refer to Figures 8-7 and 8-8.
h. Create a rib feature, refer to Figures 8-9 and 8-10.

Creating the Base Feature

1. Start SOLIDWORKS and then start a new part document from the **New SOLIDWORKS Document** dialog box.

   It is evident from the model that the sketch of its base feature is drawn on the Right Plane. Therefore, you need to select the **Right Plane** from the **FeatureManager Design Tree** to create the base feature.

2. Select the **Right Plane** from the **FeatureManager Design Tree** and choose the **Extruded Boss/Base** button from the **Features CommandManager**; the sketching environment is invoked and the **Right Plane** is oriented normal to the view, if not then orient it.

3. Draw the sketch of the base feature of the model which consists of a rectangle having dimensions 69 mm x 45 mm.

4. Add the required relations and dimensions to the sketch and exit the sketching environment.

5. Set the value of the **Depth** spinner to **10 mm** and choose the **OK** button to exit the **Boss-Extrude PropertyManager**. The base feature of the model is created, as shown in Figure 8-2.

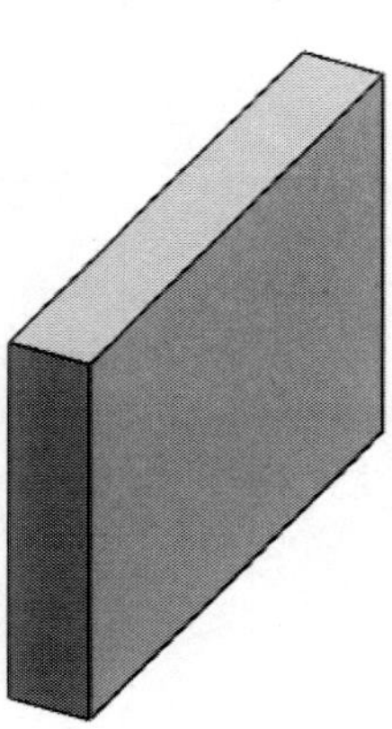

Figure 8-2 *Base feature of the model*

Creating the Second Feature of the Model

The second feature of the model is also an extruded feature. Draw the sketch for the second feature on the back face of the base feature and extrude this sketch upto the given depth.

1. Choose the **Rapid Sketch** button from the **Sketch CommandManager**, if it is not chosen by default.

2. Then, choose the **Line** button from the **Sketch CommandManager** and move the cursor on the back face of the base feature; the sketch plane is displayed in the temporary graphics. Click on the back face to invoke the sketching environment.

3. Draw the sketch of the second feature, and then add the required relations, as shown in Figure 8-3. Then, exit the sketching environment.

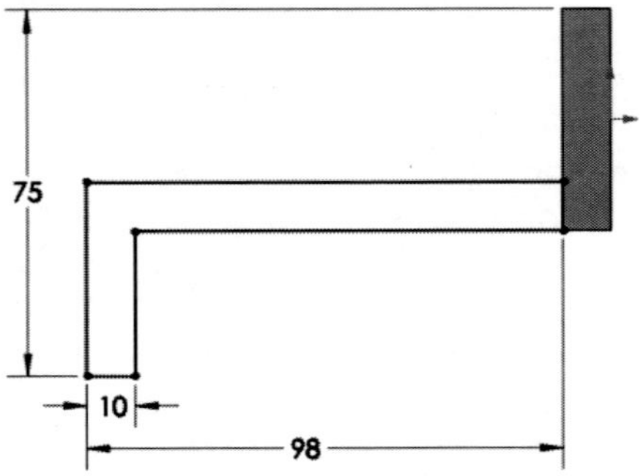

Figure 8-3 *Sketch for the second feature*

4. Make sure that the sketch is selected and then choose the **Extruded Boss/Base** tool from the **Feature CommandManager** and choose the **Reverse Direction** button in the **Direction 1** rollout from the **Boss-Extrude PropertyManager**.

5. Set the value of the **Depth** spinner to **38** and exit the feature creation.

   The model after creating the second feature is shown in Figure 8-4.

Creating the Third Feature

The third feature of this model is a circular extruded feature. The sketch for this feature will be drawn on the right planar face of the second feature and then extruded on both sides of the sketching plane.

1. Choose the **Circle** tool from the **Sketch CommandManager** and select the right planar face of the second feature. As the **Rapid Sketch** button is already chosen, the sketching environment will be displayed automatically.

2. Draw the sketch using the **Circle** tool and add the required relations and dimensions to it, as given in Figure 8-1.

3. Invoke the **Extruded Boss/Base** tool and set the value of the **Depth** spinner in the **Direction 1** rollout to **12**. Since you need to extrude the sketch in both directions with variable values, activate the **Direction 2** rollout and set the value of the **Depth** spinner in the **Direction 2** rollout to **13**, and end the feature creation. Figure 8-5 shows the model after adding the third feature.

Figure 8-4 Second feature added to the model

Figure 8-5 Third feature added to the model

Creating the Fourth Feature

The fourth feature of this model is a hole feature. You need to create a hole arbitrarily on the right face of the third feature using the **Simple Hole** tool and position it later.

1. Select the right face of the third feature and choose the **Simple Hole** button from the **Features CommandManager**, or choose **Insert > Features > Simple Hole** from the SOLIDWORKS menu to invoke the **Hole PropertyManager**.

2. Select the **Up To Next** option from the **End Condition** drop-down list and set the value of the **Hole Diameter** spinner to **16**.

3. Choose the **OK** button from the **Hole PropertyManager**; the hole feature is placed arbitrarily on the selected face.

   Now, you need to position the hole feature concentric to the circular feature.

4. Select the hole feature from the **FeatureManager Design Tree**; a pop-up toolbar is displayed.

5. Choose **Edit Sketch** from the pop-up toolbar; the sketching environment is displayed.

6. Apply concentric relation between the circle and the circular edge of the face.

7. Press CTRL+B keys on the keyboard to rebuild the model; the hole feature is positioned.

Creating the Fifth Feature

1. Follow the procedure given in the previous section to create the fifth feature, which is also a hole feature, on the specified Pitch circle, as shown in Figure 8-1. This feature will be created on the right face of the third feature. The hole feature is created using the **Up To Next** option, with the diameter of the hole as 4 mm. Next, define the placement of the feature by adding the required relations and dimensions to it.

Patterning the Hole Feature

After creating the fifth feature, which is a hole feature, you will pattern it using the **Circular Pattern** tool.

1. Choose the **Circular Pattern** button from the **Linear Pattern** flyout; the **CirPattern PropertyManager** is invoked and you are prompted to select the features to pattern.

2. Select the hole feature created in the previous step from the drawing area. Alternatively, expand the **FeatureManager Design Tree** displayed in the drawing area and then select the **Hole2** feature from it.

3. Click on the **Pattern Axis** selection box in the **CirPattern PropertyManager** and select the circular edge of the hole feature created earlier; preview of the circular pattern of the hole feature is displayed.

4. Select the **Equal spacing** check box, if it is not selected, and set the value in the **Number of Instances** spinner to **6**. Also, set the value in the **Angle** spinner to **360**. Choose the **OK** button from the PropertyManager.

Creating the Hole Feature

The next feature to be created is also a hole feature.

1. Create this hole feature by using the same procedure followed to create the fourth feature. This hole feature needs to be placed on the right planar face of the base feature. Therefore, after selecting the right planar face of the base feature, place the hole feature on it.

2. Next, define the placement of the hole feature by adding the required relations and dimensions. Figure 8-6 shows the model after adding all the hole features.

Figure 8-6 *Model after adding all the hole features*

Creating the Fillet Feature

Next, you need to create the fillet feature. It is evident from the model that the fillets to be added to it are of different radii. In SOLIDWORKS, you can specify different radii to the selected edges, faces, or loops individually in a single fillet feature.

1. Choose the **Fillet** button from the **Features CommandManager**; the **Fillet PropertyManager** is displayed. If the **FilletXpert PropertyManager** is displayed, choose the **Manual** button from it to display the **Fillet PropertyManager**.

2. Select the **Multiple Radius Fillet** check box from the **Fillet Parameters** rollout in the **Fillet PropertyManager**.

3. Select the edges to be filleted, as shown in Figure 8-7. As the **Multiple Radius Fillet** check box is selected, each selected edge has a separate Radius callout.

4. Modify the values of the radii as required in their respective Radius callouts.

5. Choose the **OK** button from the **Fillet PropertyManager**.

The isometric view of the model after adding the fillet feature is shown in Figure 8-8.

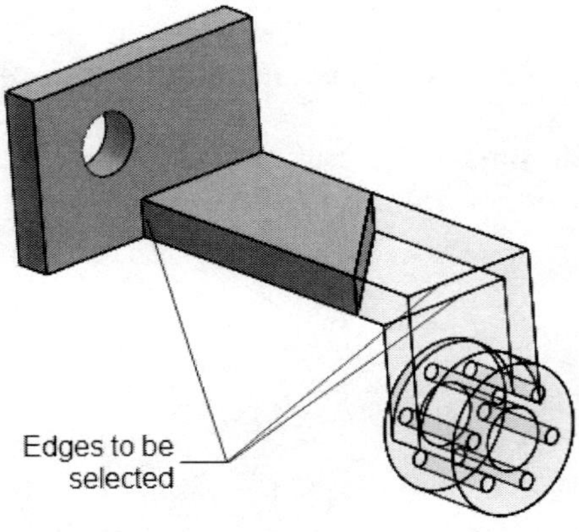

Figure 8-7 *Edges to be filleted*

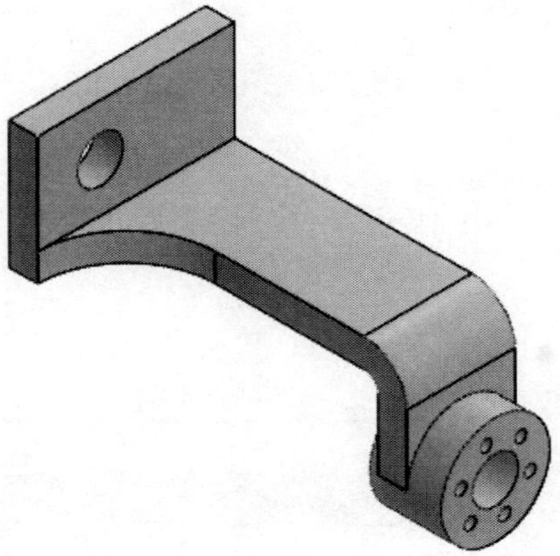

Figure 8-8 *Model after adding the fillet feature*

Creating the Rib Feature

The next feature that you need to create is a rib feature. The sketch of the rib feature needs to be drawn on a sketching plane at an offset distance from the back planar face of the model. Therefore, you first need to create a reference plane at an offset distance from the back planar face of the model.

1. Choose the **Plane** button from the **Reference Geometry** flyout; the **Plane PropertyManager** is invoked.

2. Rotate the model and select its back planar face as the first reference. Select the **Flip offset** check box below the **Offset distance** spinner and set the value in the **Offset distance** spinner to **19**.

3. Choose the **OK** button from the **Plane PropertyManager** to end the feature creation; a new plane is created at an offset distance from the back planar face of the model.

4. Choose the **Rib** button from the **Features CommandManager**.

5. Create a sketch for the rib feature on the newly created plane and add the required relations and dimensions to the sketch, as shown in Figure 8-9.

6. Exit the sketching environment; the **Rib PropertyManager** is displayed.

   Preview of the rib feature is displayed in the drawing area. You will notice that the direction of material addition is displayed by an arrow in the drawing area. The direction of material addition is opposite to the required direction. Therefore, you need to flip its direction. If the direction of material addition is the required one, skip the next step.

7. Select the **Flip material side** check box to flip the direction of the material addition.

   The **Both Sides** button is chosen by default in the **Parameters** rollout and the default value of the rib thickness is **10** which is also the required value. Therefore, you do not need to change it.

8. Choose the **OK** button from the **Rib PropertyManager**. Also, hide the newly created plane.

   The last feature of the model is a fillet feature. Add the fillet feature on the left edge of the rib using the **Fillet** tool. Figure 8-10 shows the isometric view of the final model.

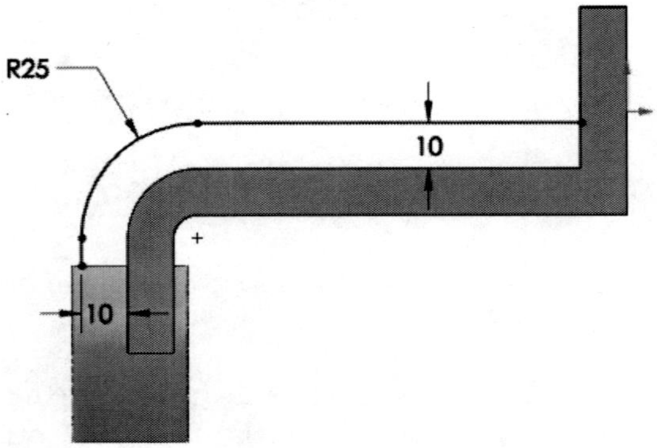

Figure 8-9 *Sketch for the rib feature*

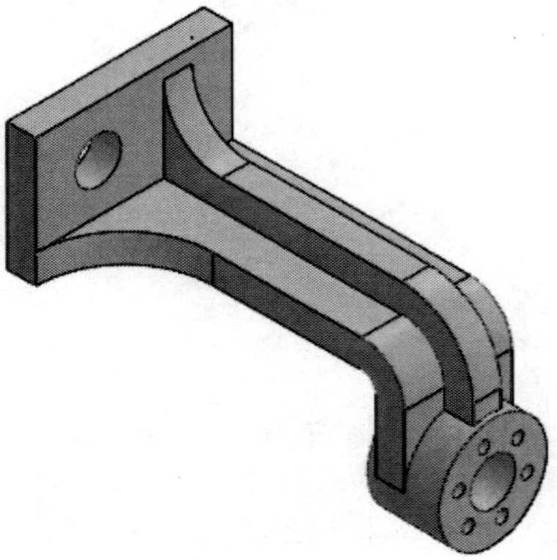

Figure 8-10 *The final solid model*

Saving the Model

1. Create a folder with the name *c08* in the *SOLIDWORKS* folder and choose the **Save** button from the Menu Bar. Save the model with the name *c08_tut01* at the location given below:

   \Documents\SOLIDWORKS\c08

2. Choose **File > Close** from the SOLIDWORKS menus to close the document.

Tutorial 2

In this tutorial, you will create the model of a cover. The dimensions of the model are shown in Figure 8-11. The model is shown in Figure 8-12. **(Expected time: 30 min)**

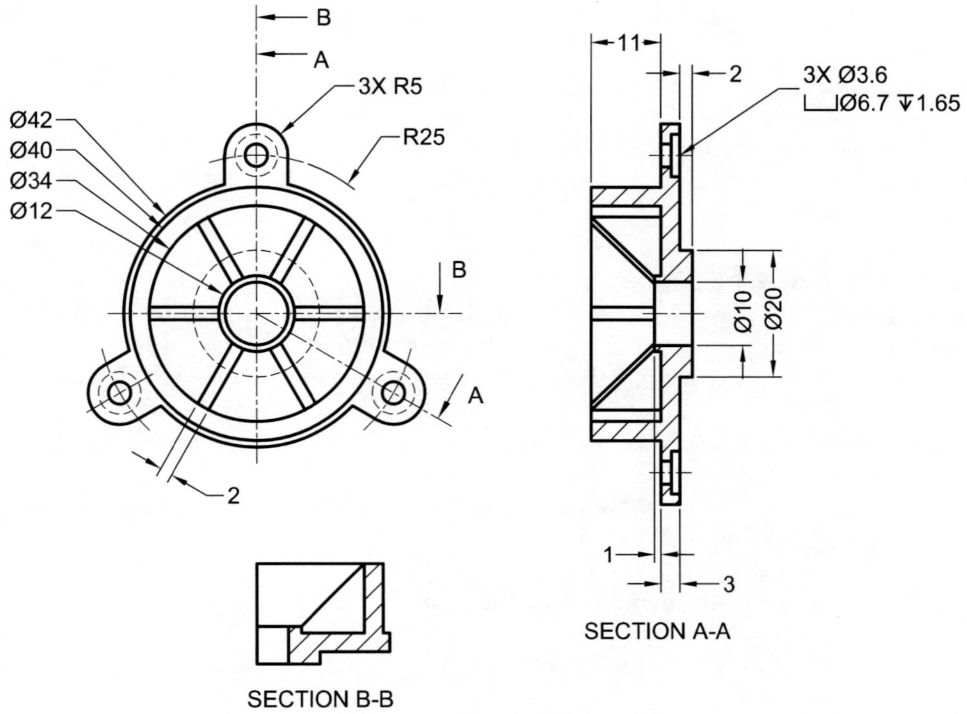

Figure 8-11 *Different views of the model with dimensions*

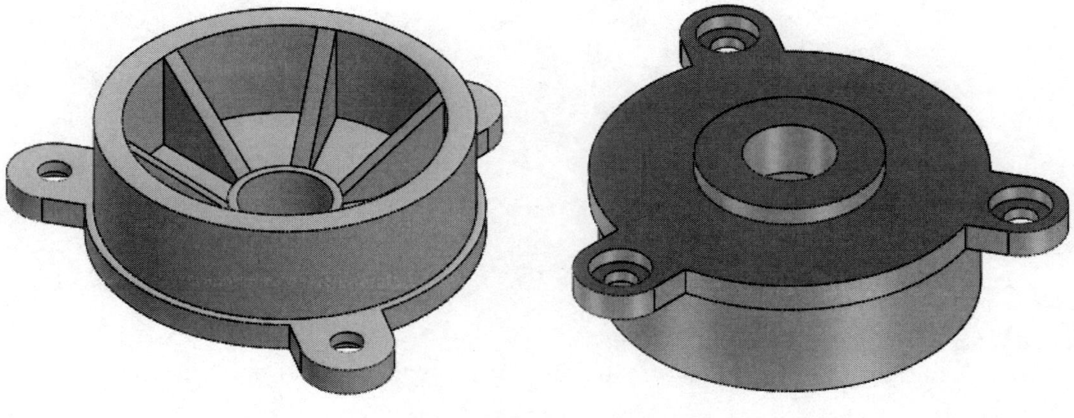

Figure 8-12 *Two different views of the model for Tutorial 2*

The following steps are required to complete this tutorial:

a. Create the base feature of the model by revolving the sketch along the centerline, refer to Figures 8-13 and 8-14.
b. Create the second feature by extruding the sketch from the sketch plane to the selected surface, refer to Figures 8-15 through 8-17.

c. Place a counterbore hole feature on the bottom face of the second feature using the **Hole Wizard** tool.
d. Pattern the second and third features along the temporary axis using the **Circular Pattern** tool, refer to Figure 8-19.
e. Create the rib feature, refer to Figure 8-20.
f. Pattern the rib feature along a temporary axis using the **Circular Pattern** tool, refer to Figure 8-21.

Creating the Base Feature

Start a new SOLIDWORKS part document. First, you need to create the base feature of the model by revolving the sketch along the axis of revolution. The axis of revolution will be a centerline and the sketch for the base feature will be drawn on the Right Plane.

1. Invoke the **Revolved Boss/Base** tool and select the **Right Plane** as the sketching plane.

2. Create the sketch for the base feature and add the required relations and dimensions to the sketch, as shown in Figure 8-13.

3. Exit the sketching environment and set the value in the **Angle** spinner to **360**.

4. Choose the **OK** button from the **Revolve PropertyManager**; the base feature is created, as shown in Figure 8-14.

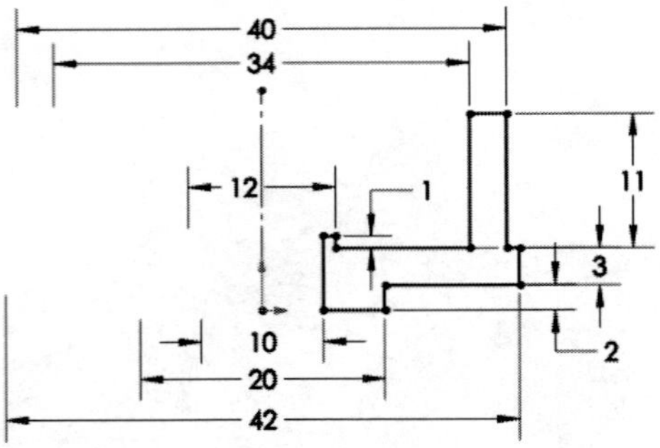

***Figure 8-13** Sketch for the base feature*

***Figure 8-14** Base feature of the model*

Creating the Second Feature

The second feature needs to be created by extruding the sketch up to the selected surface.

1. Invoke the **Extruded Boss/Base** tool and select the face as the sketching plane, as shown in Figure 8-15.

2. Create the sketch of the second feature and add the required relations and dimensions to it, as shown in Figure 8-16.

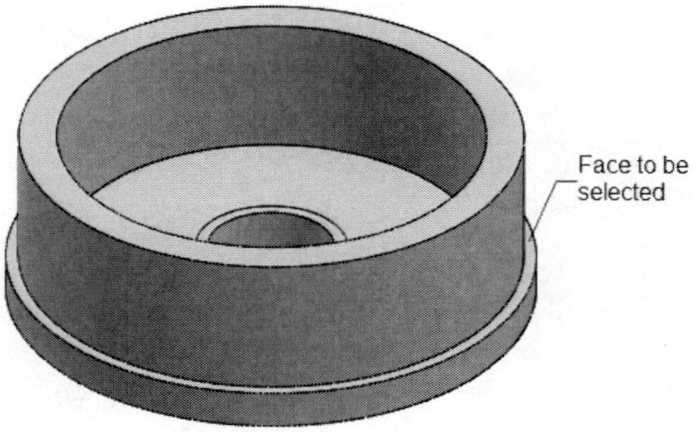

Figure 8-15 Face to be selected

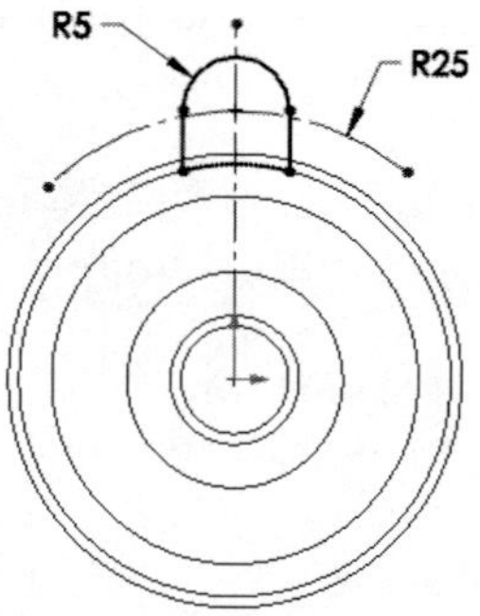

Figure 8-16 Sketch created for the second feature

3. Exit the sketching environment. Use the **Up To Surface** option to extrude the sketch. The surface to be selected is shown in Figure 8-17.

4. Choose the **OK** button from the **Boss-Extrude PropertyManager**; the second feature is created, as shown in Figure 8-18.

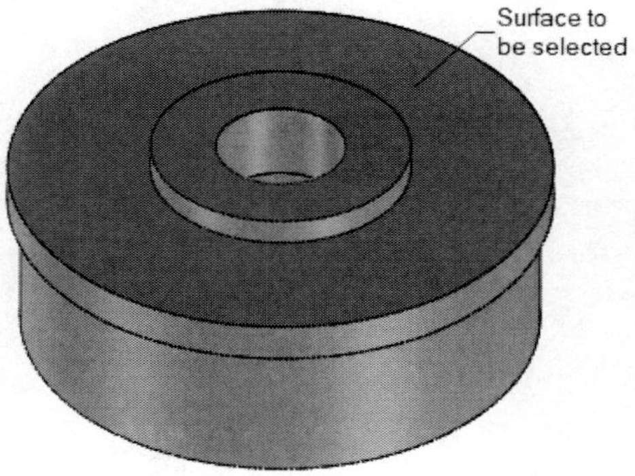

Figure 8-17 Surface to be selected

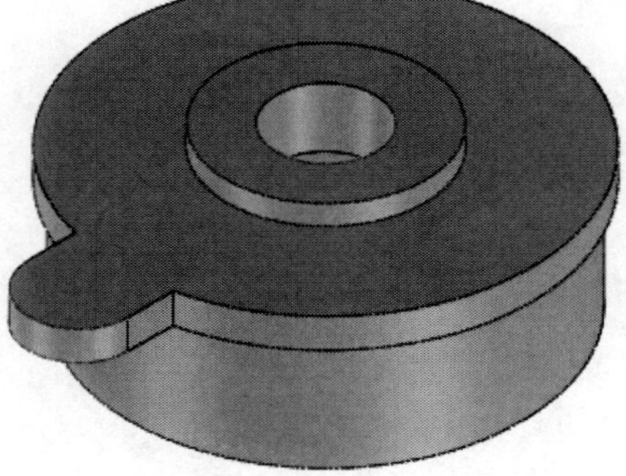

Figure 8-18 Second feature created

Creating the Hole Feature

It is evident from Figure 8-19 that a counterbore hole needs to be added to the model. It can be added by using the **Hole Wizard** tool.

1. Select the bottom face of the second feature as the placement plane for the hole feature and press the **S** key; the **Shortcut** toolbar is displayed.

2. Choose **Hole Wizard** from the **Shortcut** toolbar to invoke the **Hole Specification PropertyManager**.

3. Choose the **Counterbore** button from the **Hole Type** rollout, if it is not chosen by default, and select the **ANSI Metric** option from the **Standard** drop-down list.

4. Select the **Socket Button Head Cap Screw - ANSI B18.3.4M** option from the **Type** drop-down list. Select the **M3** option from the **Size** drop-down list. Select **Through All** in the **End Condition** drop-down list in the **End Condition** rollout.

5. Select the **Show custom sizing** check box to customize the size of the counterbore hole and then enter the following parameters:

   Through Hole Diameter: **3.6 mm** Counterbore Diameter: **6.70 mm**
   Counterbore Depth: **1.65 mm**

6. Choose the **Positions** tab from the **Hole Specification PropertyManager**; you are prompted to use dimensions and other sketching tools to position the center of the holes.

7. Specify the position of the hole and then choose the **Add Relation** button from the **Sketch CommandManager**; the **Add Relations PropertyManager** will be invoked. Now apply the concentric relation between the center point of the hole feature and the circular edge of the second feature. Choose the **OK** button to exit the **Add Relations PropertyManager**.

8. Now, choose the **OK** button from the **Hole Position PropertyManager** to end the feature creation.

Patterning the Features

After creating the second and third features, you need to pattern them using the **Circular Pattern** tool.

1. Choose the **Circular Pattern** button from the **Linear Pattern** flyout; the **CirPattern PropertyManager** is invoked.

2. Select the second and third features from the drawing area or from the **FeatureManager Design Tree** that is displayed in the drawing area.

3. Click on the **Pattern Axis** selection box in the **CirPattern PropertyManager** and select the circular edge of the base feature; preview of the circular pattern of the hole feature is displayed.

4. Set the value in the **Number of Instances** spinner to **3** and make sure the **Equal spacing** check box is selected. Clear the **Geometry pattern** check box from the **Options** rollout, if it is selected.

5. Choose the **OK** button from the **CirPattern PropertyManager**; the features are patterned, as shown in Figure 8-19.

Creating the Rib Feature

The next feature is a rib feature. The sketch for the rib feature will be created on the Front Plane.

1. Choose the **Rib** button from the **Features CommandManager** and select the **Front Plane** from the **FeatureManager Design Tree**; the sketch environment will be invoked.

2. Set the display mode to wireframe and then create the sketch for the rib feature and add the required relations, as shown in Figure 8-20.

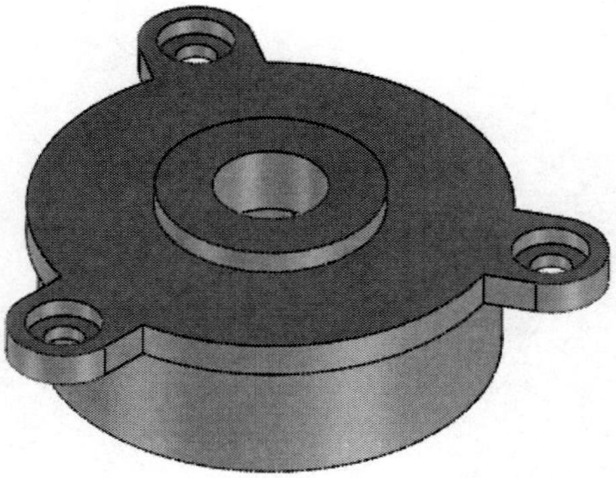

Figure 8-19 Model after patterning the features

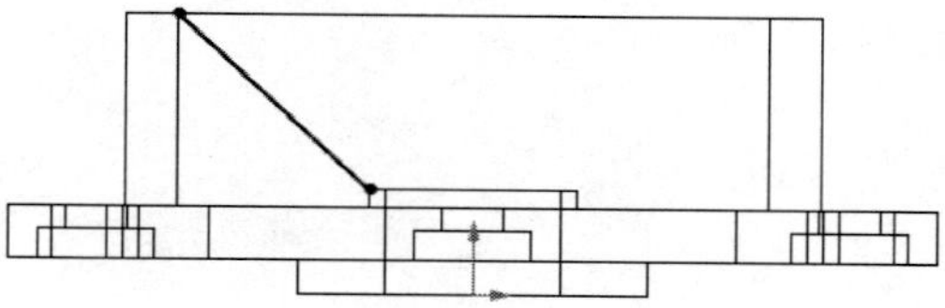

Figure 8-20 Sketch for the rib feature

3. Exit the sketching environment and set the value of the **Rib Thickness** spinner to **2**. Reverse the direction of material, if required, using the **Flip material side** check box. Use the default values for other options and choose the **OK** button from the **Rib PropertyManager**.

4. Change the display mode of the model to shaded with edges.

5. Use the **Circular Pattern** tool to create six instances of the rib feature. The final model after creating all the features is shown in Figure 8-21.

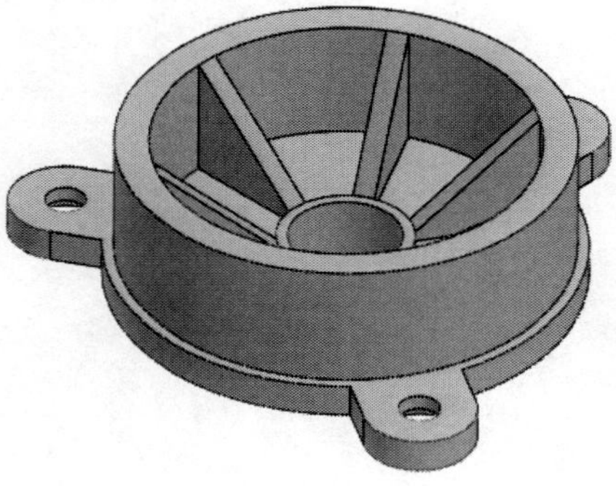

Figure 8-21 Final solid model

Saving the Model

1. Save the model with the name *c08_tut02* at the location given below:

   \Documents\SOLIDWORKS\c08

2. Choose **File > Close** from the SOLIDWORKS menus to close the document.

EXERCISES

Exercise 1

Create the model shown in Figure 8-22. The views and dimensions of the model are given in Figure 8-23.
(Expected time: 1 hr)

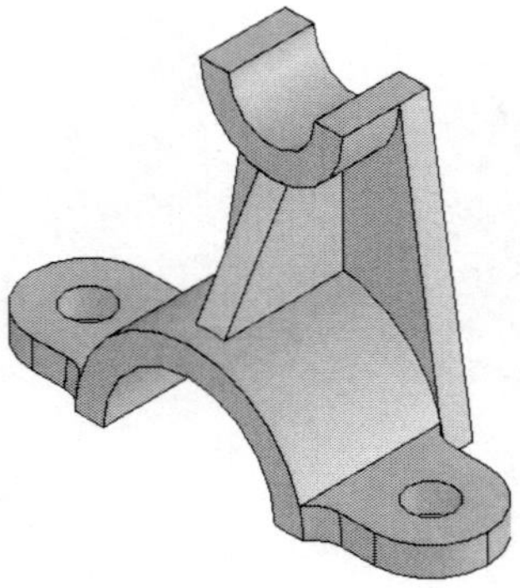

Figure 8-22 *Solid model for Exercise 1*

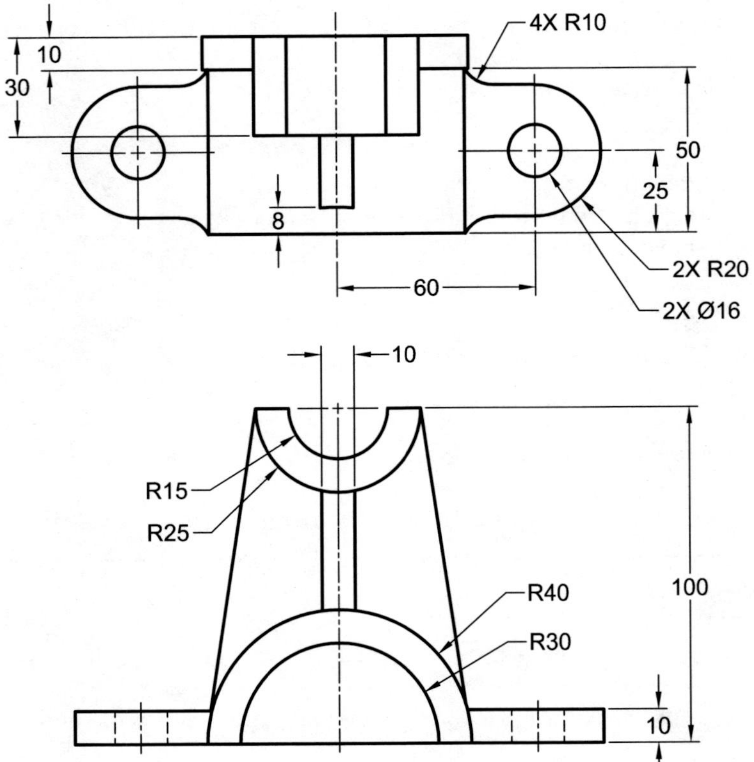

Figure 8-23 *Views and dimensions of the model for Exercise 1*

Exercise 2

Create the model shown in Figure 8-24. Next, create the section view of the model using the Front Plane. Figure 8-25 shows the section view of the model whose dimensions are given in Figure 8-26. **(Expected time: 45 min)**

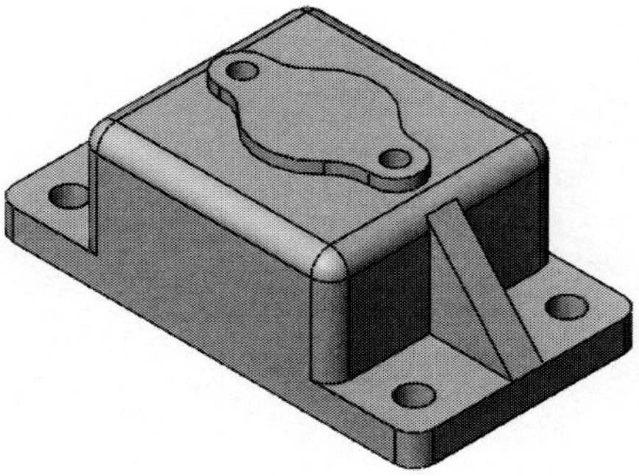

Figure 8-24 *Solid model for Exercise 2*

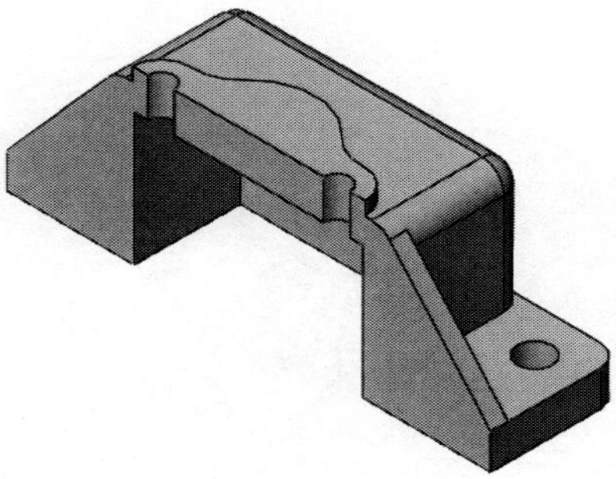

Figure 8-25 *Section view of the model*

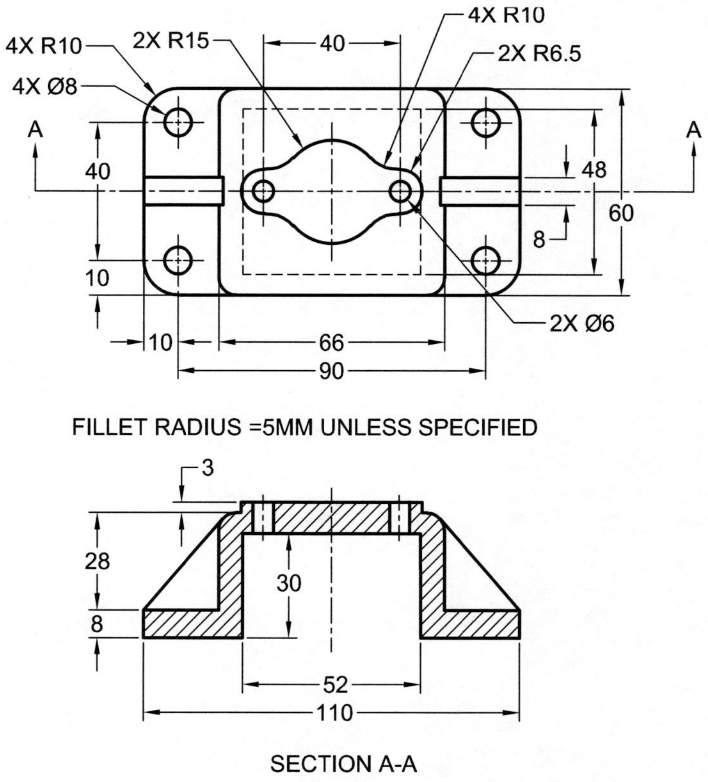

Figure 8-26 *Views and dimensions of the model for Exercise 2*

This page is intentionally left blank

Chapter 9

Editing Features

Learning Objectives

After completing this chapter, you will be able to:

- *Edit features*
- *Cut, copy, and paste features and sketches*
- *Delete features and bodies*
- *Suppress and unsuppress features*
- *Move or copy bodies*

EDITING THE FEATURES OF A MODEL

Editing is one of the most important aspects of the product design cycle. Almost all designs require editing during or after their creation. As discussed earlier, SOLIDWORKS is a feature-based parametric software. Therefore, the design created in SOLIDWORKS is a combination of individual features integrated together to form a solid model. All these features can be edited individually.

Cutting, Copying, and Pasting Features and Sketches from One Document to Another

You can also cut or copy the features and sketches from one document and paste them in another document. For example, if you need to copy a sketch created in the current document and paste it in a new document, then select the sketch and press CTRL+C to copy the item to the clipboard. Then, create a new document in the **Part** mode and select the plane on which you want to paste the sketch. Press CTRL+V to paste the sketch on the selected plane. Use the same procedure to copy features from one document to the other.

Copying Features using Drag and Drop

SOLIDWORKS also provides you with the drag and drop functionality of Windows to copy and paste an item within the document. Press and hold the CTRL key on the keyboard. Next, select and drag the item from the **FeatureManager Design Tree**. Drag the cursor to a location where you want to paste the item and release the left mouse button. If the item to be pasted is defined using the dimensions or the relations, the **Copy Confirmation** dialog box will be displayed to delete or dangle those constraints.

Dragging and Dropping Features from One Document to the Other

You can also drag and drop features as well as sketches from one document to the other. To do so, you should open both the documents in the SOLIDWORKS session. Choose **Windows > Tile Vertical/Tile Horizontal** from the SOLIDWORKS menus; both the documents are displayed at the same time in the SOLIDWORKS window. Select the feature or the sketch in the **Feature Manager Design Tree** in one document, and then drag and place it on the required entity in the other document. Note that you cannot drag and drop the base feature.

Deleting Features

You can delete the unwanted features from the model by selecting the feature from the **FeatureManager Design Tree** or from the drawing area. After selecting the feature to be deleted, press the DELETE key on the keyboard, or right-click to invoke the shortcut menu and choose **Delete** from the shortcut menu displayed.

Deleting Bodies

CommandManager: Features > Delete/Keep Body *(Customize to add)*
SOLIDWORKS menus: Insert > Features > Delete/Keep Body

As discussed earlier, SOLIDWORKS supports multibody environment. Therefore, you can create multiple disjoint bodies in SOLIDWORKS. You can also delete unwanted bodies. The bodies to be deleted can be solid bodies or surface bodies.

Suppressing Features

CommandManager: Features > Suppress *(Customize to add)*
SOLIDWORKS menus: Edit > Suppress > This Configuration

Sometimes, you do not want some feature or features to be displayed in the model or in the drawing views. Instead of deleting such features, they can be suppressed. When you suppress a feature, it is neither visible in the model nor in the drawing views. Also, if you create an assembly using that model, the suppressed feature will not be displayed even in the assembly. You can resume such suppressed features anytime by unsuppressing them. When you suppress a feature, the features that are dependent on it are also suppressed.

Unsuppressing the Suppressed Features

CommandManager: Features > Unsuppress *(Customize to add)*
SOLIDWORKS menus: Edit > Unsuppress > This Configuration

The suppressed features can be unsuppressed using the **Unsuppress** tool. To resume the suppressed feature, select the suppressed feature from the **FeatureManager Design Tree** and choose the **Unsuppress** button.

Unsuppressing Features with Dependents

CommandManager: Features > Unsuppress with Dependents *(Customize to add)*
SOLIDWORKS menus: Edit > Unsuppress with Dependents > This Configuration

As discussed earlier, when you suppress a feature, the dependent features are also suppressed. You can resume the suppressed feature along with the dependents of the suppressed parent feature in a single-click using the **Unsuppress with Dependents** tool.

Hiding Bodies

While working in the multibody environment, you can also hide the bodies. The hidden body is not displayed in the model, assembly, or in the drawing views. To hide a body, expand the **Solid Bodies** folder in the **FeatureManager Design Tree**, and select the body to be hidden. Right-click to invoke the shortcut menu and choose the **Hide** option from it. The selected body will disappear from the drawing area. The icon of the hidden body is displayed in wireframe in the **Solid Bodies** folder. To turn on the display of the hidden body, select it from the **Solid Bodies** folder and then choose the **Show** option from the shortcut menu.

Moving and Copying Bodies

CommandManager: Direct Editing > Move/Copy Bodies
SOLIDWORKS menus: Insert > Features > Move/Copy

You can move or copy bodies in the multibody environment by using the **Move/Copy Bodies** tool.

TUTORIALS

Tutorial 1

In this tutorial, you will create the model shown in Figure 9-1. After creating some of its features, you will dynamically modify it and then undo the modifications. The views and dimensions of the model are shown in Figure 9-2. **(Expected time: 30 min)**

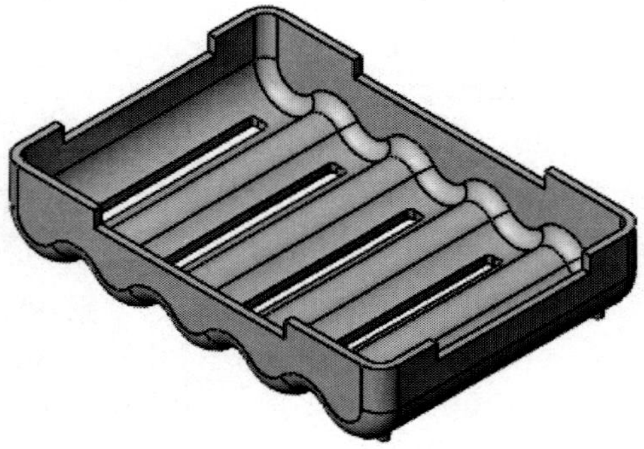

Figure 9-1 *Model for Tutorial 1*

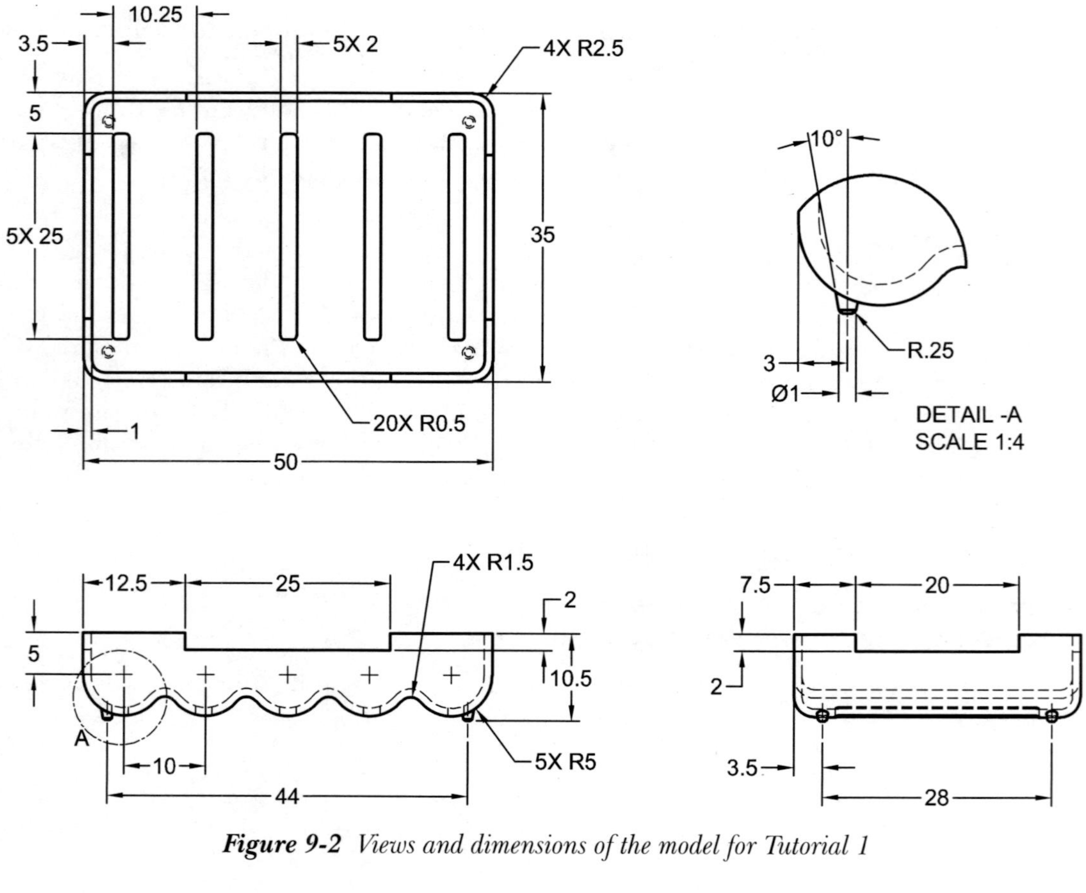

Figure 9-2 *Views and dimensions of the model for Tutorial 1*

The following steps are required to complete this tutorial:

a. Create the base feature of the model by extruding the profile to a given distance, refer to Figures 9-3 and 9-4.
b. Add fillets to the base feature, refer to Figures 9-5 and 9-6.
c. Add the shell feature to the model and remove the top face of the base feature, refer to Figures 9-7 and 9-8.
d. Dynamically modify the model, refer to Figures 9-9 through 9-11.
e. Create cuts on the sides of the model, refer to Figures 9-12 and 9-13.
f. Create slots on the lower part of the base and add fillet to the slots feature, refer to Figure 9-14.
g. Create a plane at an offset distance from the **Top Plane**.
h. Create the standoffs using the **Extrude Boss/Base** and **Fillet** tools, and pattern the standoffs, refer to Figure 9-15.
i. Save the model.

Creating the Base Feature

For creating the base feature, you need to draw the sketch of the base feature on the Front Plane and then extrude it by using the **Mid Plane** option.

1. Start SOLIDWORKS and open a new part document using the **New SOLIDWORKS Document** dialog box.

2. Invoke the **Extruded Boss/Base** tool and draw the sketch of the base feature on the **Front Plane**. Add required relations and dimensions to the sketch, as shown in Figure 9-3. Exit the sketching environment; the **Boss-Extrude PropertyManager** is invoked.

3. Extrude the sketch to a distance of 35 mm using the **Mid Plane** option.

The base feature of the model is shown in Figure 9-4.

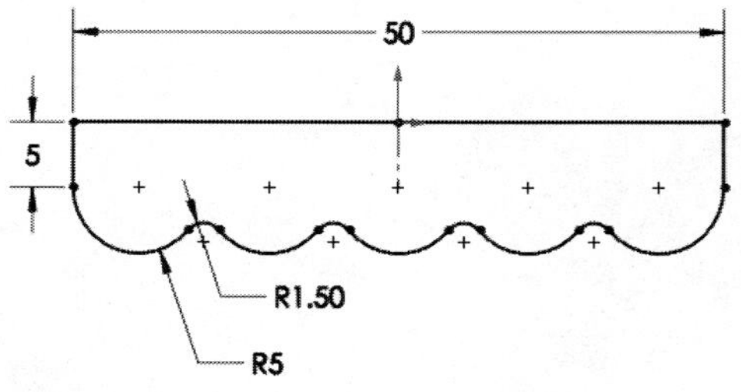

Figure 9-3 *Sketch for the base feature*

Figure 9-4 *Base feature of the model*

Adding Fillets to the Base Feature

After creating the base feature, you will fillet its lower edges.

1. Invoke the **Fillet PropertyManager** and then choose the **Constant Size Fillet** button from the **Fillet Type** rollout if it is not chosen by default.

2. Rotate the model and then select the edges of the base feature, as shown in Figure 9-5.

3. Set **2.5** as the value in the **Radius** spinner in the **Fillet Parameters** rollout and choose the **OK** button from the **Fillet PropertyManager**.

   The model after adding fillet to its edges is shown in Figure 9-6.

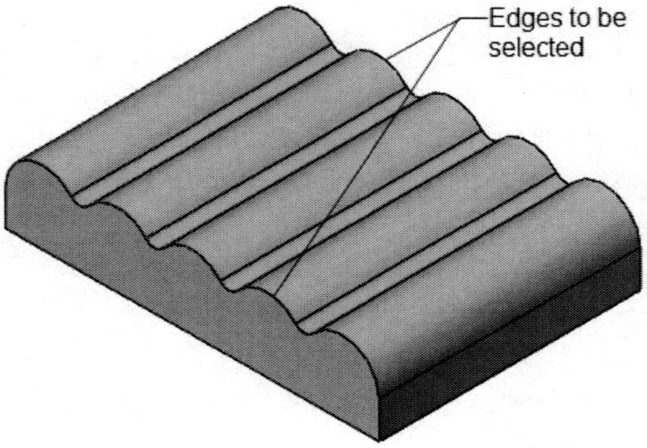

Figure 9-5 *Edges to be selected*

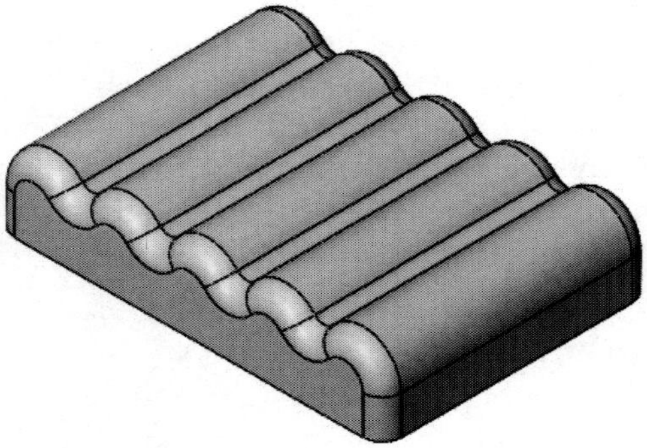

Figure 9-6 *Fillet added to the model*

Adding Shell to the Model

After creating the fillet feature, you need to shell the model using the **Shell** tool.

1. Orient the model in the isometric view and invoke the **Shell PropertyManager**.

2. Select the top planar face of the model, as shown in Figure 9-7.

3. Set the value in the **Thickness** spinner to **1**, and choose the **OK** button from the **Shell PropertyManager**.

   The model after adding the shell feature is shown in Figure 9-8.

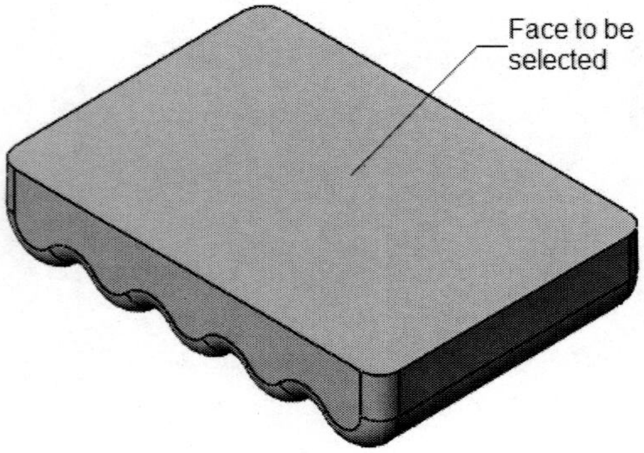

Figure 9-7 *Face to be selected*

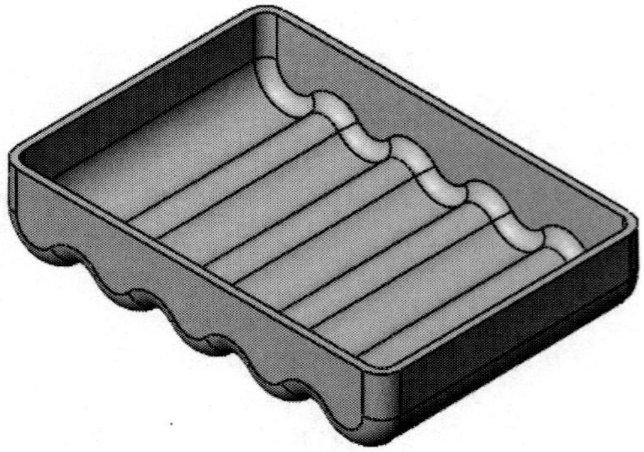

Figure 9-8 *Shell feature added to the model*

Dynamically Editing the Features

After adding the shell feature to the base of the model, you need to edit the features dynamically using the **Instant3D** tool.

1. Choose the **Instant3D** tool from the **Features CommandManager**, if not already chosen. 

2. Select the front planar face of the base feature from the drawing area; the selected face is highlighted in blue and an orange colored arrow is displayed.

3. Move the cursor to the orange colored arrow and press and hold the left mouse button; the move cursor is displayed, as shown in Figure 9-9. Drag the cursor to resize the feature.

   The preview of the resized feature and its dimensions are displayed in the drawing area.

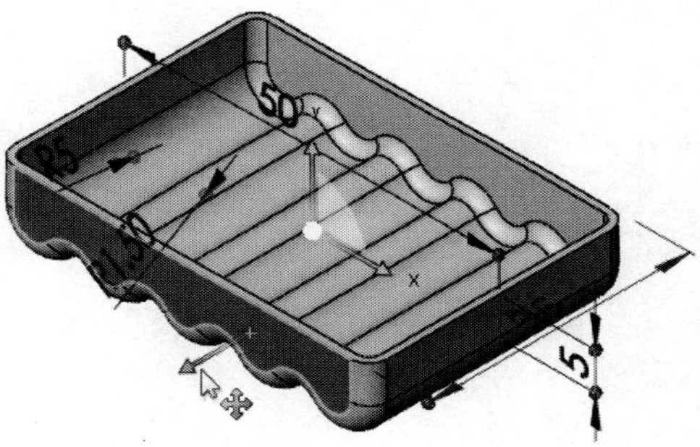

***Figure 9-9** Editing handles for editing the base feature*

As you drag the cursor, the preview and the dimensions are updated automatically.

4. Release the left mouse button after dragging the cursor to some distance. Figure 9-10 shows the preview of the feature while dragging it and Figure 9-11 shows the edited feature.

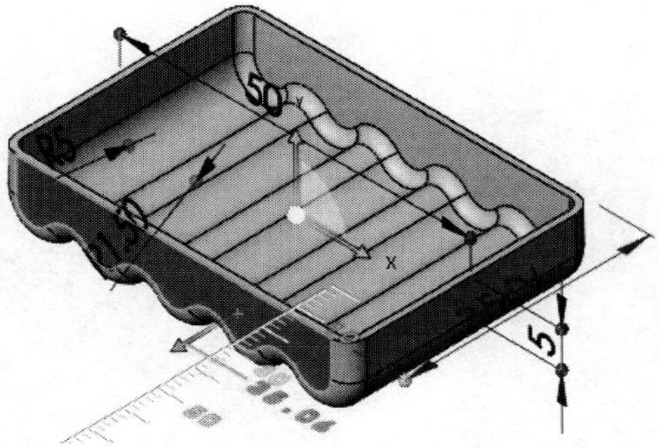

***Figure 9-10** Preview of the feature while dragging*

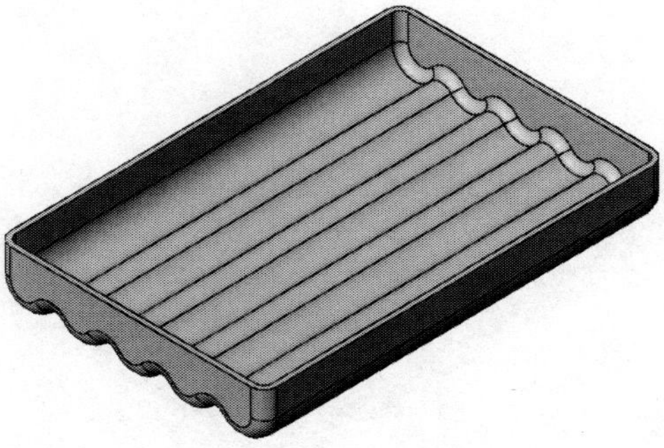

***Figure 9-11** Resultant edited feature*

After editing the model dynamically by dragging, the depth of its base feature changes. To bring the base feature back to its original size of 35 mm, you need to edit the feature again.

5. Select the base feature from the **FeatureManager Design Tree** or from the drawing area; all the dimensions of the feature are displayed in the drawing area.

6. Double-click on the dimension that is displayed in blue and reflects the depth of the base feature; the selected dimension is displayed in the edit box.

7. Set the value in the **Distance** spinner to **35** and then press the ENTER key.

8. Choose the **Rebuild** button from the Menu Bar or press CTRL+B to rebuild the model.

Creating the Cut Feature

Next, you need to create a cut feature on the front face and copy this cut feature on the right planar face.

1. Invoke the **Extruded Cut** tool and create the cut feature on the front face, as shown in Figure 9-12.

2. Select the cut feature in the **FeatureManager Design Tree**, press and hold the CTRL key, drag the selected feature, and place it on the right planar face; the **Copy Confirmation** dialog box is displayed.

3. Choose the **Delete** button from the dialog box; the cut feature is created on the right planar face.

4. Select the newly created cut feature from the **FeatureManager Design Tree**; a pop-up toolbar is displayed. Choose **Edit Sketch** from the pop-up toolbar.

5. Apply suitable constraints and dimensions to make it a fully-defined sketch.

6. Choose the **Rebuild** button from the Menu Bar or press CTRL+B to rebuild the model. The model after copying the feature is shown in Figure 9-13.

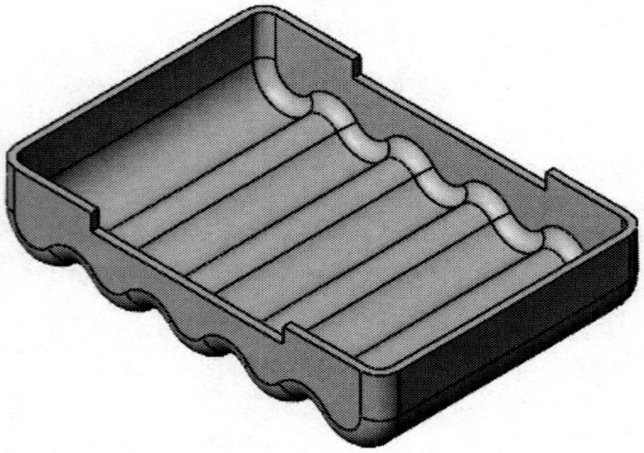

Figure 9-12 *The cut feature created on the front face*

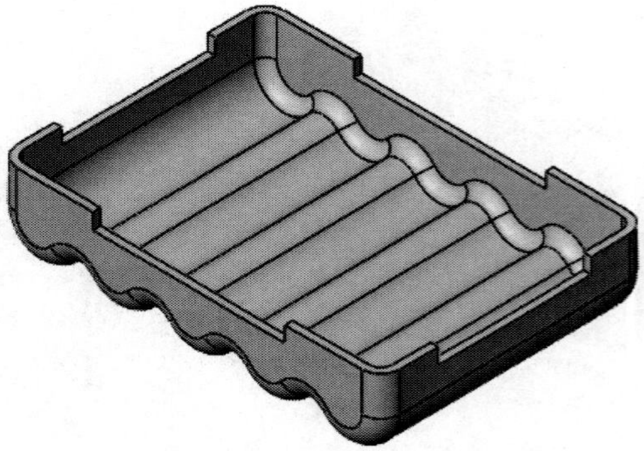

Figure 9-13 *The model after copying the feature*

7. Create slots, add fillet, and pattern the features. The model after creating these features is shown in Figure 9-14.

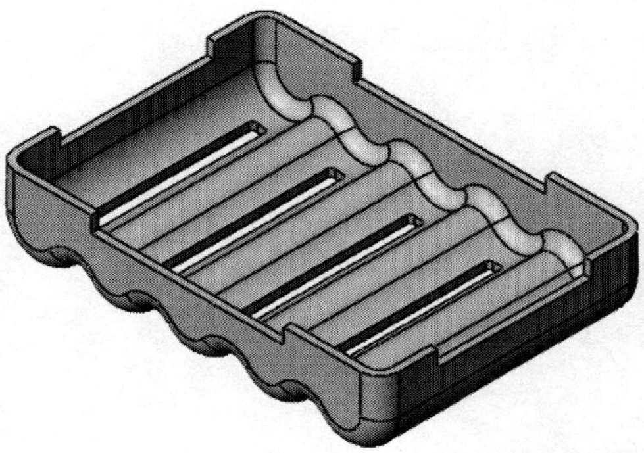

Figure 9-14 *Model after creating the features*

Tip
To create a fillet, invoke the ***Fillet*** *tool and choose the* ***Add*** *tab in the* ***FilletXpert PropertyManager****. Now, choose an edge of the slot; a pop-up toolbar will be displayed. Choose* ***Connected to start loop, 3 Edges*** *from the pop-up toolbar to select all the vertical edges of the slot.*

Creating the Standoff

Now, you need to create the standoff for the model. It is created by extruding a sketch drawn on the sketch plane at an offset distance from the Top Plane. You also need to specify a draft angle while creating this feature.

1. Create a reference plane at an offset distance of 10.5 mm from the **Top Plane**. To flip the direction of the reference plane, you need to select the **Flip offset** check box from the **Plane PropertyManager**, if required.

2. Select the newly created plane as the sketching plane, draw the sketch of the standoff, and apply the required relations and dimensions. The sketch consists of a circle of 1 mm diameter. For other dimensions, refer to Figure 9-2.

3. Extrude the sketch using the **Up To Next** option with an outward draft angle of 10-degree. Hide the newly created plane; the standoffs of the model are created.

4. Rotate the model and add a fillet of radius 0.25 mm to the base of the standoff.

   The rotated and zoomed view of the complete standoff is displayed in Figure 9-15.

5. Pattern the filleted standoff feature using the **Linear Pattern** tool. The isometric view of the final model is shown in Figure 9-16.

Saving the Model

1. Save the model with the name *c09_tut01* at the following location:

   \Documents\SOLIDWORKS\c09

2. Choose **File > Close** from the SOLIDWORKS menus to close the document.

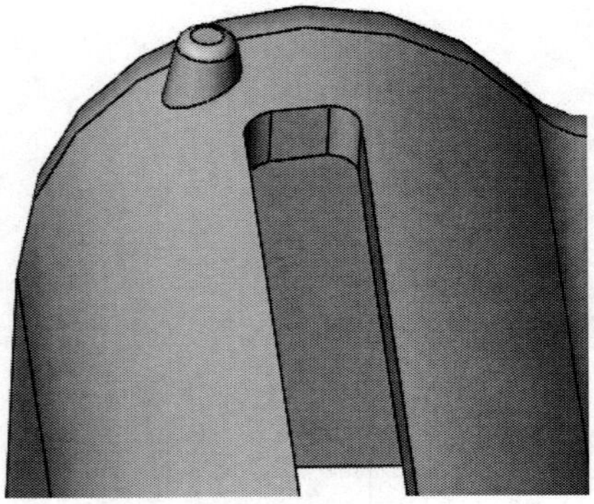

Figure 9-15 *Rotated and zoomed view of the model to show the standoff*

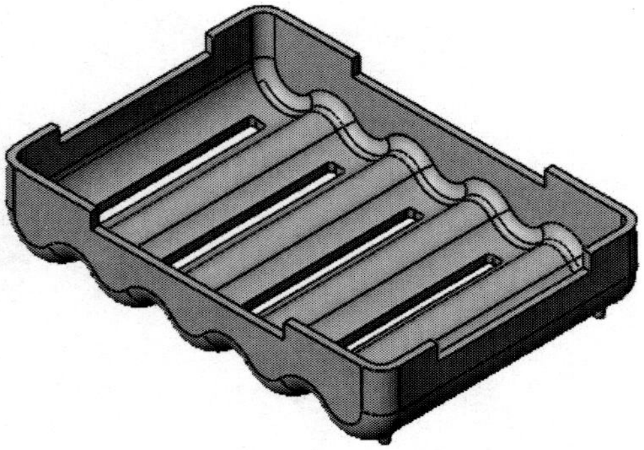

Figure 9-16 *Final model*

Tutorial 2

In this tutorial, you will create the model shown in Figure 9-17. The views and dimensions of the model are shown in the same figure. **(Expected time: 45 min)**

The following steps are required to complete this tutorial:

a. Create the base feature of the model by revolving the sketch along its central axis, refer to Figures 9-18 and 9-19.
b. Draw the sketch of the second feature on the top face of the base feature and extrude it to a given dimension, refer to Figures 9-20 and 9-21.
c. Create the revolve cut feature, refer to Figures 9-22.
d. Create the hole using the **Simple Hole** tool, and then pattern it using the **Circular Pattern** tool, refer to Figure 9-23.
e. Create a drilled hole feature using the **Hole Wizard** tool, refer to Figure 9-23.
f. Mirror the hole feature about the **Right Plane**, refer to Figure 9-23.
g. Apply the fillet, refer to Figure 9-23.
h. Perform the live sectioning of the model, refer to Figure 9-24.
i. Save the model.

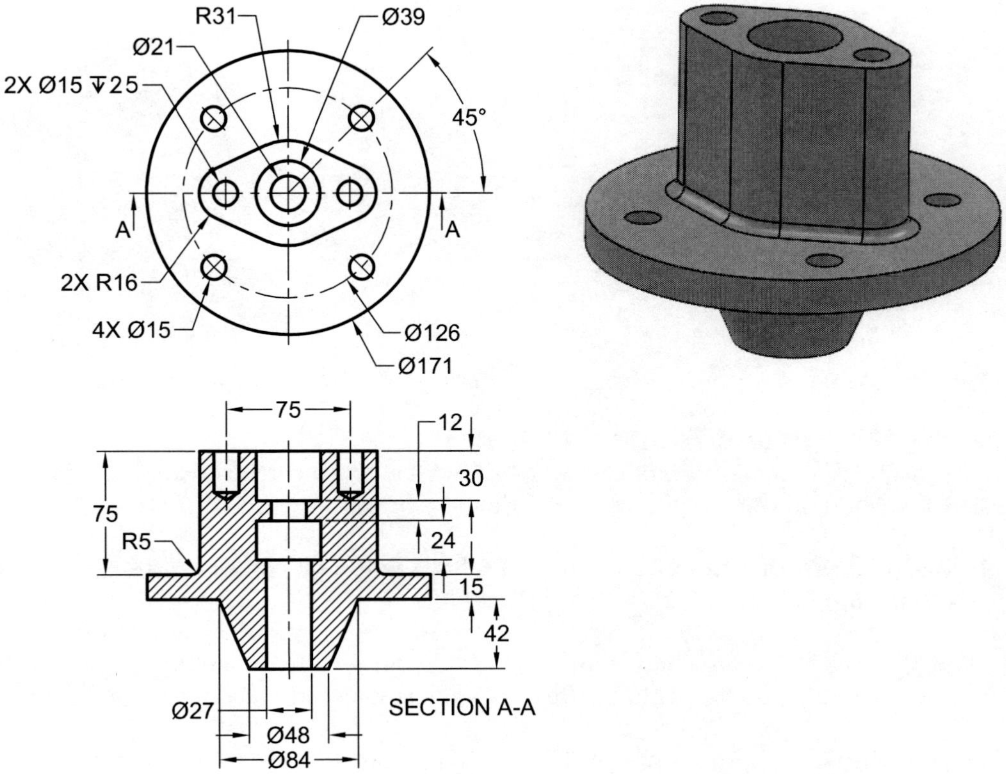

Figure 9-17 Views and dimensions of the model for Tutorial 2

Creating the Base Feature

First, you need to create the base feature of the model by revolving the sketch created on the **Front Plane**.

1. Start a new SOLIDWORKS Part document using the **New SOLIDWORKS Document** dialog box.

2. Invoke the **Revolved Boss/Base** tool and draw the sketch of the base feature on the **Front Plane**. Add the required relations and dimensions to the sketch, as shown in Figure 9-18.

3. Exit the sketching environment.

   You do not need to set any parameters in the **Revolve PropertyManager** because the default value in the **Angle** spinner is 360 degrees as required.

4. Choose the **OK** button from the **Revolve PropertyManager**. The base feature created after revolving the sketch is shown in Figure 9-19.

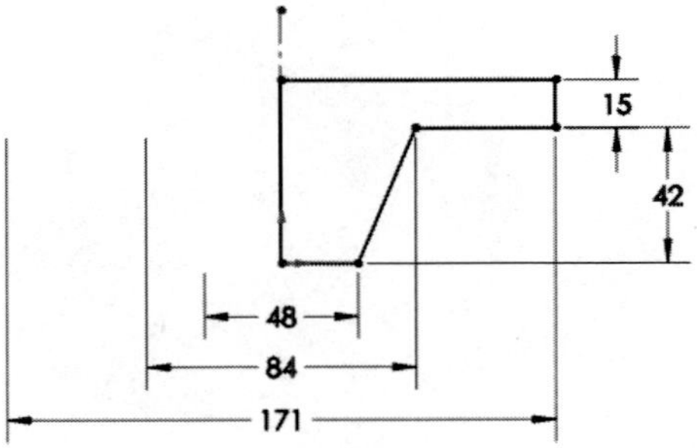

Figure 9-18 Sketch for the base feature

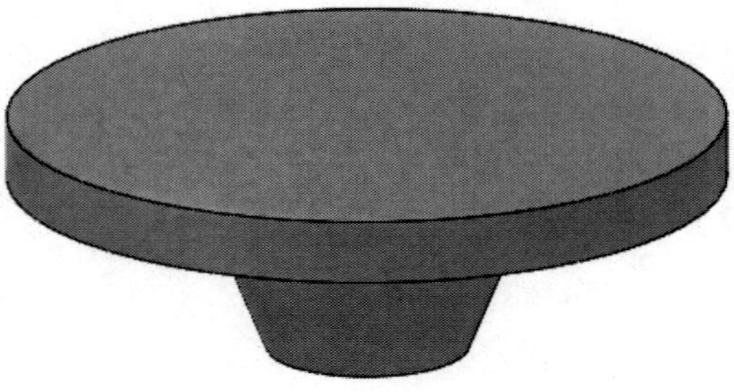

Figure 9-19 The dimetric view of the base feature

Creating the Second Feature

The second feature of this model is an extruded feature. It can be created by extruding the sketch created on the top planar face of the base feature.

1. Invoke the **Extruded Boss/Base** tool and select the top planar face of the base feature as the sketching plane.

2. Draw the sketch of the second feature and apply the required relations and dimensions to it, as shown in Figure 9-20. Make sure the sketch is symmetric about the centerline.

3. Extrude the sketch upto a distance of 75 mm. The isometric view of the model after creating the second feature is shown in Figure 9-21.

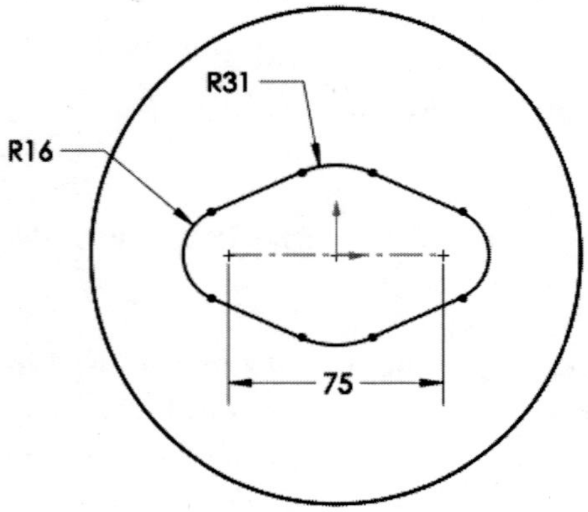

Figure 9-20 Sketch for the second feature

Figure 9-21 Model after adding the second feature

Creating the Third Feature

The third feature of the model can be created by revolving a sketch using the cut option. The sketch for this feature will be created on the **Front Plane**.

1. Invoke the **Revolved Cut** tool and select the **Front Plane** as the sketching plane.

2. Draw the sketch of the revolved cut feature, and then apply required relations and dimensions to it, as shown in Figure 9-22.

3. Exit the sketching environment and create a revolved cut feature with a default angle value of 360 degrees.

Creating the Remaining Features

1. Create the remaining features of the model using the **Fillet**, **Simple Hole**, **Mirror**, and **Hole Wizard** tools.

   The isometric view of the model after creating all the remaining features is displayed in Figure 9-23.

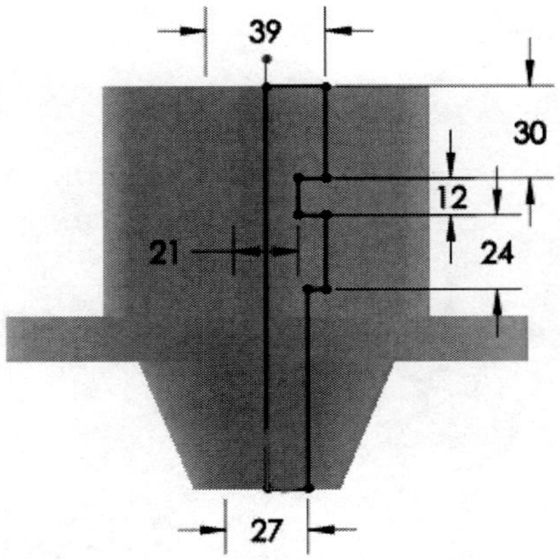

Figure 9-22 *Sketch for the revolved cut feature*

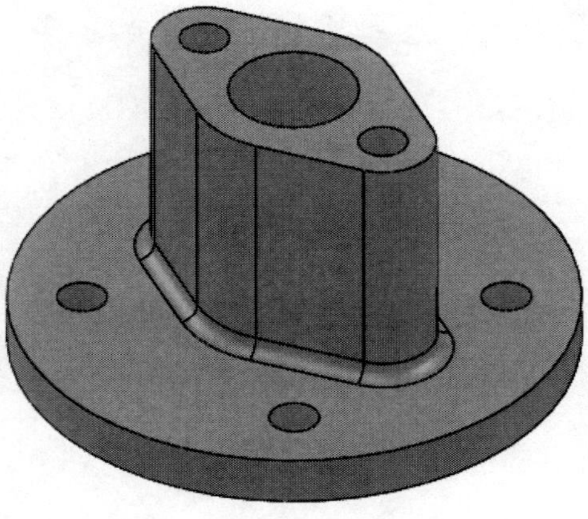

Figure 9-23 *Isometric view of the final model*

Sectioning the Model

1. Choose the top planar surface of the base feature and right-click; a shortcut menu is displayed.

2. Choose the **Live Section Plane** option from the shortcut menu; a sectioning plane is displayed with one ring and two wings at the center.

3. Right-click and choose the **Fit To Part** option.

4. Move the cursor to the vertical wing, press and hold the left mouse button, and drag the cursor to section the model along the vertical plane. Figure 9-24 shows the live section of the model when the sectioning plane is at an angle of 270 degree.

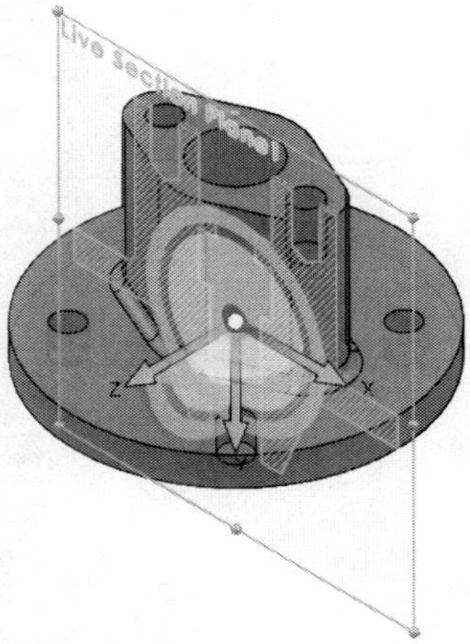

Figure 9-24 *Model with the sectioning plane*

5. Similarly, you can rotate the sectioning plane using the other wing.

6. To remove the sectioning plane, move the cursor near the edges of the section plane in the drawing area and then right-click. Next, choose the **Delete** option from the shortcut menu displayed.

Saving the Model

Now, you need to save the model.

1. Save the model with the name *c09_tut02* at the following location:

   \Documents\SOLIDWORKS\c09

2. Choose **File > Close** from the SOLIDWORKS menus to close the document.

EXERCISES

Exercise 1

Create the model, as shown in Figure 9-25. The other views and dimensions of the model are also given in the same figure. **(Expected time: 45 min)**

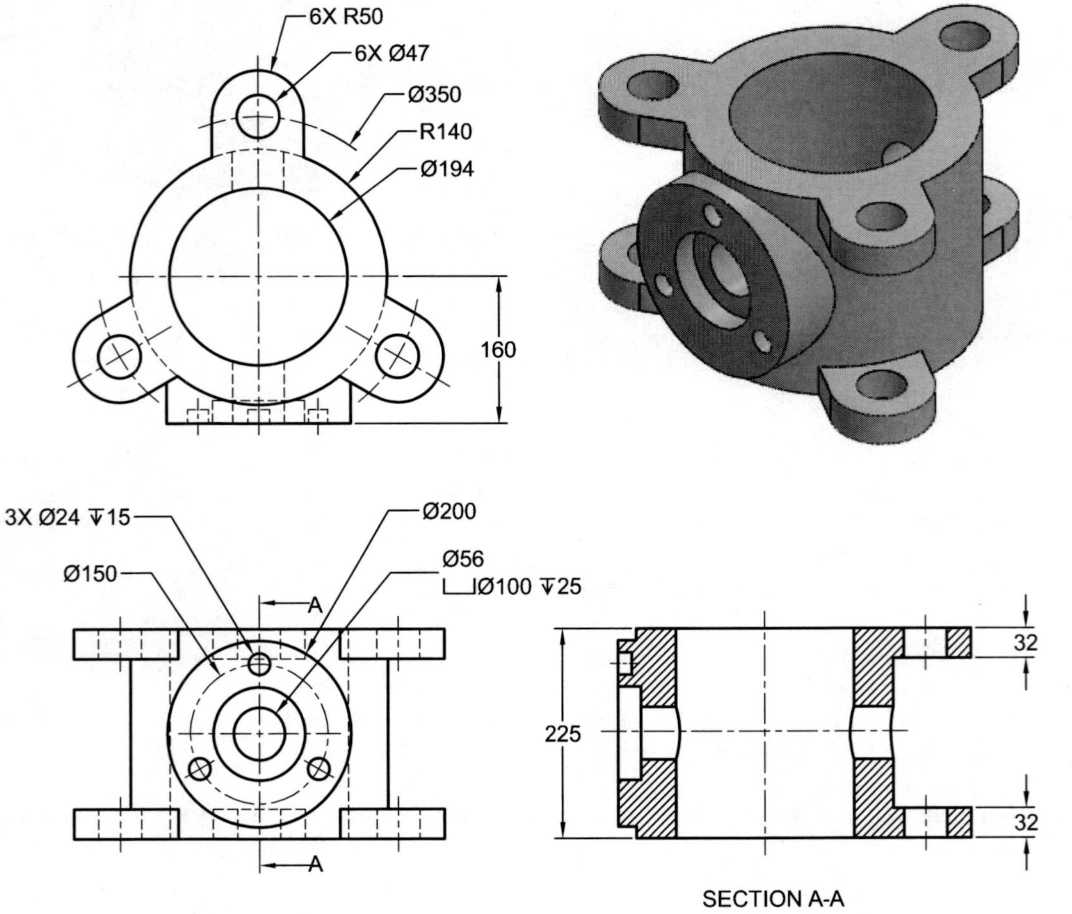

Figure 9-25 *Views and dimensions of the model for Exercise 1*

Exercise 2

Create the model, as shown in Figure 9-26. The views and dimensions of the model are shown in the same figure. **(Expected time: 30 min)**

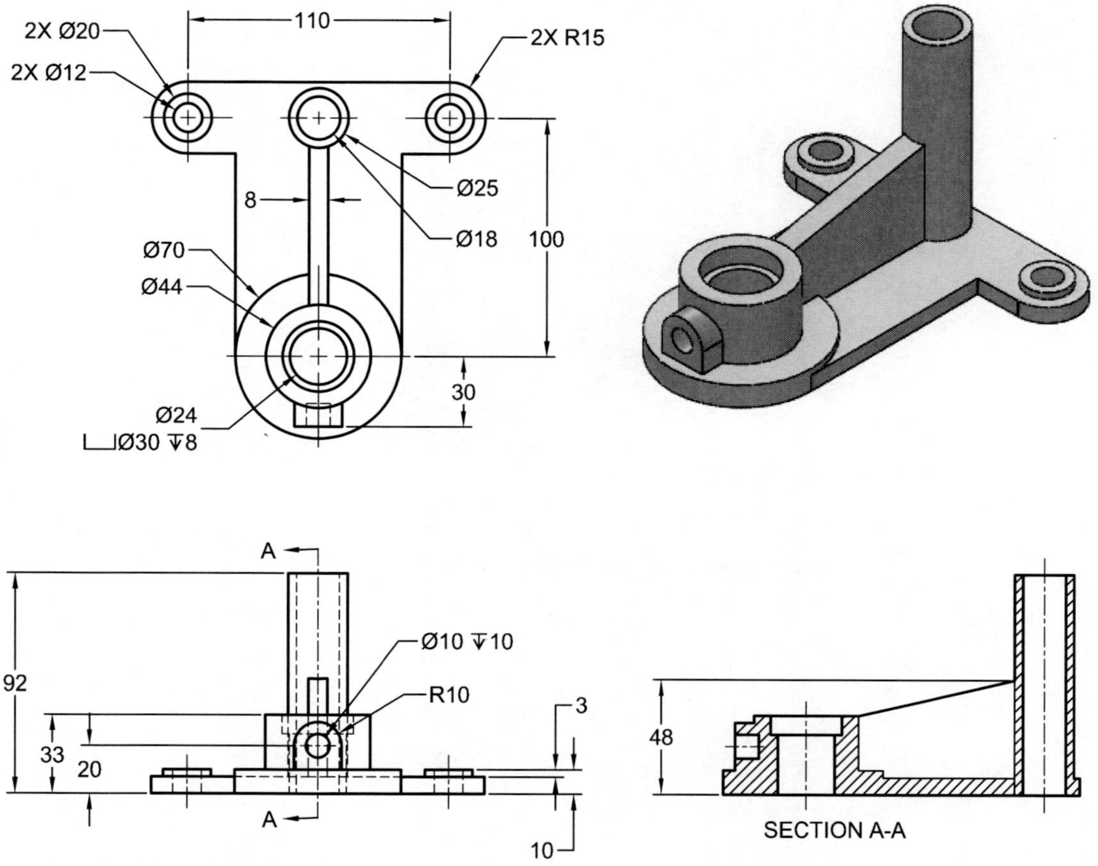

Figure 9-26 *Views and dimensions of the model for Exercise 2*

This page is intentionally left blank

Chapter 10

Advanced Modeling Tools-III

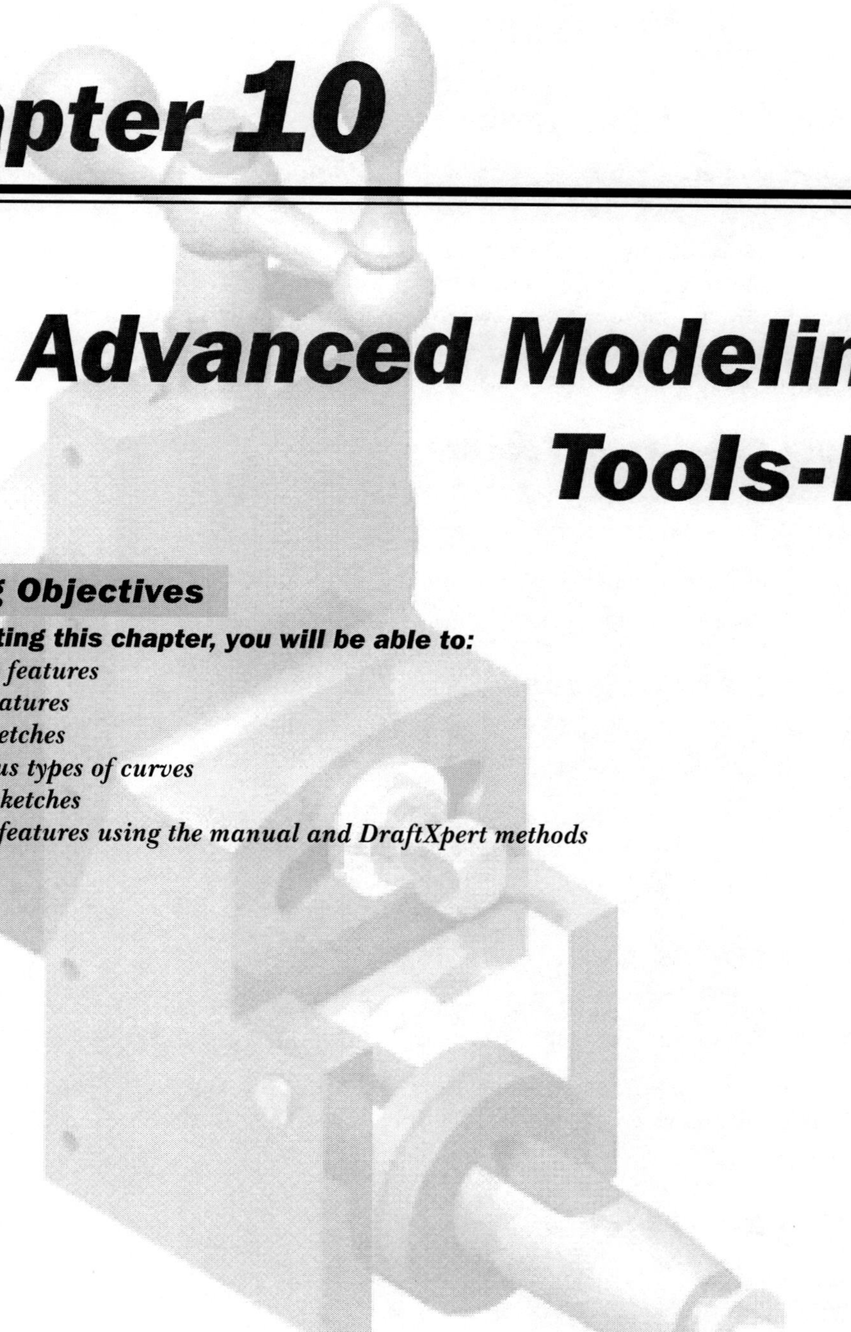

Learning Objectives

After completing this chapter, you will be able to:

- *Create sweep features*
- *Create loft features*
- *Create 3D sketches*
- *Create various types of curves*
- *Extrude 3D sketches*
- *Create draft features using the manual and DraftXpert methods*

ADVANCED MODELING TOOLS

Some of the advanced modeling tools were discussed in the earlier chapters. In this chapter, you will learn about some more advanced modeling tools such as sweep, loft, draft, curves, 3D sketches, and so on.

Creating Sweep Features

CommandManager:	Features > Swept Boss/Base
SOLIDWORKS menus:	Insert > Boss/Base > Sweep

One of the most important advanced modeling tools is the **Swept Boss/Base** tool. This tool is used to extrude a closed profile along an open or a closed path. Therefore, you need a profile and a path to create a sweep feature. A profile is a section for the sweep feature and a path is the course taken by the profile while creating the sweep feature. The profile has to be a sketch, but the path can be a sketch, curve, or an edge.

Creating Cut-Sweep Features

CommandManager:	Features > Swept Cut
SOLIDWORKS menus:	Insert > Cut > Sweep

You can also remove material from an existing feature or a model by creating a cut-sweep feature.

Creating Loft Features

CommandManager:	Features > Lofted Boss/Base
SOLIDWORKS menus:	Insert > Boss/Base > Loft

The loft features are created by blending more than one similar or dissimilar sections together to get a free form shape. These similar or dissimilar sections may or may not be parallel to each other. Note that the sections for the solid lofts should be the closed sketches.

Creating Lofted Cuts

CommandManager:	Features > Lofted Cut
SOLIDWORKS menus:	Insert > Cut > Loft

You can remove the material in a part by using the **Cut Loft PropertyManager**. This PropertyManager is invoked by choosing **Insert > Cut > Loft** from the SOLIDWORKS menus.

Creating 3D Sketches

CommandManager:	Sketch > Sketch > 3D Sketch
SOLIDWORKS menus:	Insert > 3D Sketch

In the earlier chapters, you have learned to draw 2D sketches in the sketching environment. In this chapter, you will learn to draw 3D sketches. 3D sketches are mostly used to create 3D paths for the sweep features, 3D curves, and so on.

Mirroring Sketch Entities in a 3D Sketch

In SOLIDWORKS 2018, you can mirror the sketch entities of a 3D sketch. To do so, start a 3D sketch and then create sketch entities. In the open sketch, choose the **Mirror Entities** button from the **Sketch CommandManager**; the **Mirror PropertyManager** will be displayed. Select the entities from the drawing area and then select a plane or a planar entity from the drawing area as the mirror line; the selected entities will be mirrored.

Creating Curves

In SOLIDWORKS, you can create various types of curves that can be used to create complex shapes. The tools to create such curves are grouped together in the **Curves** flyout of the **Features CommandManager**. The curve types that can be created by using these tools are discussed next.

Creating a Projected Curve

CommandManager:	Features > Curves > Project Curve
SOLIDWORKS menus:	Insert > Curve > Projected

The **Project Curve** tool is used to project a closed/open sketched entity on one or more than one planar or curved faces. You can also project a sketched entity on another sketched entity to create a 3D curve.

Creating Split Lines

CommandManager:	Features > Curves > Split Line
SOLIDWORKS menus:	Insert > Curve > Split Line

The **Split Line** tool is generally used to project a sketch on a planar or curved face. This sketch in turn splits or divides the single face into two or more faces.

Creating a Composite Curve

CommandManager:	Features > Curves > Composite Curve
SOLIDWORKS menus:	Insert > Curve > Composite

The **Composite Curve** tool is used to create a curve by combining 2D or 3D curves, sketched entities, and part edges into a single curve. You need to ensure that the selected entities form a continuous chain. A composite curve is mainly used while creating a sweep or loft feature.

Creating a Curve Through XYZ Points

CommandManager:	Features > Curves > Curve Through XYZ Points
SOLIDWORKS menus:	Insert > Curve > Curve Through XYZ Points

The **Curve Through XYZ Points** tool is used to create a curve by specifying the coordinate points.

Creating a Curve Through Reference Points

CommandManager:	Features > Curves > Curve Through Reference Points
SOLIDWORKS menus:	Insert > Curve > Curve Through Reference Points

The **Curve Through Reference Points** tool enables you to create a curve by selecting the sketched points, vertices, origin, endpoints, or center points.

Creating a Helical/Spiral Curve

CommandManager:	Features > Curves > Helix and Spiral
SOLIDWORKS menus:	Insert > Curve > Helix/Spiral

The **Helix and Spiral** tool is used to create a helical curve or a spiral curve. This curve is generally used as the sweep path to create springs, threads, spiral coils, and so on.

Extruding a 3D Sketch

You can also extrude a 3D sketch drawn in the 3D sketching environment. A 3D sketch is always extruded along a direction vector, which can be a line, edge, planar face, or a plane.

Creating Draft Features

CommandManager:	Features > Draft
SOLIDWORKS menus:	Insert > Features > Draft

The **Draft** tool is used to add taper to the selected faces of a model. Draft feature is mostly added to the models that are to be molded or cast as it will be easier to remove them from the mold or die.

TUTORIALS

Tutorial 1

In this tutorial, you will create the model shown in Figure 10-1. The dimensions of the model are shown in the same figure. **(Expected time: 45 min)**

The following steps are required to complete this tutorial:

a. The base feature of the model is a sweep feature. First, you need to create a path for the sweep feature on the Front Plane. Next, you need to create a thin circular sweep feature because the base feature of the model is a hollow feature, refer to Figures 10-2 through 10-4.
b. Create the remaining features, refer to Figure 10-5.
c. Save the model.

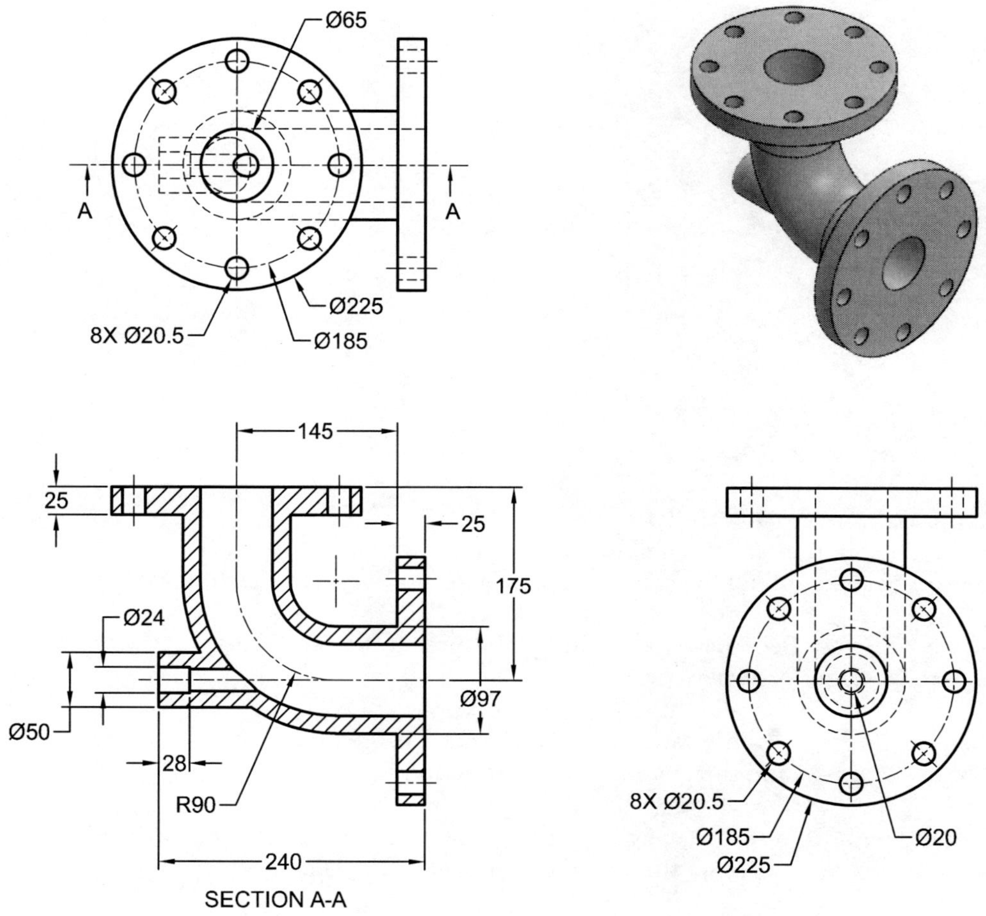

Figure 10-1 *Views and dimensions of the model for Tutorial 1*

Creating the Path for the Sweep Feature

As discussed earlier, the base feature of the model is a sweep feature. To create the sweep feature, you first need to create its path on the Front Plane.

1. Start a new SOLIDWORKS part document using the **New SOLIDWORKS Document** dialog box.

2. Draw the sketch of the path for the sweep feature on the **Front Plane** and add required relations and dimensions to the sketch, as shown in Figure 10-2. Exit the sketching environment and change the view to isometric, as shown in Figure 10-3.

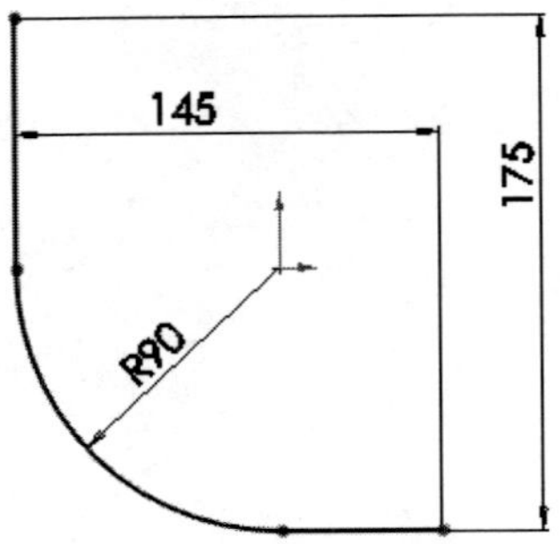

Figure 10-2 Sketch of the path

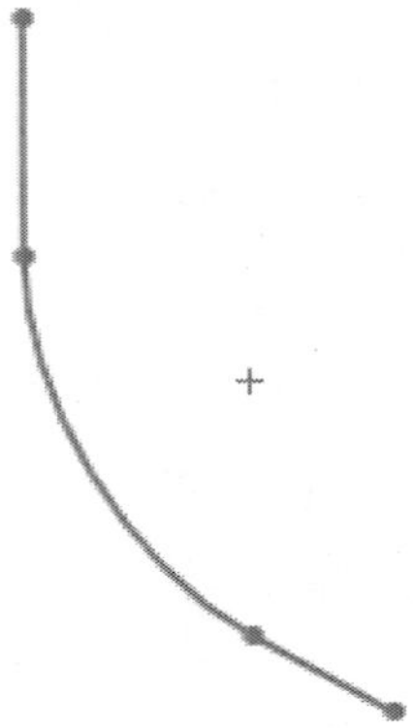

Figure 10-3 Isometric view of the sketch

Creating the Sweep Feature

The sweep feature that you need to create is a thin circular sweep feature. You need to use the **Thin Feature** rollout to specify the parameters of the sweep feature.

1. Choose the **Swept Boss/Base** tool from the **Features CommandManager**; the **Sweep PropertyManager** is displayed.

2. Select the **Circular Profile** radio button from the **Profile and Path** rollout. As a result, the **Path** selection box and the **Diameter** spinner is displayed.

3. Select the path from the drawing area, the name of the sketch will be displayed in the **Path** selection box. Also, set the value of diameter in the **Diameter** spinner to 97.

4. Select the check box on the left of the **Thin Feature** rollout; the **Thin Feature** rollout is invoked.

5. Set the value of thickness in the **Thickness** spinner to **16**.

   Since the direction of thickness addition is opposite to the required direction, therefore you need to reverse the direction.

6. Choose the **Reverse Direction** button from the **Thin Feature** rollout. Choose the **OK** button from the **Sweep PropertyManager** or choose **OK** from the confirmation corner.

   The base feature created by sweeping a circular profile along a path is shown in Figure 10-4.

Note

Instead of creating a thin sweep feature, you can also create a solid sweep feature and then add a shell feature to hollow the base feature.

Creating Remaining Features

1. Create the remaining features (flanges) on both ends of the sweep feature using the **Extruded Boss/Base** tool, refer to Figure 10-5.

2. Create the hole using the **Simple Hole** tool and create circular pattern of the hole feature.

3. Create a plane at an offset distance of **240** mm from the right face of the model. Make sure that the **Flip offset** check box is selected while creating reference plane. Then, create the circular feature by extruding it using the **Up To Next** option. For dimensions refer to Figure 10-1.

4. Create the custom-sized counterbore hole using the **Hole Wizard** tool on the extruded feature created in the previous step, refer to Figure 10-1. The final solid model for Tutorial 1 is shown in Figure 10-5.

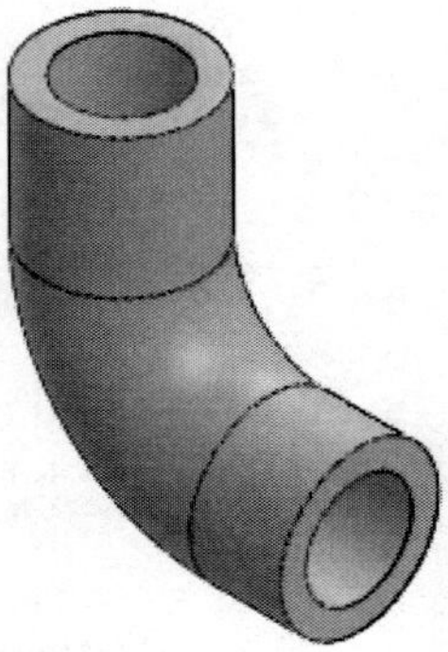

Figure 10-4 *Base feature of the model*

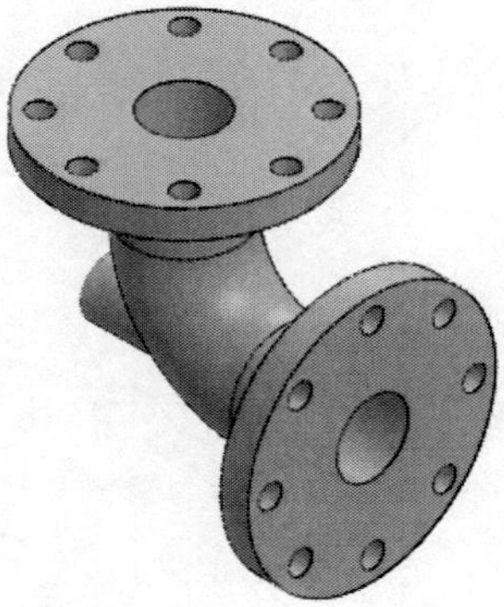

Figure 10-5 *Final model for Tutorial 1*

Saving the Model

1. Save the model with the name *c10_tut01* at the following location:

   \Documents\SOLIDWORKS\c10

2. Choose **File > Close** from the SOLIDWORKS menus to close the file.

Tutorial 2

In this tutorial, you will create the chair frame shown in Figure 10-6. The dimensions of the chair frame are also shown in the same figure. **(Expected time: 30 min)**

The following steps are required to complete this tutorial:

a. Invoke the 3D sketching environment and then draw the sketch of a 3D path. Create only the left half of the 3D path in the 3D sketching environment, refer to Figure 10-7.
b. Sweep the profile along the 3D path using the **Thin Feature** option, refer to Figure 10-8.
c. Mirror the sweep feature using the Front Plane, refer to Figure 10-9.
d. Save the model.

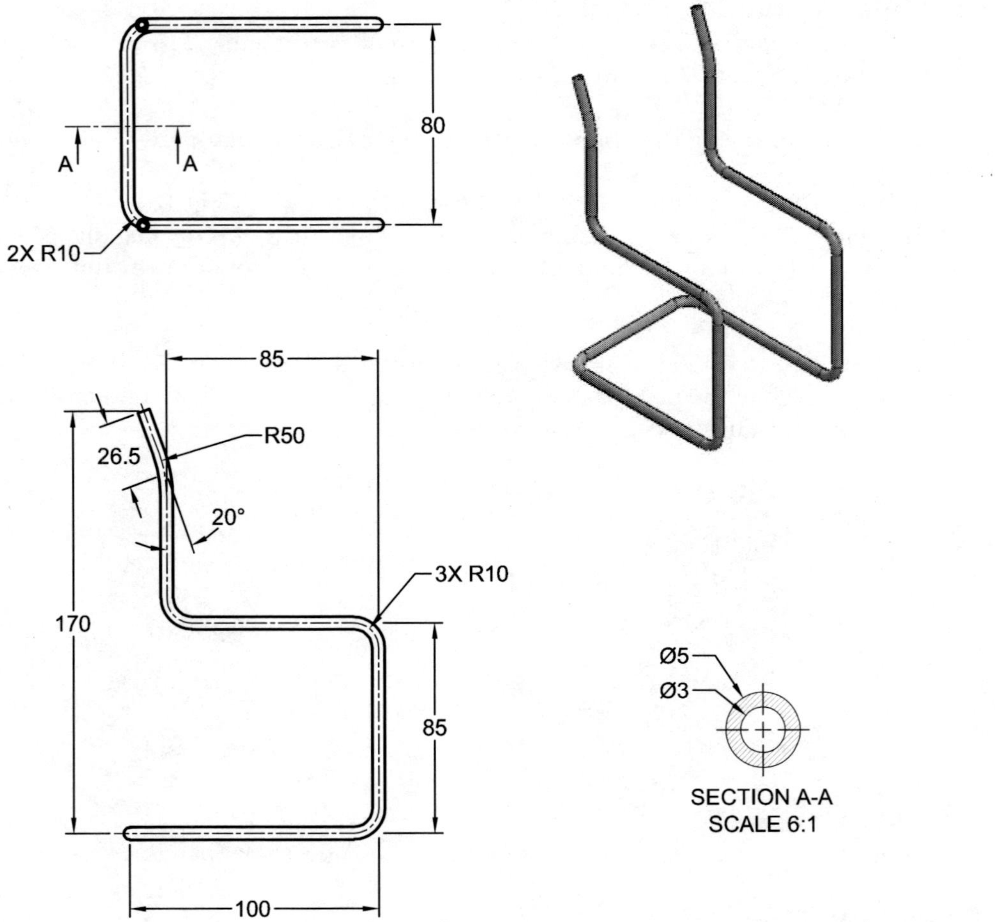

Figure 10-6 *Views and dimensions of the model for Tutorial 2*

Creating the Path for the Sweep Feature Using the 3D Sketching Environment

It is evident from Figure 10-6 that the model needs to be created by sweeping a profile along the 3D path. Therefore, you need to create a path for the sweep feature in the 3D sketching environment.

1. Start a new SOLIDWORKS part document.

2. Change the current view to isometric.

3. Choose **Insert > 3D Sketch** from the SOLIDWORKS menus or choose the **3D Sketch** button from the **Sketch** flyout in the **FeatureManager Design Tree** to invoke the 3D sketching environment.

On doing so, the 3D sketching environment is invoked and the origin is displayed in red. Also, the confirmation corner is displayed on the top right corner of the drawing area. The sketching tools that can be used in the 3D sketching environment are also highlighted.

You need to draw the left half of the sketch as the path of the sweep profile.

4. Invoke the **Line** tool; the select cursor is replaced by the line cursor. The **XY** symbol displayed below the line cursor indicates that the line will be sketched in the XY plane by default.

   You need to draw the first line in the ZX plane. Therefore, you need to toggle the plane before you start creating the sketch.

5. Press the TAB key twice to switch to the ZX plane.

6. Move the line cursor to the origin. When an orange dot is displayed, click to specify the start point of the line.

7. Move the cursor in positive Z direction of the triad; a small triad with Z appears below the cursor indicating that you are drawing the line along the **Z** direction. Click to specify the endpoint of the line when a value close to **40** is displayed above the cursor; a rubber-band line is attached to the cursor.

8. Move the cursor toward the right along the infinite line indicating the X-axis. Click to specify the endpoint of the line when a value close to **100** is displayed above the cursor.

9. Press the TAB key to switch to the XY plane. Next, move the cursor vertically upward.

10. Specify the endpoint of the line when the value above the cursor is close to **85**.

11. Similarly, draw the remaining part of the sketch, and then add required relations and dimensions to it.

12. Add the Parallel relation between the last line drawn in the 3D sketch and the Front plane. The final sketch is displayed in Figure 10-7.

13. Exit the 3D sketching environment by using the confirmation corner.

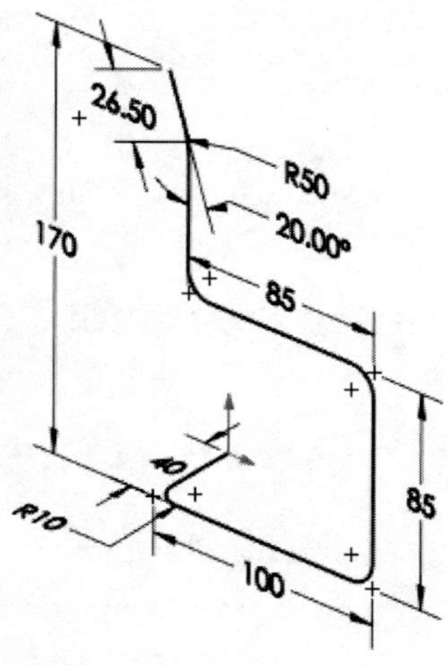

Figure 10-7 Sketch of the 3D path

Creating the Sweep Feature along the 3D Path

After creating the 3D path for the sweep feature, you need to sweep a circular profile along the 3D path using the **Swept Boss/Base** tool.

1. Choose the **Swept Boss/Base** button from the **Features CommandManager**; the **Sweep PropertyManager** is displayed.

2. Select the **Circular Profile** radio button from the **Profile and Path** rollout. As a result, the **Path** selection box and the **Diameter** spinner is displayed.

3. Select the path from the drawing area if not selected by default, the name of the path will be displayed in the **Path** selection box. Also set the value of diameter in the **Diameter** spinner to 5.

   As evident from Figure 10-6, the frame of the chair is made up of a hollow pipe. Therefore, you need to create a thin sweep feature to create a hollow chair frame.

4. Expand the **Thin Feature** rollout and set the value in the **Thickness** spinner to **1**.

5. Choose the **Reverse Direction** button from the **Thin Features** rollout to reverse the direction of the thin feature creation.

6. Choose the **OK** button from the **Sweep PropertyManager** to end the creation of the feature.

   The model after creating the sweep feature is displayed in Figure 10-8.

7. Mirror the sweep feature about the Front Plane using the **Mirror** tool. The model after mirroring is shown in Figure 10-9.

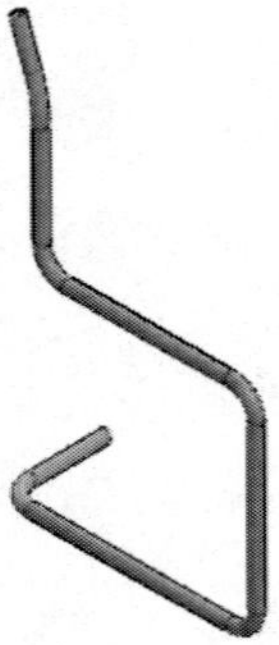

Figure 10-8 *Sweep feature created by sweeping a profile along a 3D path*

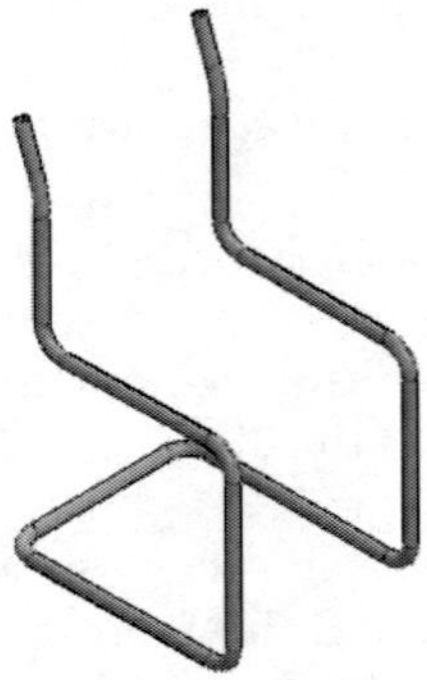

Figure 10-9 *The final model*

Saving the Model

1. Save the model with the name *c10_tut02* at the following location:

   \Documents\SOLIDWORKS\c10

2. Close the document.

EXERCISES

Exercise 1

Create the model of the Upper Housing shown in Figure 10-10. The dimensions of the model are shown in Figure 10-11. **(Expected time: 1 hr)**

Tip

This model is divided into three major parts. The first part is the base and is created by extruding the sketch upto a distance of 80 mm using the ***Mid Plane*** *option.*

The second part of this model is the right portion of the discharge venturi and is created by using the sweep feature. The path of the sweep feature will be created on the Right Plane.

The third part of this model is the left portion of the discharge venturi created using the loft feature. You need to create the first section of the loft feature on the planar face of the sweep feature created earlier. The second section will be created on a plane at an offset distance from the planar face of the sweep feature created earlier. Create a loft feature using the two sections created earlier. The other features needed to complete the model are fillets, hole, circular pattern, and so on.

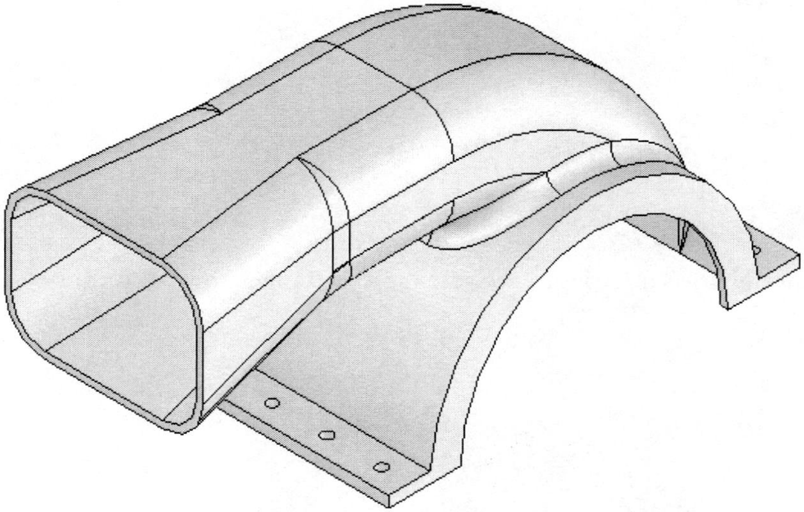

Figure 10-10 *Model of the Upper Housing*

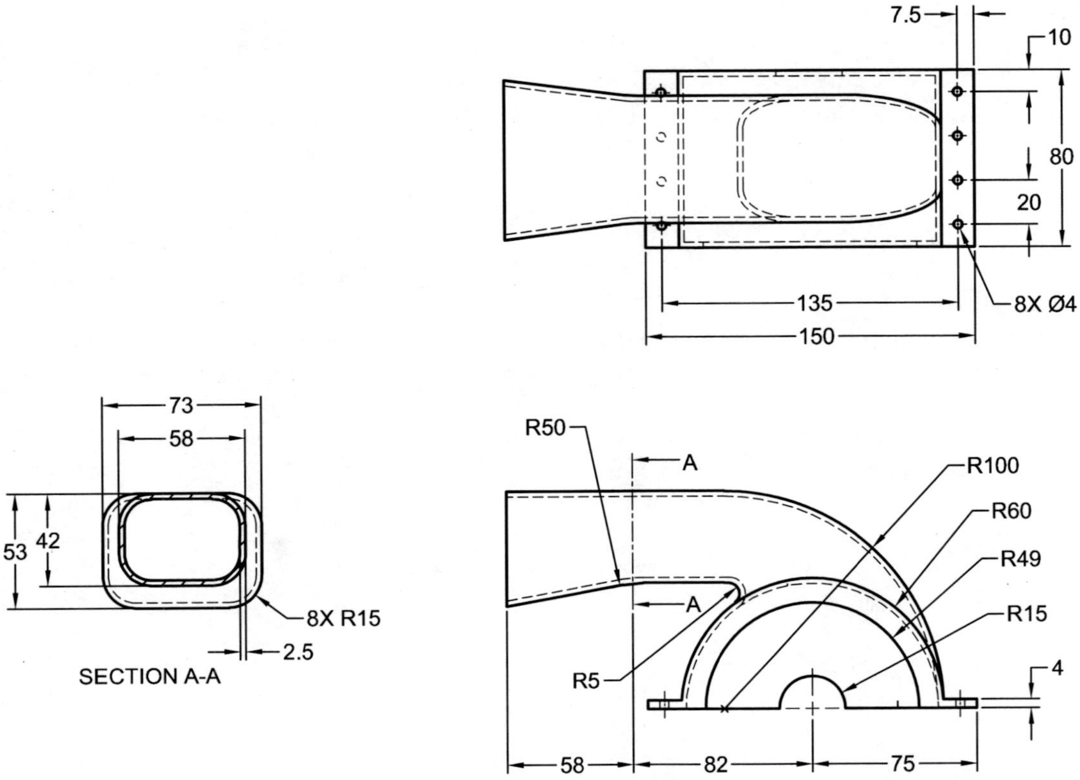

Figure 10-11 *Dimensions of the Upper Housing*

Exercise 2

Create the model shown in Figure 10-12. The dimensions of the model are shown in Figure 10-13. **(Expected time: 1 hr)**

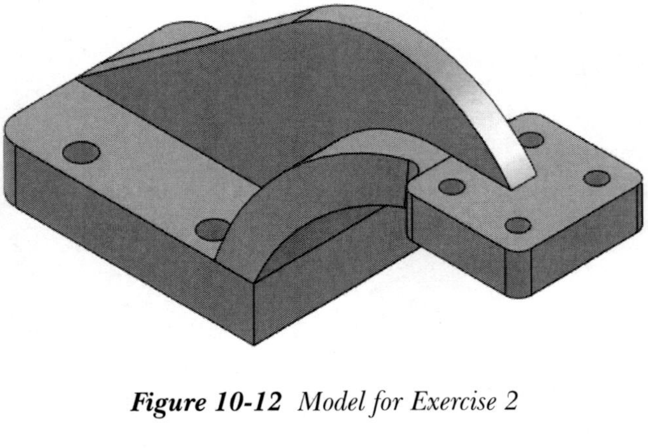

Figure 10-12 *Model for Exercise 2*

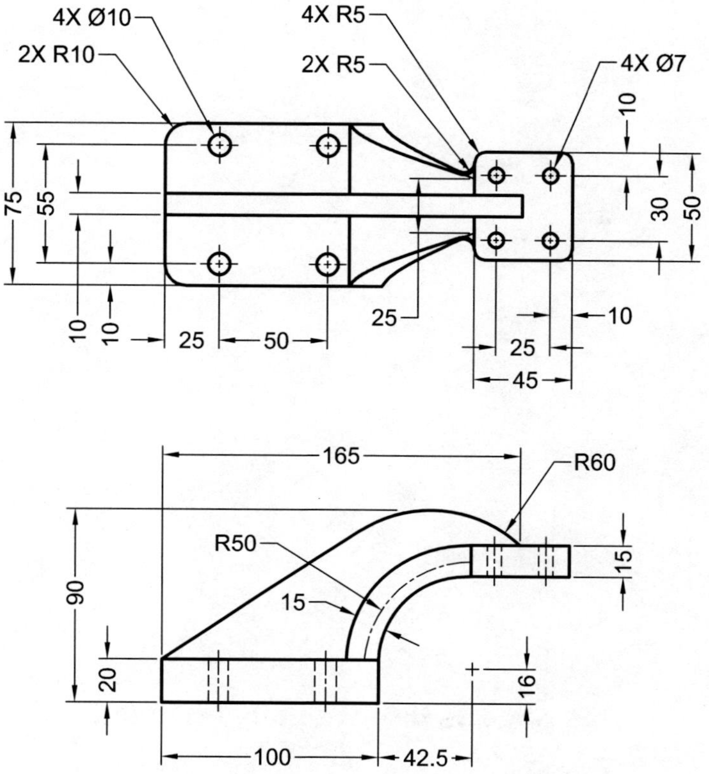

Figure 10-13 Dimensions of the model

This page is intentionally left blank

Chapter 11

Advanced Modeling Tools-IV

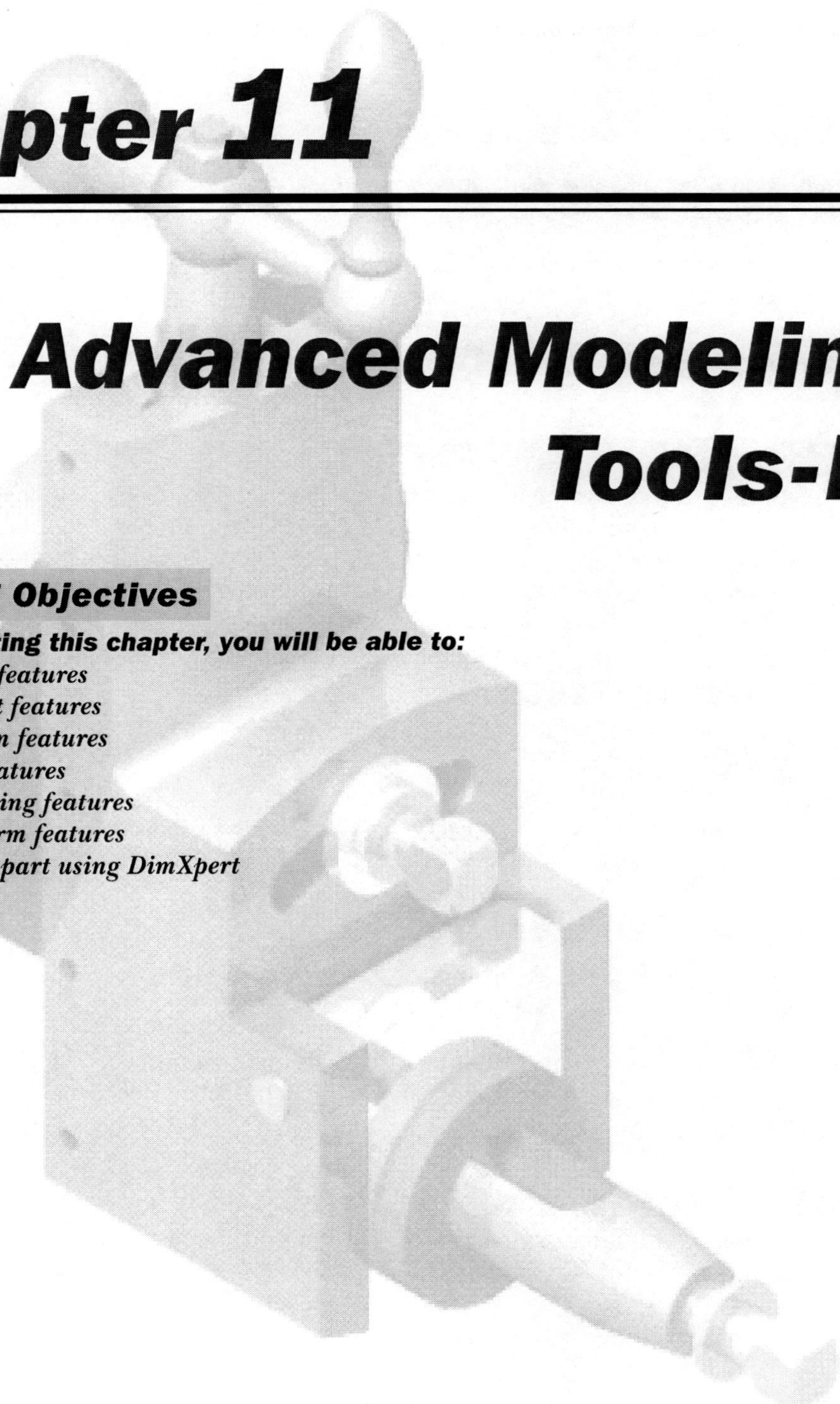

Learning Objectives

After completing this chapter, you will be able to:

- *Create dome features*
- *Create indent features*
- *Create deform features*
- *Create flex features*
- *Create fastening features*
- *Create freeform features*
- *Dimension a part using DimXpert*

ADVANCED MODELING TOOLS

Some of the advanced modeling tools have already been discussed in Chapters 7, 8, and 10. In this chapter, you will learn about some more advanced modeling tools that can be used to enhance your modeling ability.

Creating Dome Features

CommandManager: Features > Dome *(Customize to add)*
SOLIDWORKS menus: Insert > Features > Dome

The **Dome** tool is used to create a dome feature on the selected face. Depending on the direction along which a feature is created, a dome can be of convex or concave shape.

Creating Indents

CommandManager: Features > Indent *(Customize to add)*
SOLIDWORKS menus: Insert > Features > Indent

The **Indent** tool is extremely useful for the packaging industries, industrial designers, and product designers. The **Indent** tool is used in the multibody environment. This tool is used to add or scoop out material from a target body. The body that is used to add or scoop out the material is known as the tool body. Depending on the geometry, the tool is used to add or scoop out material from the target body.

Creating Deform Features

CommandManager: Features > Deform *(Customize to add)*
SOLIDWORKS menus: Insert > Features > Deform

The **Deform** tool is used to create free-style designs by manipulating the shape of the entire model or a particular portion of it.

Creating Flex Features

CommandManager: Features > Flex *(Customize to add)*
SOLIDWORKS menus: Insert > Features > Flex

The **Flex** tool is used to perform free form bending, twisting, tapering, and stretching of a selected body. The portion of the bodies to be deformed is defined by two planes known as trimming planes.

CREATING FASTENING FEATURES

SOLIDWORKS has a set of tools that are used to design the plastic products. These tools are known as fastening feature tools and are discussed next.

Creating the Mounting Boss

CommandManager: Features > Mounting Boss *(Customize to add)*
SOLIDWORKS menus: Insert > Fastening Feature > Mounting Boss

The **Mounting Boss** tool is used to create mounting boss features that are used in plastic components to accommodate fasteners while assembling the components.

Creating Snap Hooks

CommandManager: Features > Snap Hook *(Customize to add)*
SOLIDWORKS menus: Insert > Fastening Feature > Snap Hook

The snap hooks are generally created in small plastic boxes to create a push fit type arrangement to close or open a box.

Creating Snap Hook Grooves

CommandManager: Features > Snap Hook Groove *(Customize to add)*
SOLIDWORKS menus: Insert > Fastening Feature > Snap Hook Groove

The snap hook grooves are the cut features that are created to accommodate a snap hook in order to create a push fit type arrangement to close or open a box. Note that this feature works only in the case of multibody models. This is because you need to take reference from the snap hook created in one of the bodies to create the snap hook groove in the other body.

Creating Vents

CommandManager: Features > Vent *(Customize to add)*
SOLIDWORKS menus: Insert > Fastening Feature > Vent

The **Vent** tool is used to create vents in the existing models. This tool uses a closed sketch as the boundary of the vent and the open or closed sketched segments inside the closed sketch as the ribs and spars of the vent.

Creating a Lip/Groove Feature

CommandManager: Features > Lip / Groove *(Customize to add)*
SOLIDWORKS menus: Insert > Fastening Feature > Lip / Groove

The **Lip/Groove** tool is used to create lip or groove features on a component. These features are mainly used in plastic boxes in which they create a push-fit type arrangement. This arrangement is used for opening and closing of the plastic boxes. In the groove feature, material is removed from a component whereas in the lip feature material is added to the component.

CREATING FREEFORM FEATURES

CommandManager: Features > Freeform *(Customize to add)*
SOLIDWORKS menus: Insert > Features > Freeform

The **Freeform** tool is used to deform a face to get the required shape. This tool allows you to create the freeform features using the n-sided face of an existing feature. You can select the face of an existing feature and then deform it to any freeform shape.

DIMENSIONING A PART USING DimXpert

The set of tools in the **DimXpert CommandManager** are used to specify dimensions and tolerances on a part. Practice of using such type of dimensioning is termed as Mode-based definition. This helps in checking the design intent and providing exact information to the production team about the parts without the need of engineering drawings. The tolerances created using the DimXpert tools can be further used for tolerance analysis. Note that the dimensions created using the DimXpert tools are not parametric in nature. You can create the dimensions on the parts manually or automatically. The **DimXpertManager** keeps a track of the dimensions placed on the model.

TUTORIALS

Tutorial 1

In this tutorial, you will create the plastic cover shown in Figure 11-1. The dimensions of this cover are shown in Figure 11-2. Note that a draft angle of 1-degree has to be applied to the side faces of the cover. The parameters of the mounting boss are: Boss height = **14** mm, Boss diameter = **4.8** mm, Diameter of the boss step = **4** mm, Height of the boss step = **1.5** mm, Draft angle of the main boss = **2**-degree, Diameter of the inside hole = **3** mm, Diameter of the inside counter bore = **4** mm, Depth of the inside counter bore = **2** mm, Draft angle of the inside hole = **1**-degree, Length of fins = **4** mm, Width of fins = **1.5** mm, Height of fins = **12** mm, Draft angle of fins = **1**-degree, Distance for the fin chamfer from the edge of fin = **1.5** mm, Angle for the fin chamfer = **45**-degrees, and Number of Fins = **4**.

(Expected time: 30 min)

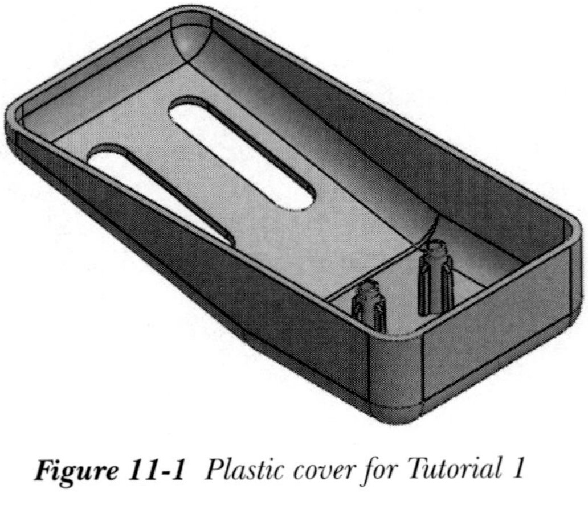

Figure 11-1 *Plastic cover for Tutorial 1*

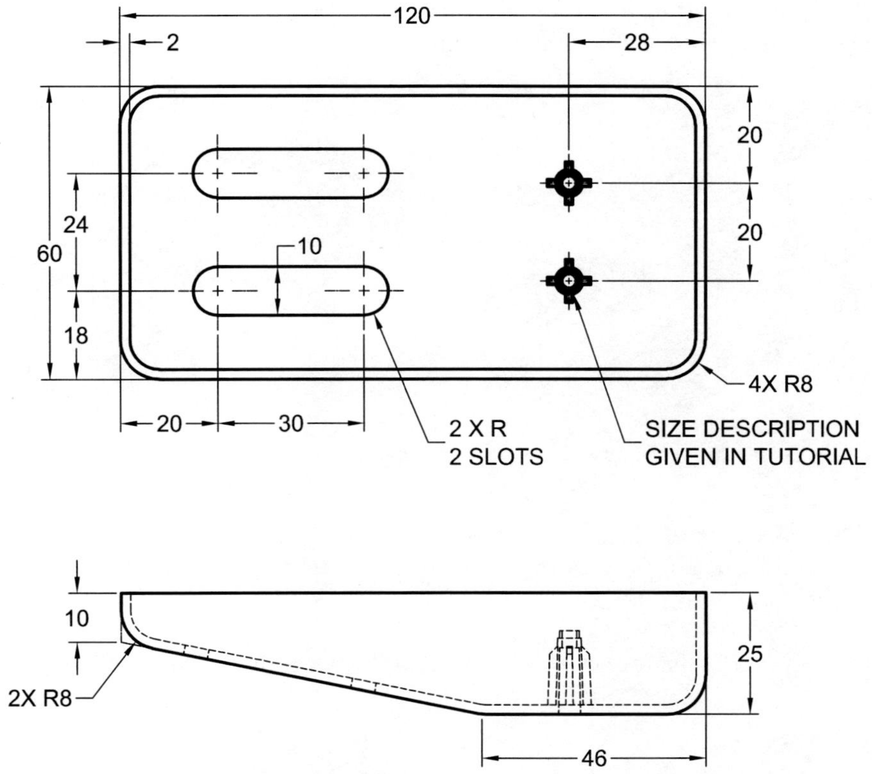

Figure 11-2 *Dimensions of the plastic cover*

The following steps are required to complete this tutorial:

a. Create the base feature of the model on the **Front Plane** and extrude it using the **Mid Plane** option, refer to Figure 11-3.
b. Add the face draft to the side faces of the model and then create the fillets, refer to Figure 11-4.
c. Create the shell feature by removing the top face, refer to Figure 11-5.
d. Create the cut feature, refer to Figure 11-6.
e. Add one of the mounting bosses and then edit its sketch to locate it by using dimensions, refer to Figure 11-8.
f. Mirror the mounting boss feature on the other side, refer to Figure 11-9.

Creating the Base Feature

1. Start SOLIDWORKS and then start a new part file.

2. Create the extruded base feature of the model on the Front Plane using the **Mid Plane** option. The base feature is shown in Figure 11-3.

Adding Draft and Creating Fillets

1. Invoke the **Draft** tool and add a draft of 1-degree to all the four vertical side faces of the model by selecting the top planar face as the neutral face.

2. Add a fillet of **8** mm radius to all the sharp edges of the model excluding those of the top planar face of the base feature, refer to Figure 11-4.

Figure 11-3 *Base feature of the model*

Figure 11-4 *Base feature with draft and fillet*

Creating the Shell and Cut Features

1. Create the shell feature for a wall thickness **2** mm by removing the top planar face of the model, as shown in Figure 11-5.

2. Create the extruded cut feature, as shown in Figure 11-6. For dimensions, refer to Figure 11-2.

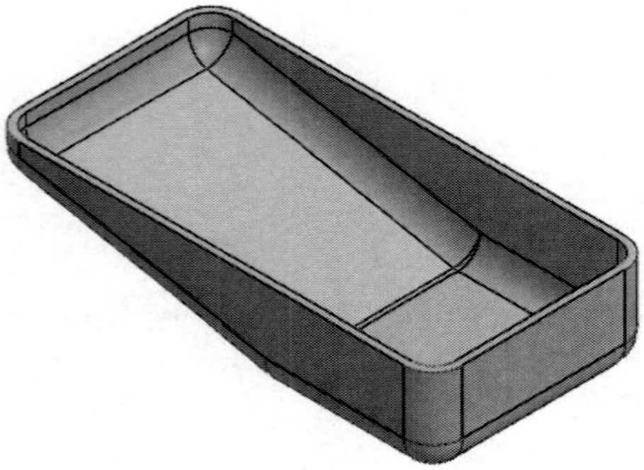

Figure 11-5 *Model after creating the shell*

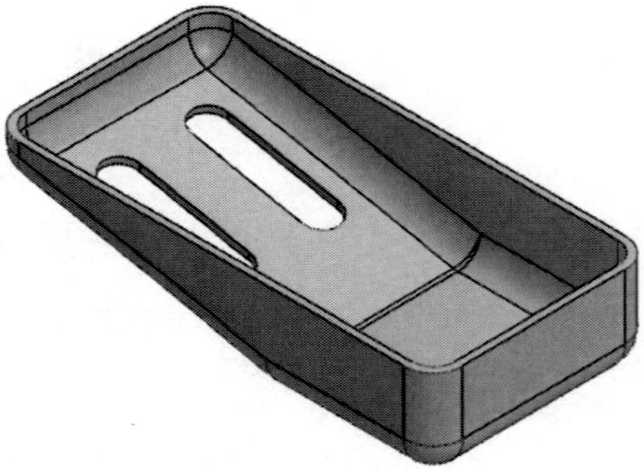

Figure 11-6 *Model after creating the cut feature*

Creating the Mounting Boss

Next, you need to create two mounting bosses. First you need to create only one mounting boss and then mirror it to create another instance. After creating the mounting bosses, you need to edit its sketch to place it at the exact location using the given dimensions.

1. Choose **Insert > Fastening Feature > Mounting Boss** from the SOLIDWORKS menus; the **Mounting Boss PropertyManager** is displayed and the **Select a face or a 3D point** selection box in the **Position** rollout is highlighted in it.

2. Click on the horizontal face close to the top edge as the face to place the mounting boss; a preview of the mounting boss feature is displayed, as shown in Figure 11-7.

3. Modify the mounting boss parameters based on the values given in the tutorial description.

4. Click in the **Select a vector to define orientation of the fins** selection box of the **Fins** rollout and then select one of the bottom horizontal edges to reorient the fins in the mounting boss.

5. Choose the **OK** button from the **Mounting Boss PropertyManager**; the mounting boss is created.

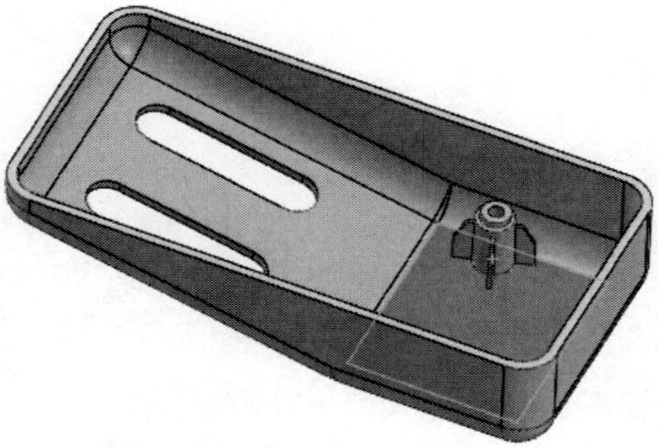

Figure 11-7 *Preview of the mounting boss*

You will notice that by default, the mounting boss is created at a point where you have selected the plane. So, you need to change the position of the mounting boss and move it to the required location. This is done by editing the sketch of the mounting boss which is automatically created when you create the mounting boss.

6. Click on the (▸) sign located on the left of the **Mounting Boss** in the **FeatureManager Design Tree** to expand it. Now, right-click on the 3D Sketch and choose **Edit Sketch** from the shortcut menu.

7. Position the 3D Sketch to its actual location using the **Smart Dimension** tool. Refer to Figure 11-2 for dimensions.

8. Exit the sketching environment; the mounting boss is created and gets placed at its proper location, as shown in Figure 11-8.

9. Use the **Mirror** tool to create mirror image of the mounting boss. Select **Front Plane** as the mirroring plane. The final model of the plastic cover is shown in Figure 11-9.

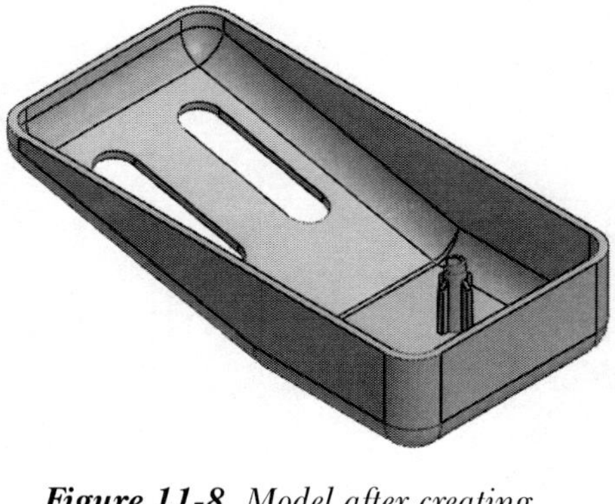

Figure 11-8 *Model after creating and positioning the mounting boss*

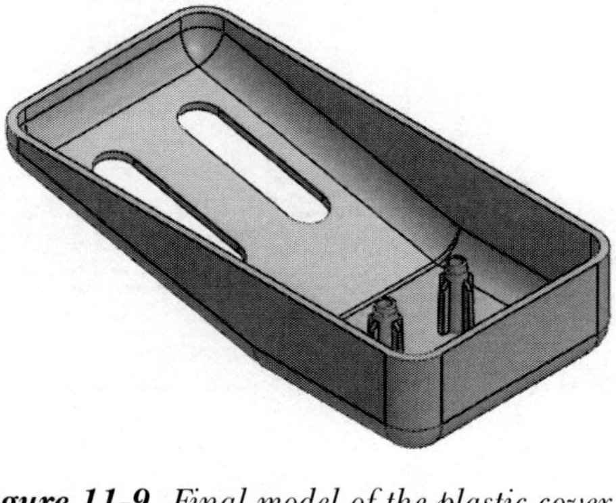

Figure 11-9 *Final model of the plastic cover*

Saving the Model

1. Save the model with the name *c11_tut01* at the following location:

\Documents\SOLIDWORKS\c11

2. Choose **File > Close** from the SOLIDWORKS menus to close the document.

Tutorial 2

In this tutorial, you will create the plastic cover shown in Figure 11-10. The dimensions of this cover are shown in Figure 11-11. In the lower vent, the vertical lines are ribs and the horizontal lines are spars. The other parameters of this vent are given below:

Depth of ribs = 1 mm, Width of ribs = 2 mm, Depth of spars = 0.5 mm, Width of spars = 2 mm, Offset of spars from surface = 0.5 mm. **(Expected time: 30 min)**

Figure 11-10 *Model for Tutorial 2*

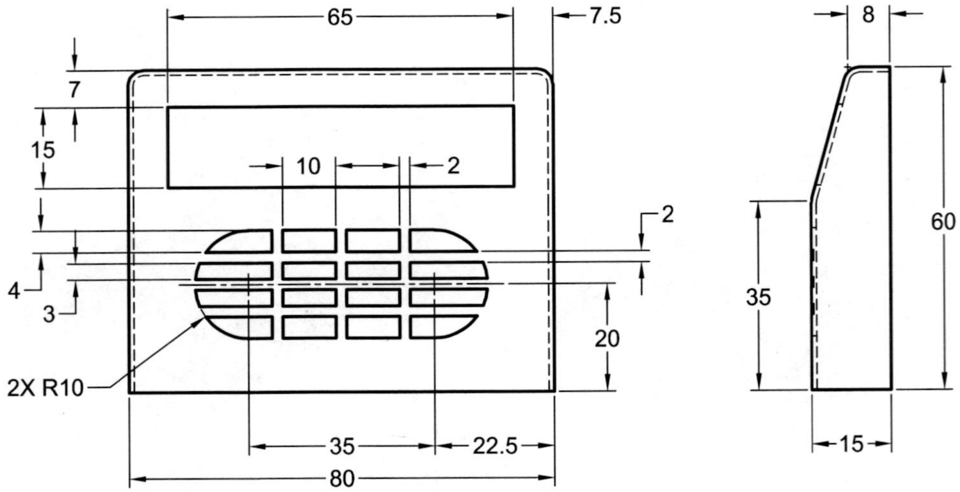

NOTE:
FILLET ALL SHARP EDGES, EXCEPT THE EDGES ON BOTTOM FACE, TO 3MM.
SHELL THICKNESS 1MM
HIDDEN LINES ARE SUPPRESSED FOR CLARITY

Figure 11-11 *Dimensions of the cover for Tutorial 2*

The following steps are required to complete this tutorial:

a. Create the base feature of the model on the **Right Plane** and extrude it using the **Mid Plane** option, refer to Figure 11-12.
b. Add the required fillets, refer to Figure 11-13.
c. Create the shell feature by removing the back and bottom faces, refer to Figure 11-14.
d. Create the first vent, refer to Figure 11-16.
e. Create the second vent, refer to Figure 11-18.
f. Save the model.

Creating the Base Feature

1. Start a new part file using the **New SOLIDWORKS Document** dialog box.

2. Create the base feature of the model using the **Right Plane** as the sketching plane, refer to Figure 11-12.

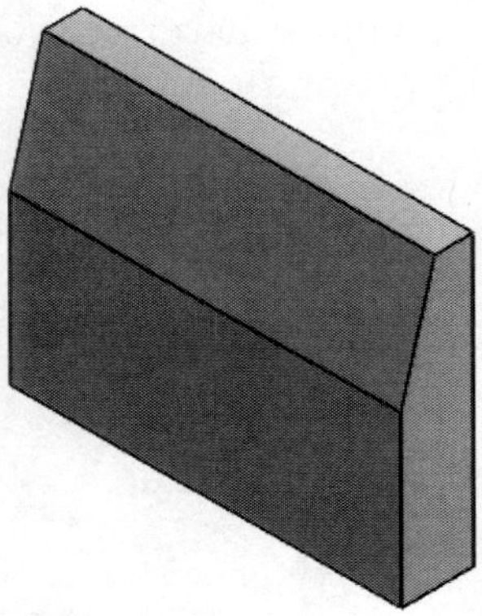

Figure 11-12 *Base feature of the model*

Adding Fillets

1. Add fillets of radius **3** mm to all the sharp edges of the model, except for the edges on the bottom and back faces. The model after adding the fillets is shown in Figure 11-13.

Creating the Shell Feature

1. Create the shell feature with a shell thickness value of **1** mm. Remove the back face and the bottom face of the model, as shown in Figure 11-14.

Figure 11-13 *Model after adding fillets*

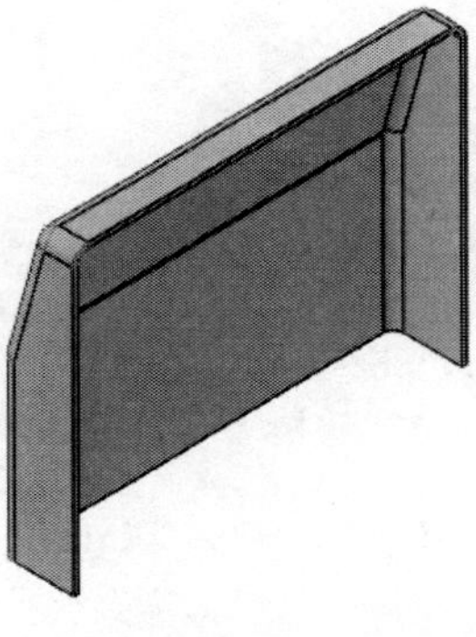

Figure 11-14 *Model after creating the shell feature*

Creating Vents

In the given model, you need to create two vents. The lower vent, which has arcs at the two ends, also has fins and spars. However, the upper vent consists of only the boundary. You will first create the lower vent.

1. Select the front face of the model as the sketching plane and draw the sketch of the vent, as shown in Figure 11-15.

2. Exit the sketching environment and then invoke the **Vent** tool; the **Vent PropertyManager** is displayed and the selection box in the **Boundary** rollout is activated.

3. Select the outer loop as the boundary of the vent. Make sure that the front face of the model is selected as the face on which the vent will be placed.

4. Select the vertical lines as ribs and then set the parameters of the ribs based on the information given in the tutorial statement.

5. Select the horizontal lines as spars and then set the parameters of the spars based on the information given in the tutorial statement. Reverse the direction of spars offset if the preview disappears.

6. Choose **OK** to exit the **Vent PropertyManager**. The model after creating the lower vent is shown in Figure 11-16.

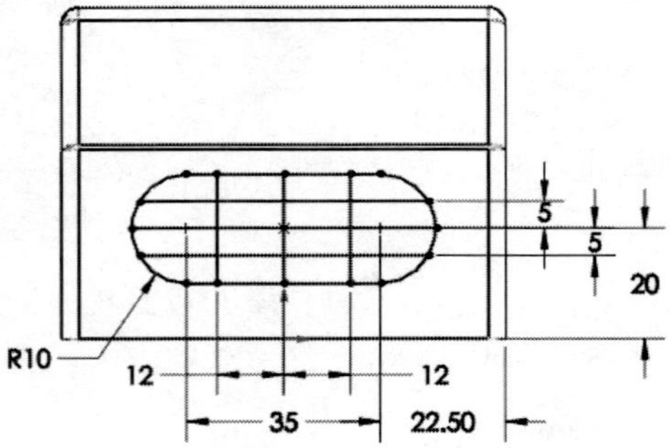

Figure 11-15 *Sketch for the vent*

Figure 11-16 *Model after creating the lower vent*

7. Draw the sketch for the upper vent on the front face of the model, as shown in Figure 11-17.

8. Exit the sketching environment and then invoke the **Vent** tool. Select the sketch as the boundary of the vent to create the vent.

9. Exit the **Vent** tool. The final model after creating the vent is shown in Figure 11-18.

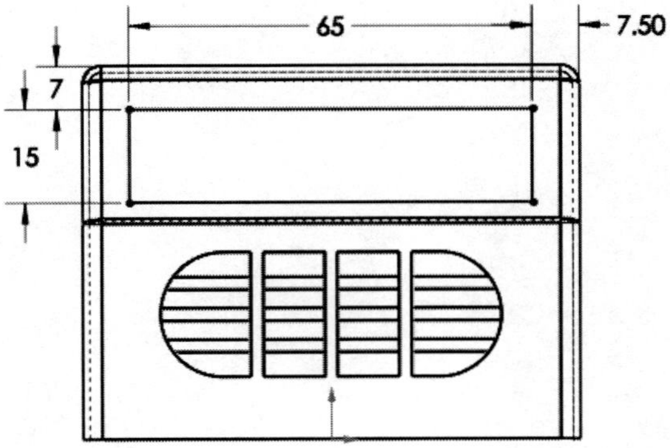

Figure 11-17 Sketch for the upper vent

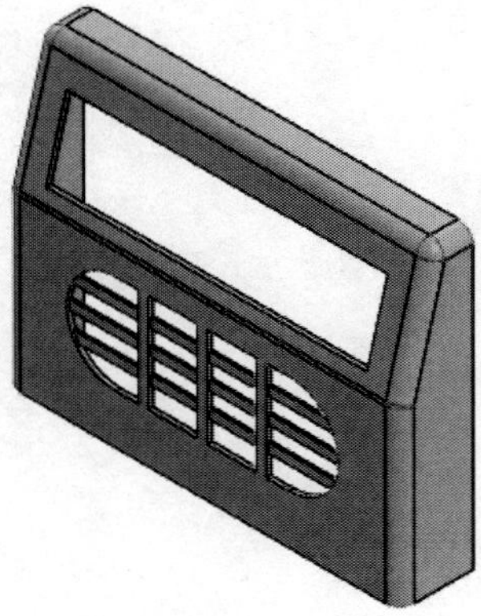

Figure 11-18 Final model

Saving the Model

1. Save the model with the name *c11_tut02* at the following location:

   \Documents\SOLIDWORKS\c11

2. Choose **File > Close** from the SOLIDWORKS menus to close the document.

EXERCISE

Exercise 1

Open the model created in Tutorial 1 of this chapter and then use the **Flex** tool to twist, bend, and stretch it. Figure 11-19 shows the model after twisting the body of the model.

(Expected time: 15 min)

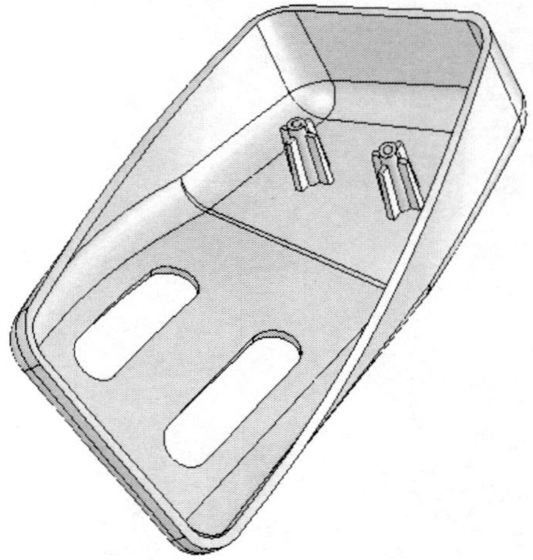

Figure 11-19 Twisted body of the plastic cover

This page is intentionally left blank

Chapter 12

Assembly Modeling-I

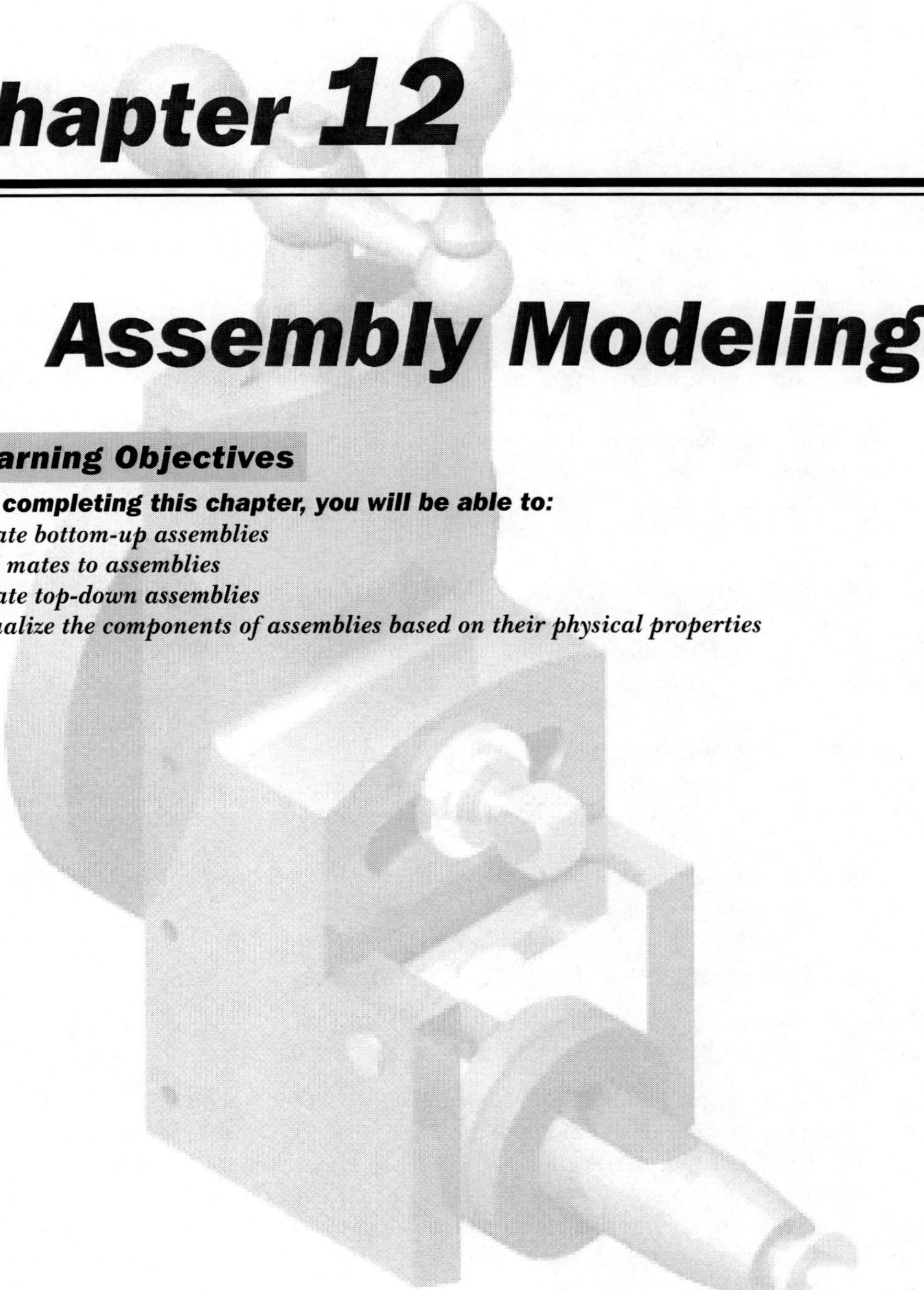

Learning Objectives

After completing this chapter, you will be able to:

- *Create bottom-up assemblies*
- *Add mates to assemblies*
- *Create top-down assemblies*
- *Visualize the components of assemblies based on their physical properties*

ASSEMBLY MODELING

An assembly design consists of two or more components assembled together at their respective work positions by using parametric relations. In SOLIDWORKS, these relations are called mates. Mates allow you to constrain the degrees of freedom of components at their respective work positions. To start the assembly mode of SOLIDWORKS, invoke the **New SOLIDWORKS Document** dialog box and then choose the **Assembly** button or from the **Welcome - SOLIDWORKS 2018** dialog box.

Types of Assembly Design Approach

In SOLIDWORKS, assemblies are created using two types of design approach: bottom-up approach and top-down approach. These approaches are discussed next.

Bottom-up Assembly Design Approach

The bottom-up assembly design approach is the traditional and the most widely preferred approach of assembly design. In this assembly design approach, all components are created as separate part documents, and then they are placed and referenced in the assembly as external components. In this type of approach, components are created in the **Part** mode and saved as the *.sldprt* documents. After creating and saving all components of the assembly, start a new assembly document (*.sldasm*) and insert the components in it using the tools provided in the **Assembly** mode. After inserting the components, assemble them using the assembly mates. The main advantage of this assembly design approach is that the view of the part is not restricted because there is only a single part in the current file. Therefore, this approach allows you to concentrate on the complex individual features. This approach is preferred while handling large assemblies or assemblies with complex parts.

Top-down Assembly Design Approach

In the top-down assembly design approach, the components are created in the same assembly document and can be saved within the assembly or as separate part files. Therefore, the top-down assembly design approach is entirely different from the bottom-up design approach. In this approach, you will start your work in the assembly document and the geometry of one part will help in defining the geometry of the other.

Starting an Assembly from the Part Document

SOLIDWORKS menus:	File > Make Assembly from Part
Menu Bar:	New flyout > Make Assembly from Part/Assembly

You can also start an assembly document from the part document. If the part document of the base component of the assembly is opened, choose **New > Make Assembly from Part/Assembly** from the Menu Bar or choose **File > Make Assembly from Part** from the SOLIDWORKS menus. If the **New SOLIDWORKS Document** dialog box is invoked, choose the **OK** button from this dialog box; an assembly document will be started and the **Begin Assembly PropertyManager** will be invoked. Choose the **OK** button from this PropertyManager to place the component at the origin.

Assembling Components

After placing the components in the assembly document, you need to assemble them. By assembling the components, you can constrain their degrees of freedom. As mentioned earlier, the components are assembled using mates. Mates help you to precisely place and position the component with respect to the other components and surroundings in the assembly. You can also define the linear and rotatory movements of the component with respect to the other components. Additionally, you can create a dynamic mechanism and check its stability by precisely defining the combination of mates. There are two methods for adding mates to the assembly, by using the **Mate PropertyManager** and by using the **Smart Mates**.

ASSEMBLY VISUALIZATION

In SOLIDWORKS, you can display the components of an assembly based on their physical properties. You can also group and display the components in same colors based on a property.

TUTORIALS

Tutorial 1

In this tutorial, you will create all components of a Bench Vice assembly and then assemble them. The Bench Vice assembly is shown in Figure 12-1. The dimensions of various components of the assembly are given in Figure 12-2 through 12-6. **(Expected time: 2 hr 45 min)**

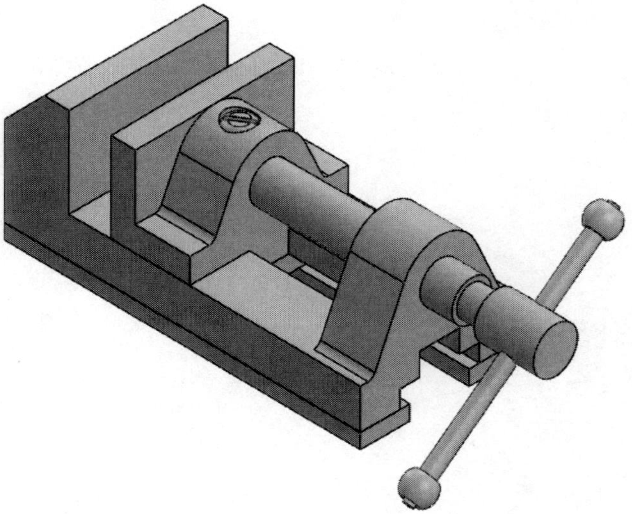

Figure 12-1 *Bench Vice assembly*

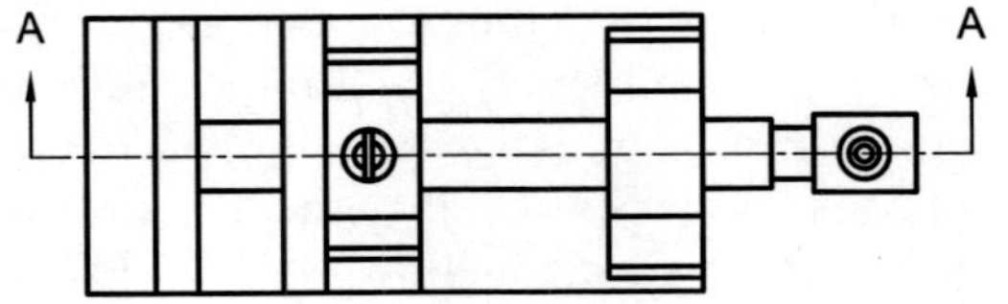

ITEM NO.	PART NO.	DESCRIPTION	QTY.
1	Vice Body		1
2	Vice Jaw		1
3	Jaw Screw		1
4	Clamping Plate		1
5	Base Plate		2
6	Screw Bar		1
7	Bar Globes		2
8	Oval Fillister		1
9	Set Screw1		4
10	Set Screw2		2

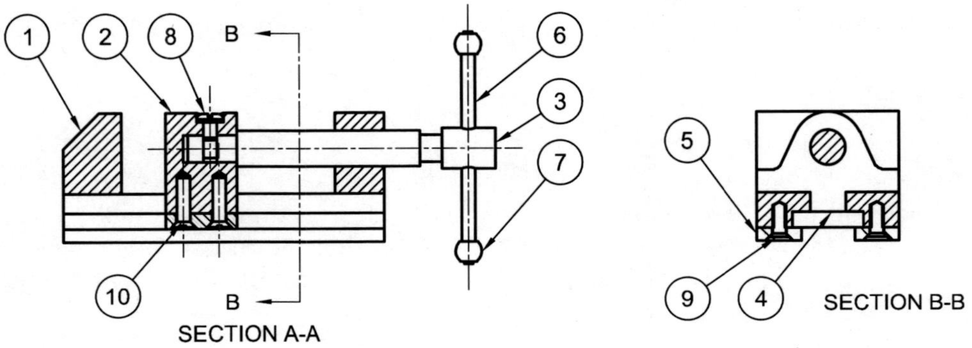

Figure 12-2 *Bench Vice assembly*

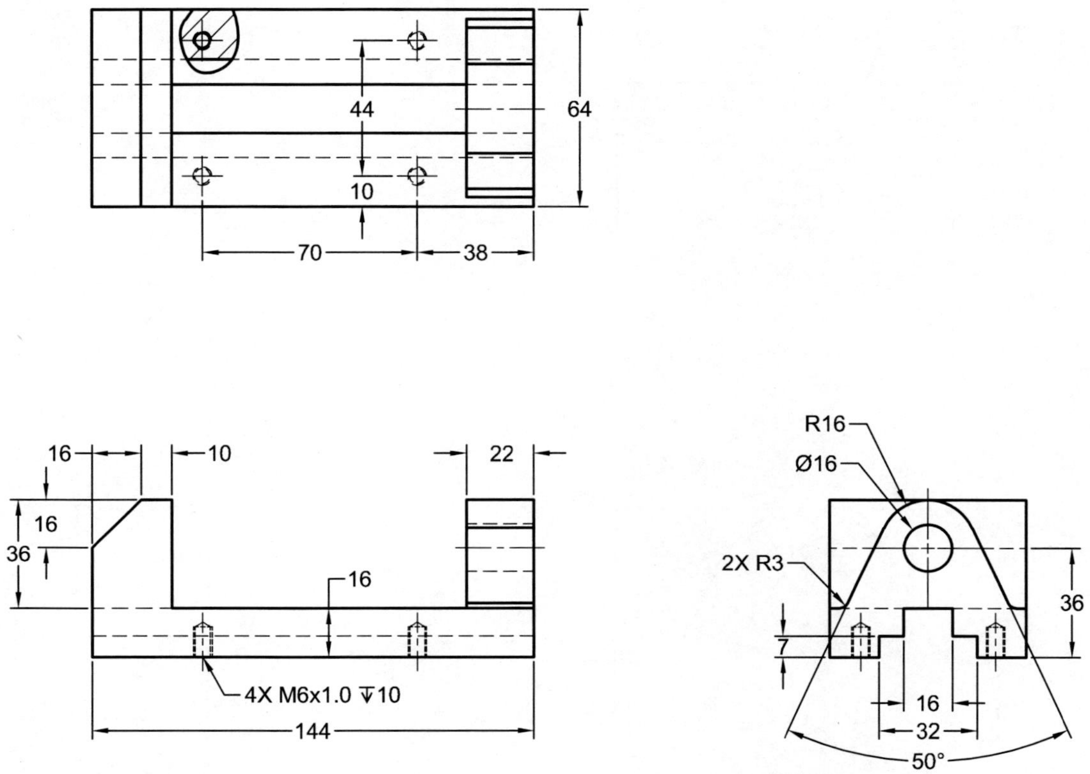

Figure 12-3 Orthographic views and dimensions of the Vice Body

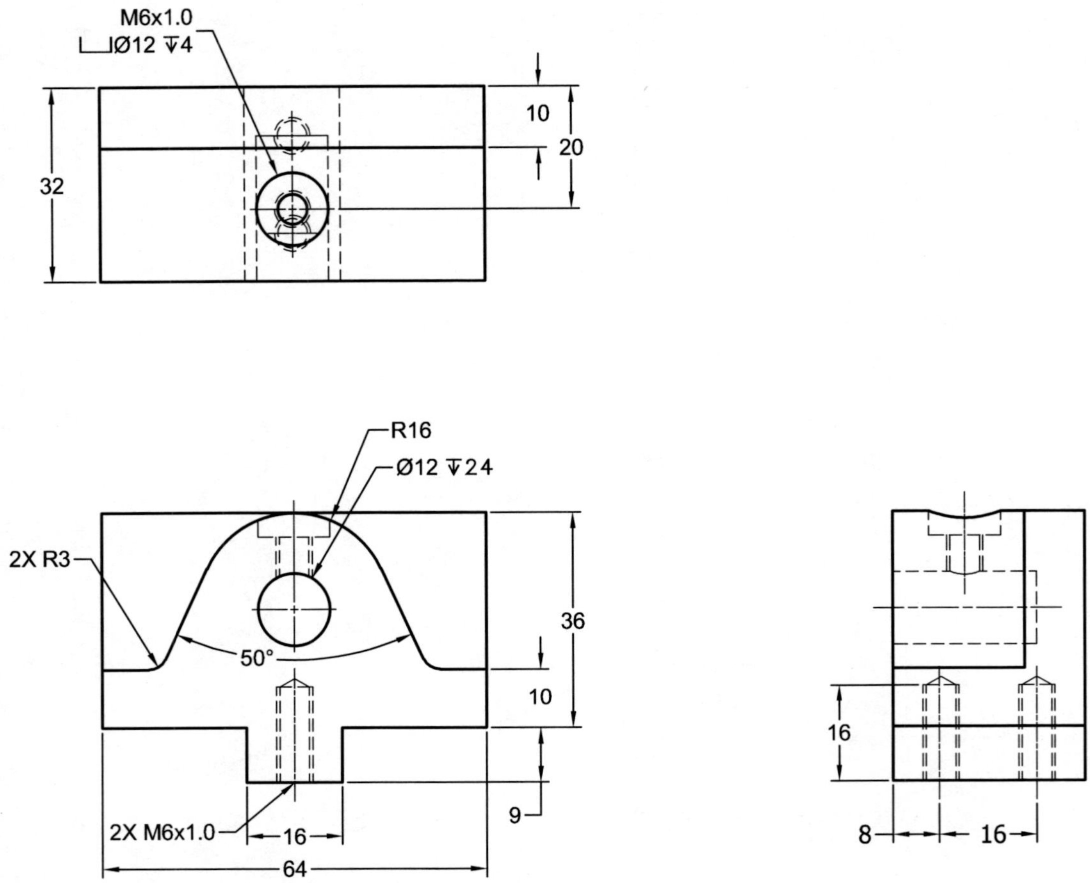

Figure 12-4 *Orthographic views and dimensions of the Vice Jaw*

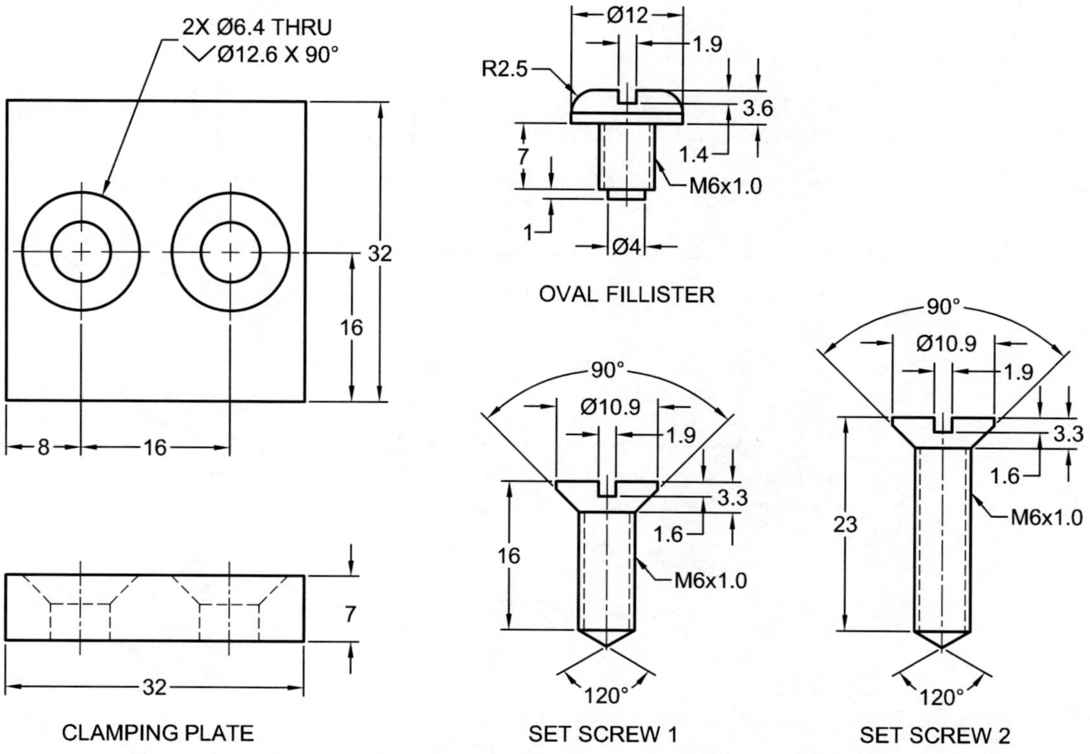

Figure 12-5 *Orthographic views and dimensions of various components of the Bench Vice assembly*

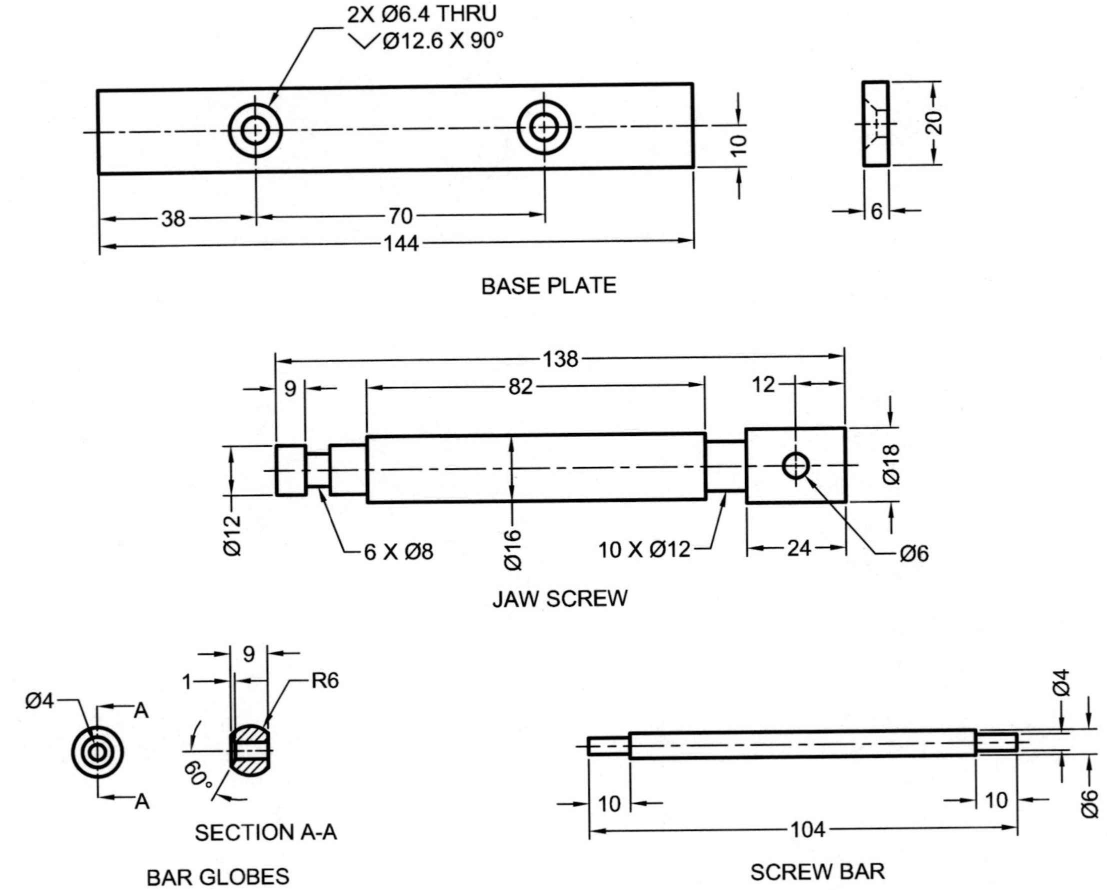

Figure 12-6 *Orthographic views and dimensions of various components of the Bench Vice assembly*

The following steps are required to complete this tutorial:

a. Create all components as individual part documents and save them at the location *\Documents\SOLIDWORKS\c12\Bench Vice.*
b. Open the Vice Body and Vice Jaw part documents and define mate references in both the part documents, refer to Figures 12-7 and 12-8.
c. Create a new assembly document and open all part documents. Place the first component, which is the Vice Body. Next, drag and drop the Vice Jaw in the assembly document. It will automatically be assembled with the Vice Jaw because mate references have already been defined in both the part documents, refer to Figures 12-9 and 12-10.
d. Drag and drop the Jaw Screw in the assembly document and apply mates, refer to Figures 12-11 through 12-17.
e. Apply mates for constraining the required degrees of freedom , refer to Figures 12-18 and 12-19.
f. Assemble the Clamping Plate, refer to Figures 12-20 through 12-23.
g. Assemble the Oval Fillister and Set Screws by using the feature-based mates, refer to Figures 12-24 and 12-25.

h. Similarly, assemble other components.
i. Save the assembly.

Creating Components

1. Create all the components of the Bench Vice assembly as separate part documents. Specify the names of the documents, as shown in Figures 12-2 through 12-6. Save the files at the location *\Documents\SOLIDWORKS\c12\Bench Vice*. Make sure that *Bench Vice* is your current folder.

Creating Mate References

In this tutorial, you will assemble the first two components of the assembly by using mate references. For assembling components using mate references, first you need to create mate references. To do so, you need to open the part documents in which the mate references will be added.

1. Choose the **Open** button from the Menu Bar; the **Open** dialog box will be displayed. Browse to the location of *Bench Vice* folder.

2. Double-click on the Vice Body; the Vice Body part document is opened in the SOLIDWORKS window.

3. Choose the **Mate Reference** tool from the **Reference Geometry** flyout in the **Features CommandManager**; the **Mate Reference PropertyManager** is invoked. Also, the **Primary Reference Entity** selection box becomes active.

4. Select the planar face of the model as the primary reference, refer to Figure 12-7; the selected planar face is highlighted and the **Secondary Reference Entity** selection box becomes active.

5. Select the **Coincident** option from the **Mate Reference Type** drop-down list in the **Primary Reference Entity** rollout.

6. As the **Secondary Reference Entity** selection box is active, select the planar face of the model as the secondary reference, refer to Figure 12-7; the selected face is highlighted and the **Tertiary Reference Entity** selection box becomes active.

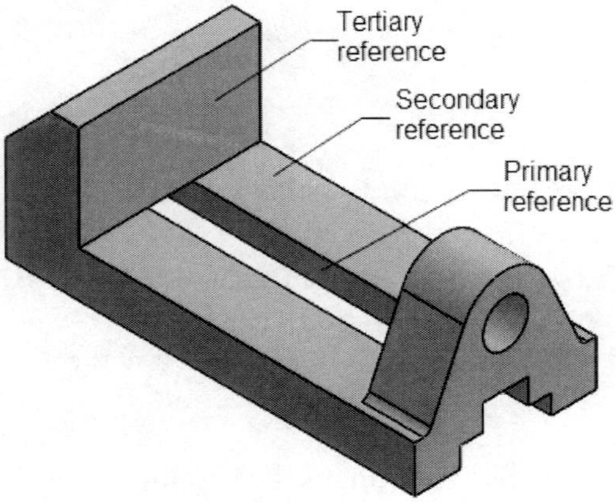

Figure 12-7 *Faces to be selected as mate references*

7. Select the **Coincident** option from the **Mate Reference Type** drop-down list in the **Secondary Reference Entity** rollout.

8. Select the planar face of the model as the tertiary reference, as shown in Figure 12-7; the selected face is highlighted.

9. Select the **Parallel** option from the **Mate Reference Type** drop-down list in the **Tertiary Reference Entity** rollout.

10. Enter **Vice Mate Reference** as the name of the mate reference in the **Mate Reference Name** edit box available in the **Reference Name** rollout.

11. Choose the **OK** button from the **Mate Reference PropertyManager**.

12. Open the Vice Jaw part document.

13. Create the mate reference in the Vice Jaw part document. The faces to be selected as references are shown in Figure 12-8. For the Vice Jaw part, enter **Vice Mate Reference** as the name of the mate reference in the **Mate Reference Name** edit box.

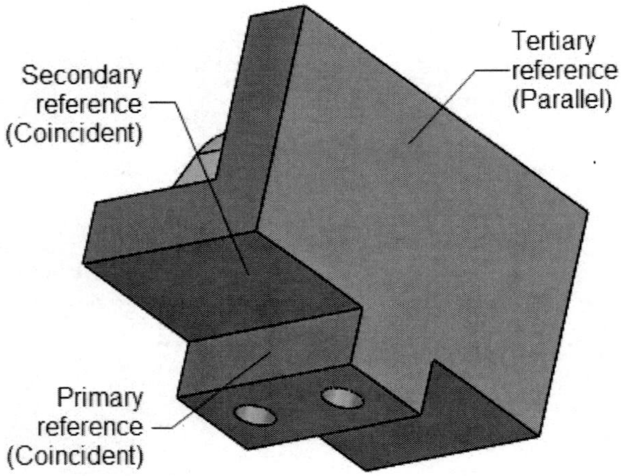

Figure 12-8 *Faces to be selected as mate references*

14. Close all the part documents except *Vice Body.sldprt* and *Vice Jaw.sldprt* if they are opened.

Assembling the First Two Components of the Assembly

After creating mate references in the part documents, you need to assemble the components. To do so, you need to start a new SOLIDWORKS assembly document.

1. Start a new SOLIDWORKS assembly document; the **Begin Assembly PropertyManager** is invoked automatically and the names of the opened components are displayed in the **Open documents** selection box.

2. Select the Vice Body from the **Open documents** selection box; a preview of the Vice Body is displayed along with the component cursor.

   It is recommended to place the first component of the assembly at the assembly origin.

3. Choose the **OK** button from the **Begin Assembly PropertyManager**; the first component is placed coincident to the origin.

4. Change the current view to isometric.

Next, you need to place the second component in the assembly. As mentioned earlier, the second component of the assembly, which is the Vice Jaw, is assembled with the Vice Body by using mate references.

5. Choose the **Insert Components** button from the **Assembly CommandManager**; the **Insert Component PropertyManager** is invoked and the names of the components opened in other windows are displayed in the **Open documents** selection box. 

6. Select the Vice Jaw from the **Open documents** selection box; a preview of the Vice Jaw is displayed in the drawing area.

When you move the cursor close to the Vice Body in the assembly document, a preview of the Vice Jaw after applying mates with the Vice Body is displayed in the assembly document.

7. Place the component at the required location. The mates specified in the mate references are applied between the Vice Jaw and the Vice Body automatically.

Figure 12-9 shows the second component placed in the assembly document and Figure 12-10 shows the isometric view of the Vice Jaw assembled with the Vice Body.

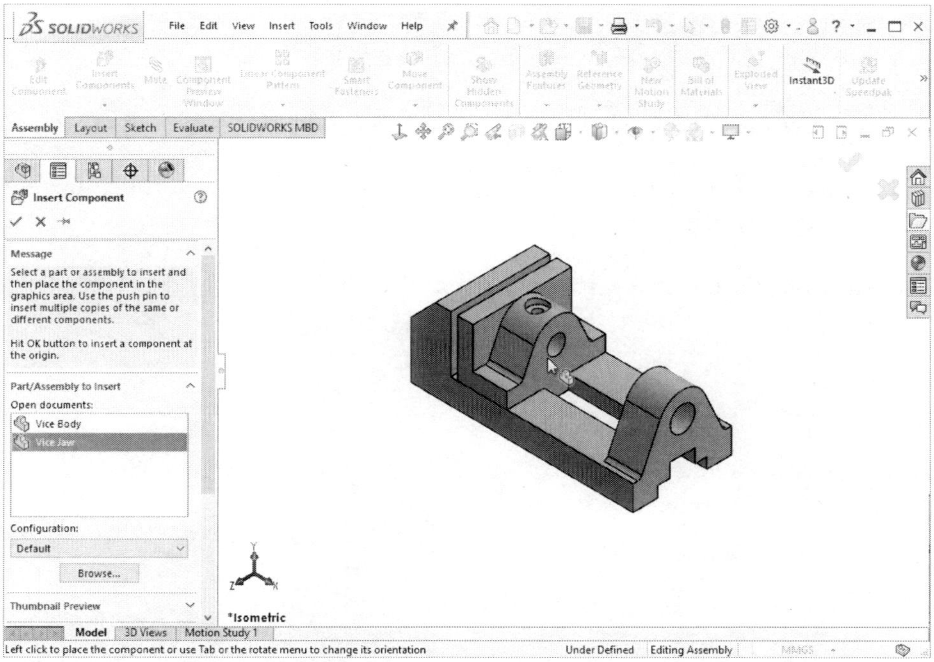

Figure 12-9 *The second component placed in the assembly document*

8. Before proceeding further, close the part document windows of all the parts that are placed in the assembly document.

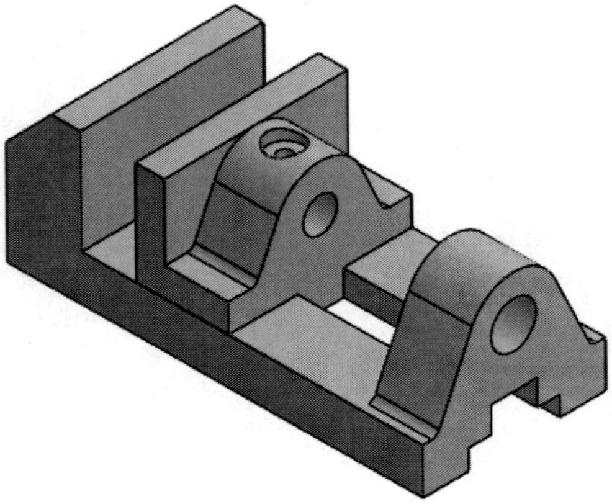

Figure 12-10 *Vice Jaw assembled with the Vice Body*

Tip
The mates defined in the mate references are applied to the components when the components are placed in the assembly. You can view the mates applied to both the components by expanding the ***Mates*** *node from the* ***FeatureManager Design Tree*** *of the assembly document.*

Assembling the Jaw Screw

Next, you need to place the Jaw Screw in the assembly document.

1. Choose the **Insert Components** button from the **Assembly CommandManager**.

2. Choose the **Browse** button in the **Part/Assembly to Insert** rollout if the **Open** dialog box is not displayed by default.

3. Double-click on the Jaw Screw to open its part document; a preview of the Jaw Screw is displayed in the drawing area.

4. Click anywhere in the drawing area to place the Jaw Screw, as shown in Figure 12-11.

   Next, you need to add assembly mates to assemble the components placed in the assembly.

5. Press and hold the ALT key and select the face of the Jaw Screw, refer to Figure 12-12.

6. Drag the Jaw Screw to the hole in the Vice Body, as shown in Figure 12-13; the Jaw Screw appears transparent and the select cursor is replaced by the smart mates cursor. Also, a symbol of the concentric mate is displayed below the cursor.

Figure 12-11 *Jaw Screw placed in the assembly*

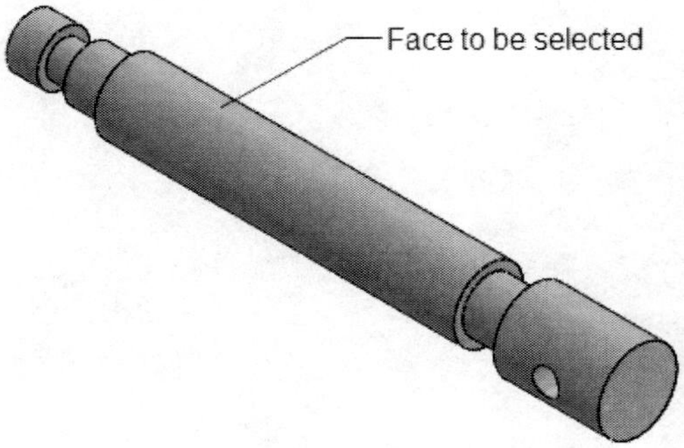

Figure 12-12 Face to be selected

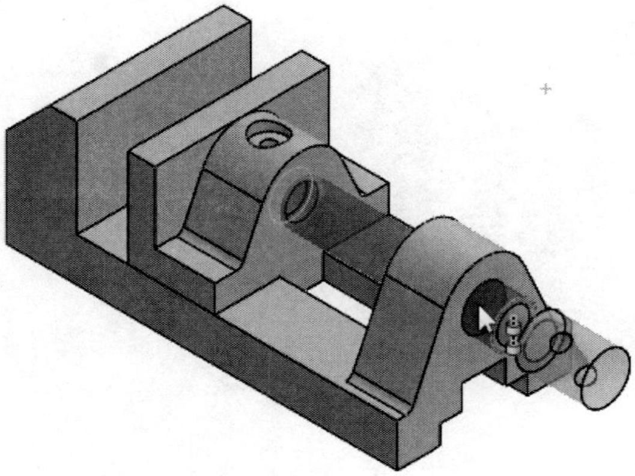

Figure 12-13 Jaw Screw being dragged

7. Release the left mouse button and the ALT key at this location; the **Mate** pop-up toolbar is displayed with the **Concentric** button chosen by default indicating that the concentric mate is the most appropriate mate to be applied. Choose the **Add/Finish Mate** button from the **Mate** pop-up toolbar. Figure 12-14 shows the Jaw Screw assembled in the Vice Body.

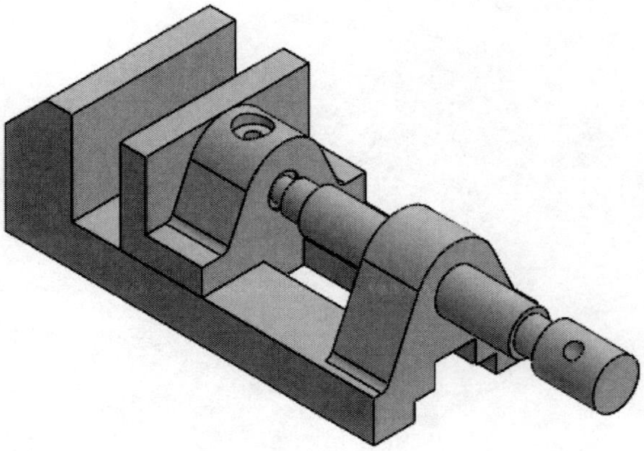

Figure 12-14 The concentric mate applied between the Jaw Screw and the Vice Body

Next, you need to apply the coincident mate between the planar faces of the Jaw Screw and the Vice Jaw.

8. Choose the **Mate** button from the **Assembly CommandManager** to invoke the **Mate PropertyManager**.

9. Rotate the assembly view and select the face of the Vice Jaw, as shown in Figure 12-15. Next, select the face of the Jaw Screw, as shown in Figure 12-16.

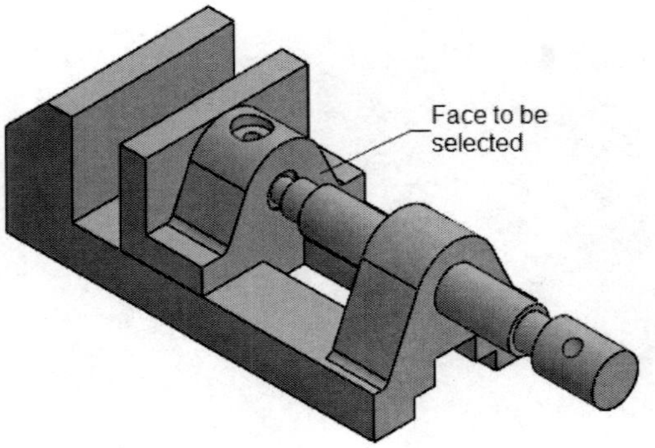

Figure 12-15 *Face of the Vice Jaw to be selected*

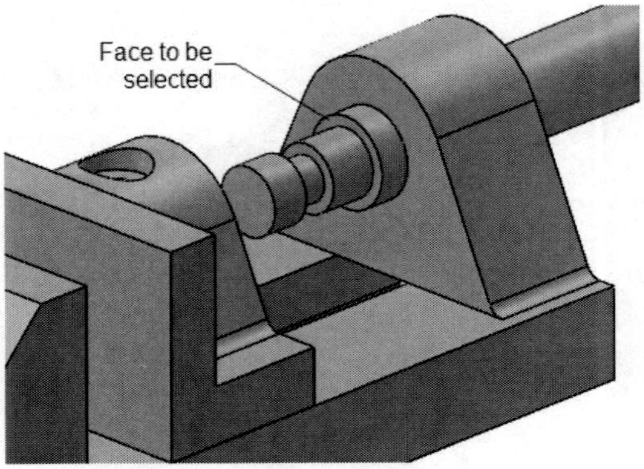

Figure 12-16 *Face of the Jaw Screw to be selected*

As soon as you select the faces, the **Mate** pop-up toolbar is invoked with the **Coincident** button chosen and a preview of the assembly with the coincident mate is displayed in the drawing area.

10. Choose the **Add/Finish Mate** button from the **Mate** pop-up toolbar. The assembly after adding the coincident mate is displayed in Figure 12-17.

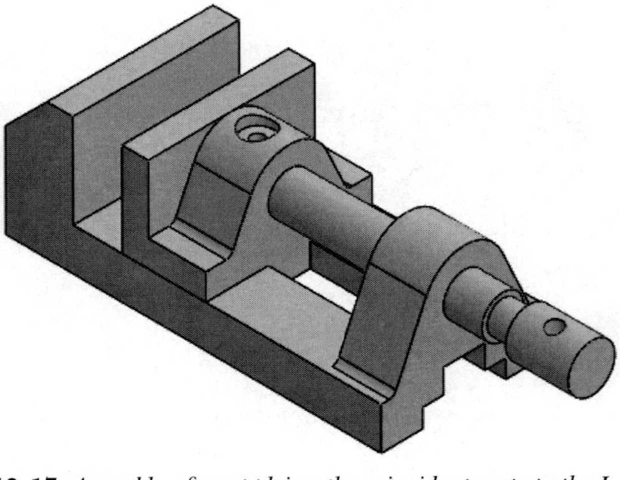

Figure 12-17 *Assembly after applying the coincident mate to the Jaw Screw*

Next, you need to apply the distance mate to the Vice Body and the Vice Jaw. Make sure you apply the distance mate with maximum and minimum limits so that the Vice Jaw can slide over the Vice Body between these limits. You can do so by using the distance mate.

11. Select the two faces, one of the Vice Body and the other of the Vice Jaw, refer to Figures 12-18 and 12-19.

As soon as you select the faces, the **Mate** pop-up toolbar is invoked with the **Coincident** button chosen and a preview of the assembly with the coincident mate is displayed in the drawing area.

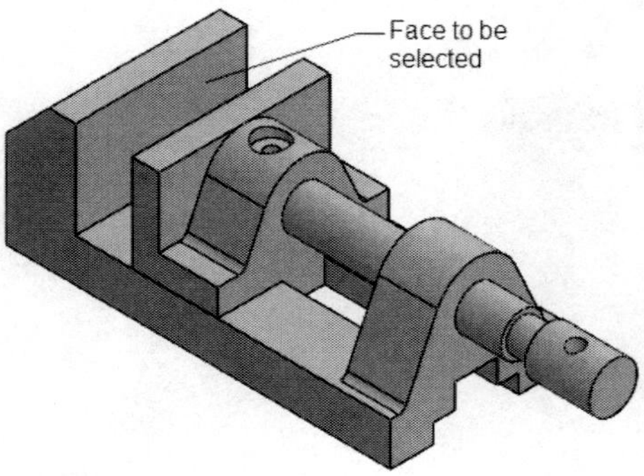

Figure 12-18 Face of the Vice Body to be selected

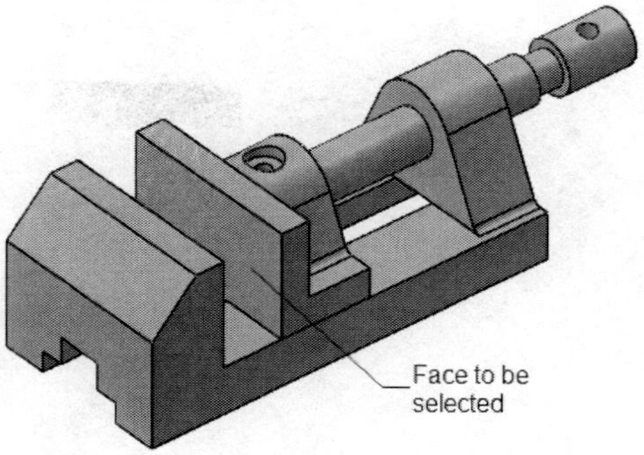

Figure 12-19 Face of the Vice Jaw to be selected

12. Next, expand the **Advanced Mates** rollout and then choose the **Distance** button from it; the **Maximum Value** and **Minimum Value** edit boxes become available.

13. Enter **60** and **0** in the **Maximum Value** and **Minimum Value** edit boxes, respectively. Next, choose the **OK** button twice from the PropertyManager.

    You will learn more about the advanced mates in the next chapter.

    To view the effect after the distance mate is applied with the specified limits, drag the Jaw Screw. You will notice that the Jaw Screw along with the Vice Jaw moves between these limits. Also, the Jaw Screw is free to rotate about its own axis.

Assembling the Clamping Plate

Next, you need to assemble the Clamping Plate with the assembly.

1. Invoke the **Insert Component PropertyManager** and choose the **Browse** button if the **Open** dialog box is not invoked by default. Open the Clamping Plate and place it in the drawing area.

2. Rotate the assembly such that the bottom face of the assembly is displayed, as shown in Figure 12-20.

3. Select the Clamping Plate by using the right mouse button and drag the cursor to rotate it. Figure 12-21 shows the rotated Clamping Plate.

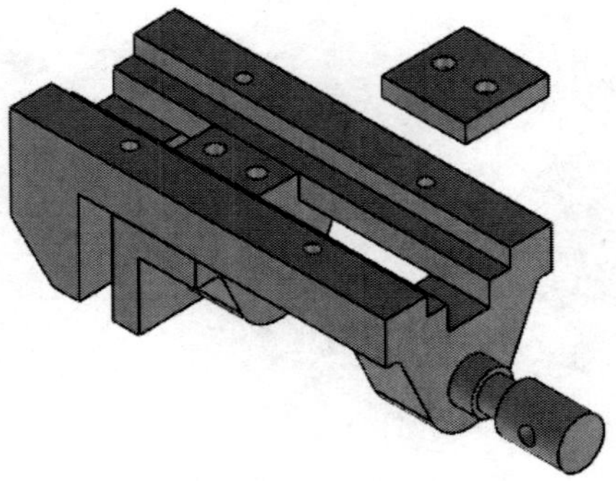

Figure 12-20 Rotated assembly showing the bottom face

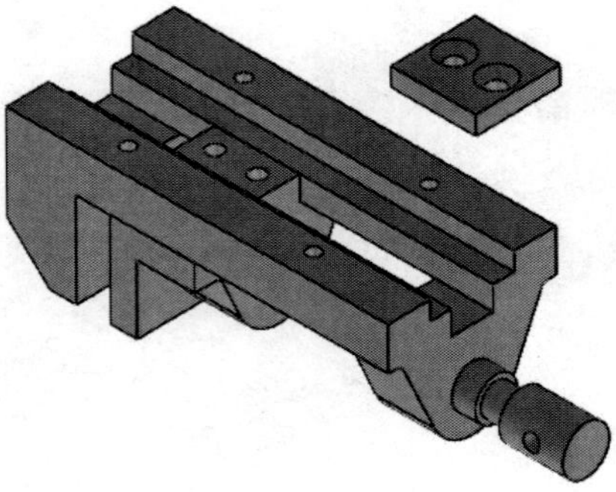

Figure 12-21 Rotated Clamping Plate

4. Apply the concentric mate between two cylindrical faces of the Clamping Plate and the two holes of the Vice Jaw, refer to Figure 12-22. You may need to move the Clamping Plate after applying the first mate.

5. Rotate the Clamping Plate by dragging it. Next, apply the coincident mate between the faces of the Clamping Plate and the Vice Jaw, as shown in Figure 12-23.

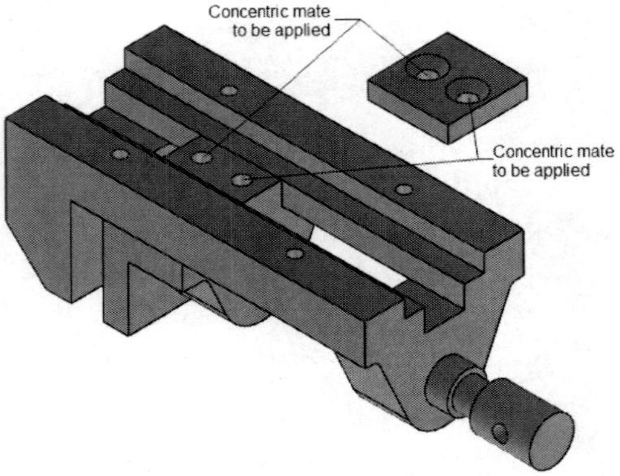

Figure 12-22 Faces to be selected to apply mate

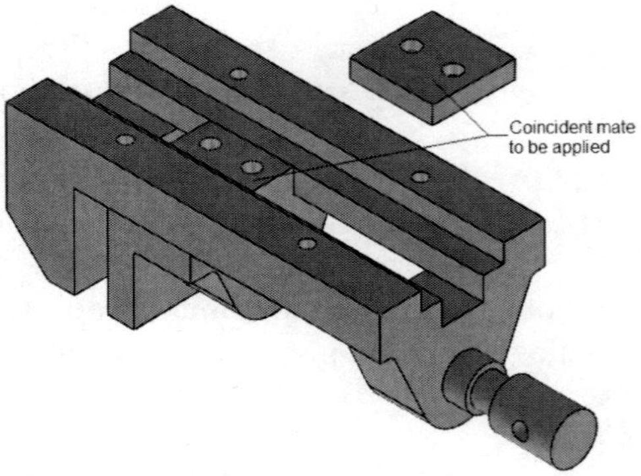

Figure 12-23 Faces to be selected to apply mate

Tip

*Whenever you invoke the **Insert Component PropertyManager**, the **Open** dialog box is also invoked simultaneously. However, this dialog box will only appear when there is no other SOLIDWORKS open document.*

Assembling the Remaining Components

After assembling the Clamping Plate, you need to assemble the Screw Bar, Base Plates, Bar Globes, Oval Fillister, Set Screw 1, and Set Screw 2.

1. Follow the same procedure for assembling the Screw Bar, Base Plates, Bar Globes, Oval Fillister, Set Screw 1, and Set Screw 2 with the assembly. Make sure that you specify the

required mating references on all parts for assembling them using the Mate References. The assembly after assembling all these components is shown in Figure 12-24. The rotated view of the assembly after assembling the Set Screw 1 and Set Screw 2 is shown in Figure 12-25.

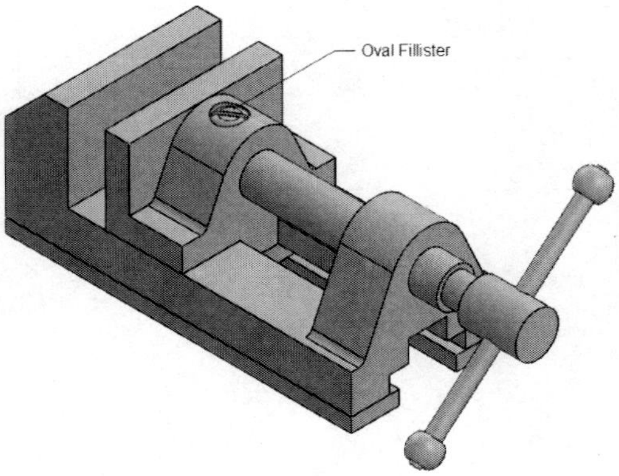

Figure 12-24 *Assembly after assembling the remaining parts*

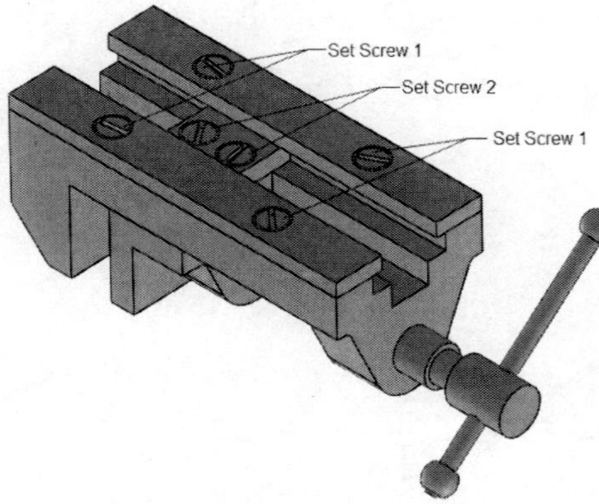

Figure 12-25 *Rotated view of the final assembly*

2. Choose the **Save** button to save the assembly with the name *Bench Vice* at the location *\Documents\SOLIDWORKS\c12\Bench Vice.*

Tutorial 2

In this tutorial, you will create all the components of the Pipe Vice assembly and then assemble them. The Pipe Vice assembly is shown in Figure 12-26. The dimensions of various components of this assembly are given in Figures 12-27 through 12-29. **(Expected time: 2 hr 45 min)**

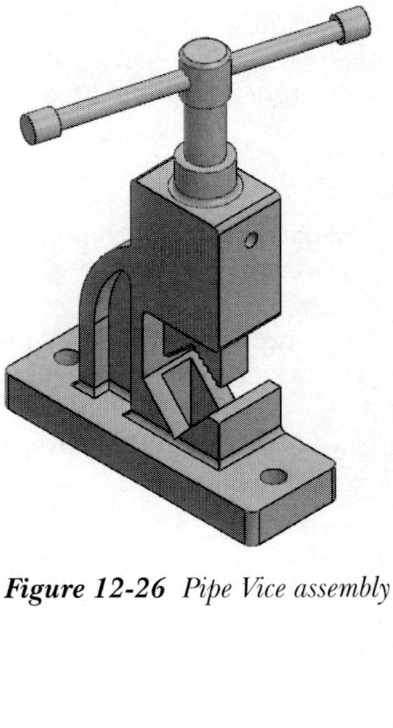

Figure 12-26 *Pipe Vice assembly*

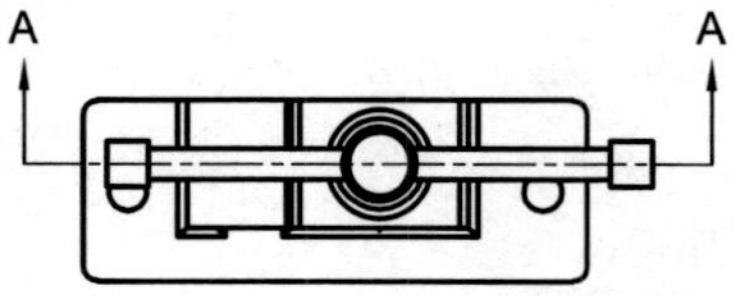

ITEM NO.	PART NO.	DESCRIPTION	QTY.
1	Base		1
2	Moveable Jaw		1
3	Screw		1
4	Handle		1
5	Handle Stop		2

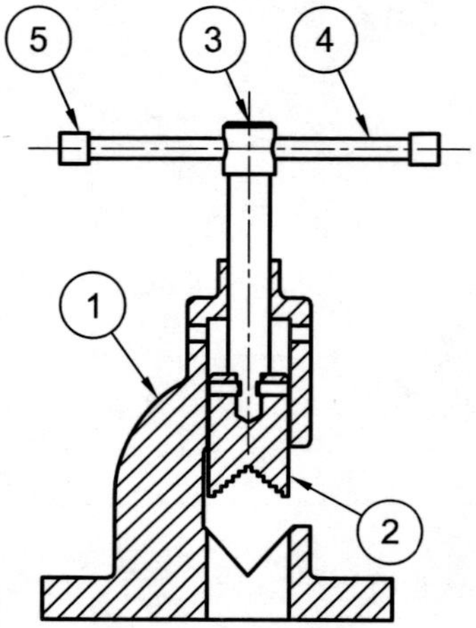

SECTION A-A

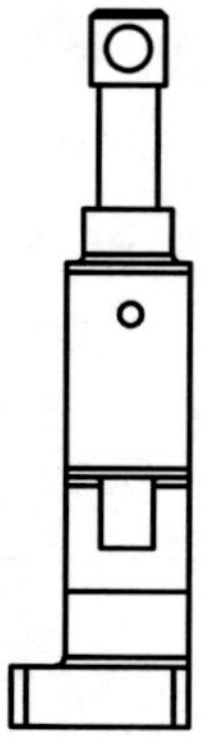

Figure 12-27 *Pipe Vice assembly*

You will create all components of the Pipe Vice assembly as separate part documents. After creating the parts, you will assemble them in the assembly document. In this tutorial, you need to use the bottom-up approach for creating the assembly.

The following steps are required to complete this tutorial:

a. Create all the components as separate part documents and then save them. The part documents will be saved at *\Documents\SOLIDWORKS\c12\Pipe Vice*.
b. Place the Base at the origin of the assembly.
c. Place the Moveable Jaw and the Screw in the assembly. Apply mates between the Moveable Jaw and the Screw, refer to Figures 12-30 through 12-32.
d. Assemble the Screw with the Base.
e. Place other components in the assembly and apply mates to the assembly, refer to Figure 12-33.

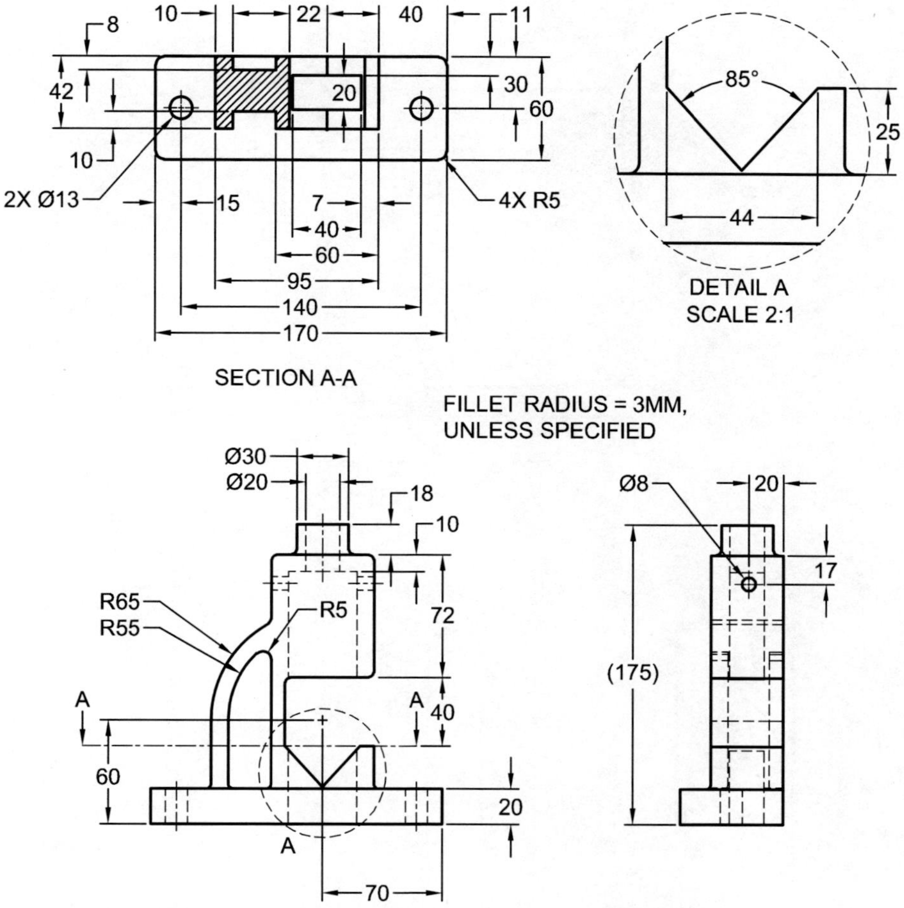

Figure 12-28 *Views and dimensions of the Base*

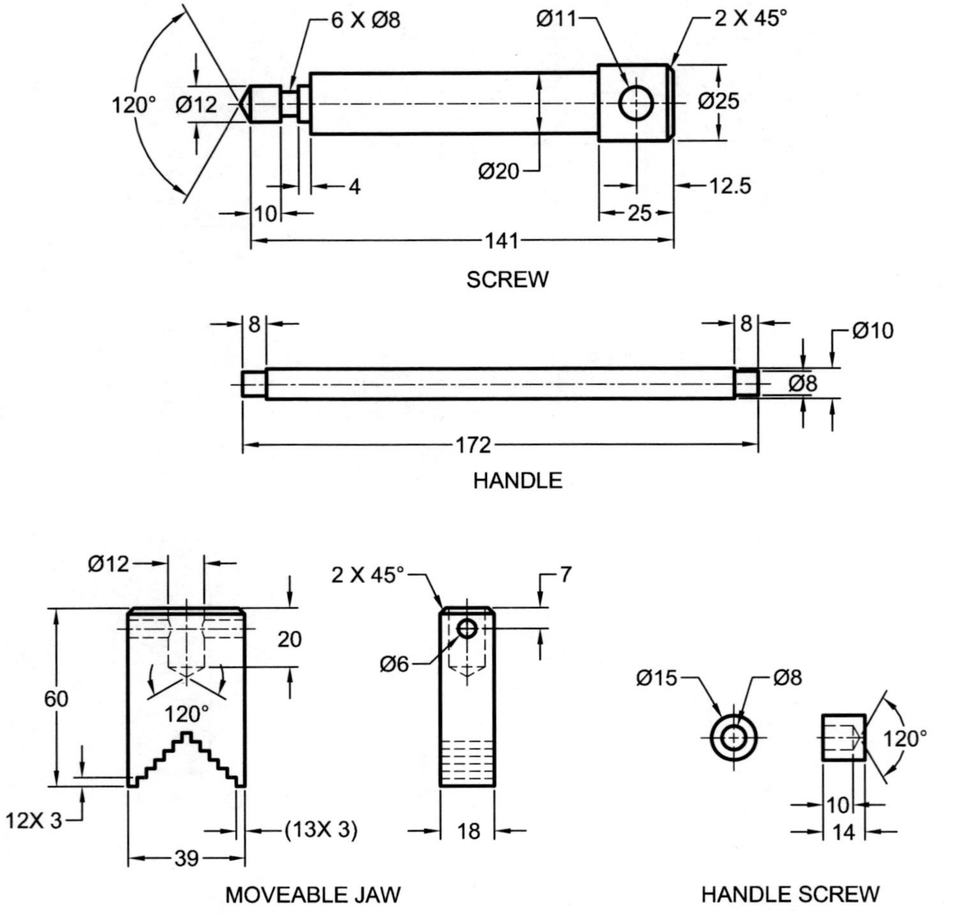

Figure 12-29 *Views and dimensions of the Screw, Handle, Moveable Jaw, and Handle Screw*

Creating Components

1. Create all the components of the Pipe Vice assembly as separate part documents. Specify the names of the files, as shown in Figures 12-27 to 12-29. Save the documents at the location *\Documents\SOLIDWORKS\c12\Pipe Vice*.

Inserting the First Component into the Assembly

After creating all the components of the Pipe Vice assembly, you need to start a new SOLIDWORKS assembly document.

1. Start a new SOLIDWORKS assembly document; the **Begin Assembly PropertyManager** is displayed by default.

2. Choose the **Browse** button from the **Part/Assembly to Insert** rollout to display the **Open** dialog box. Open the **Pipe Vice** folder and double-click on the Base.

3. Choose the **OK** button from the **Begin Assembly PropertyManager** to place the Base part origin coincident at the origin of the assembly document.

4. Change the view orientation to isometric.

Inserting and Assembling the Moveable Jaw and the Screw

After placing the first component in the assembly document, you need to place the Moveable Jaw and the Screw in the assembly document. After placing these components, you need to apply the required mates.

1. Choose the **Insert Components** button from the **Assembly CommandManager**. Then, choose the **Keep Visible** button from the PropertyManager to keep it visible. Next, invoke the **Open** dialog box by choosing the **Browse** button from the **Part/Assembly to Insert** rollout.

2. Double-click on the Moveable Jaw. Place the component anywhere in the assembly document such that it does not interfere with the existing component.

3. Similarly, place the Screw in the assembly document and choose the **OK** button from the **Insert Component PropertyManager**. Figure 12-30 shows the Moveable Jaw, Screw, and Base placed in the assembly document.

   First, you need to assemble the Screw with the Moveable Jaw. Therefore, you need to fix the Moveable Jaw.

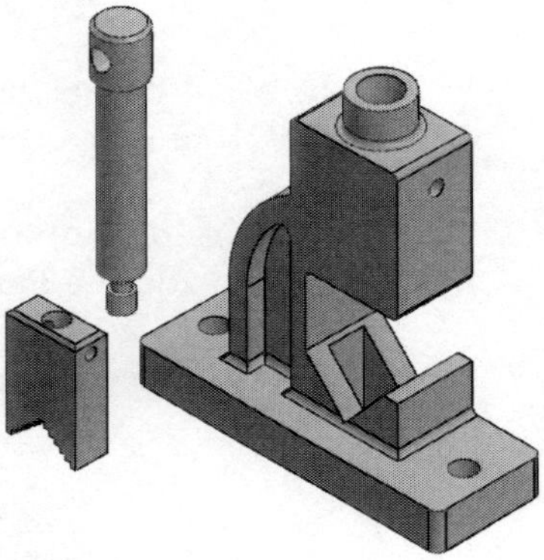

Figure 12-30 *The Moveable Jaw, Screw, and Base placed in the assembly document*

4. Select the Moveable Jaw from the drawing area or from the **FeatureManager Design Tree**. Right-click to invoke the shortcut menu.

5. Choose the **Fix** option from the shortcut menu; the Moveable Jaw becomes fixed and you cannot move or rotate it.

6. Invoke the **Move Component PropertyManager** and choose the **SmartMates** button from the **Move** rollout. Double-click on the lowermost cylindrical face of the Screw; the Screw appears transparent.

7. Drag the cursor to the hole located at the top of the Moveable Jaw. Release the left mouse button as soon as the concentric symbol is displayed below the cursor. Next, choose the **Add/Finish Mate** button from the **Mate** pop-up toolbar.

8. Select the Screw and move it up so that it is not inside the Moveable Jaw.

9. Right-click in the drawing area and then choose **Clear Selections** from the shortcut menu to clear the current selection.

10. Rotate the assembly and double-click on the lower flat face of the Screw; the Screw appears transparent.

11. Rotate the model again and select the top planar face of the Moveable Jaw. The coincident mate is applied between the two selected faces. Choose the **Add/Finish Mate** button from the **Mate** pop-up toolbar.

12. Choose the **OK** button from the **SmartMates PropertyManager**. Figure 12-31 shows the Screw after applying mates.

    Next, you need to assemble the Screw and the Moveable Jaw with the Base.

13. Select the Moveable Jaw and right click on it; a shortcut menu will be displayed. Choose the **Float** option from this shortcut menu. Now, you can move the Moveable Jaw and the Screw assembled to it.

14. Press the ALT key, select the cylindrical face of the Screw and move the screw toward the hole created on the top face of the Base to add the concentric mate.

15. Invoke the **Mate PropertyManager** and then select the front planar face of the Moveable Jaw and the front planar face of the Base.

16. Choose the **Parallel** button from the **Mate** pop-up toolbar and then choose the **Add/Finish Mate** button to add the **Parallel** mate between the selected faces.

17. Next, select the faces, as shown in Figure 12-32.

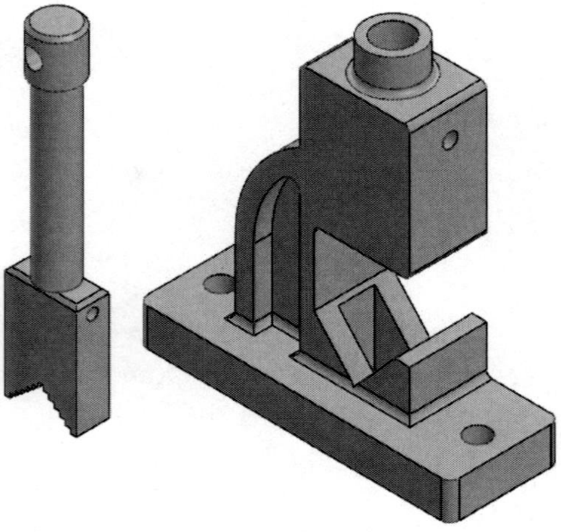

Figure 12-31 *The Screw assembled with the Moveable Jaw*

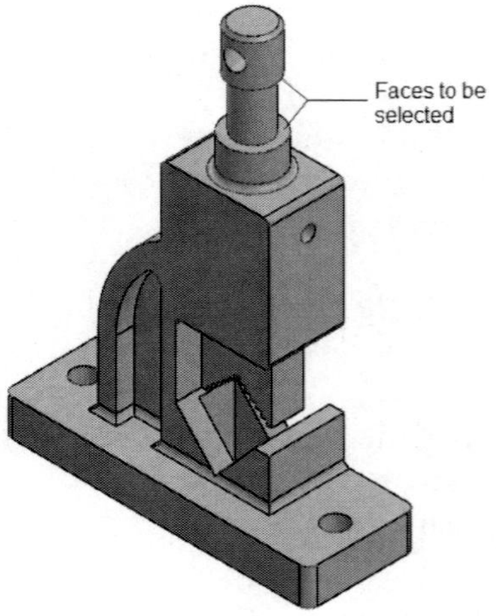

Figure 12-32 *Faces to be selected*

18. Expand the **Advanced Mates** rollout and then choose the **Distance** button from it; the **Maximum Value** and **Minimum Value** edit boxes become available.

19. Enter **65** and **5** in the **Maximum Value** and **Minimum Value** edit boxes respectively. Next, choose **OK** from the **Mate PropertyManager**.

20. Similarly, assemble the other components of the Pipe Vice assembly. Figure 12-33 shows the final Pipe Vice assembly.

Figure 12-33 *Final Pipe Vice assembly*

21. Choose the **Save** button to save the assembly document at the location *\Documents\SOLIDWORKS\c12\Pipe Vice*.

EXERCISES

Exercise 1

Create the Plummer Block assembly, as shown in Figure 12-34. The views and dimensions of various components of the assembly are shown in Figures 12-35 through 12-38.

(Expected time: 1 hr)

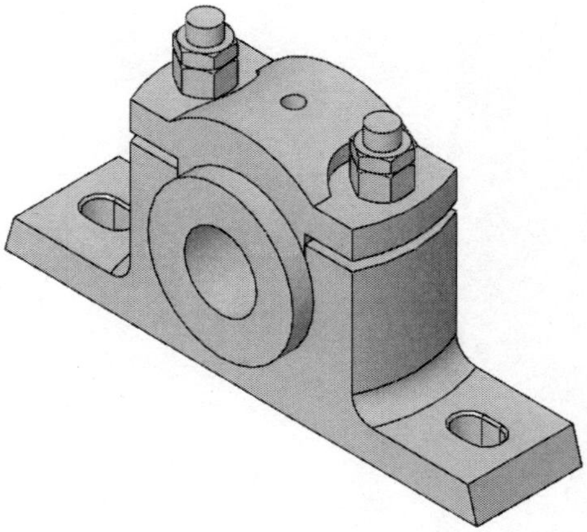

Figure 12-34 *Plummer Block assembly*

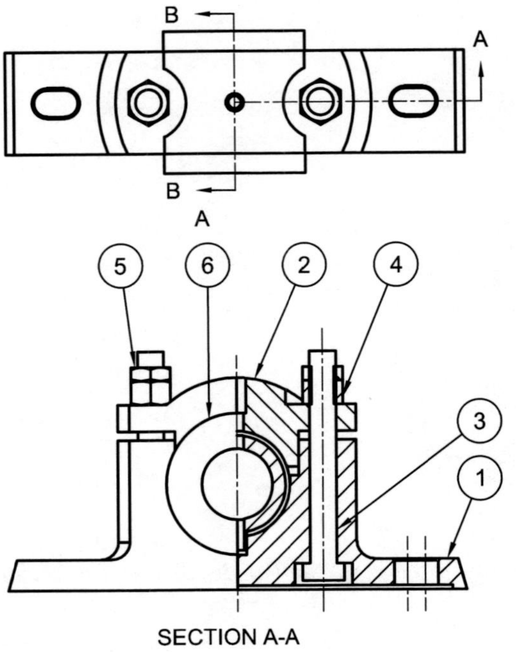

ITEM NO.	PART NO.	DESCRIPTION	QTY.
1	Casting		1
2	Cap		1
3	Bolt		2
4	Nut		2
5	Lock Nut		2
6	Brasses		1

Figure 12-35 *Plummer Block assembly*

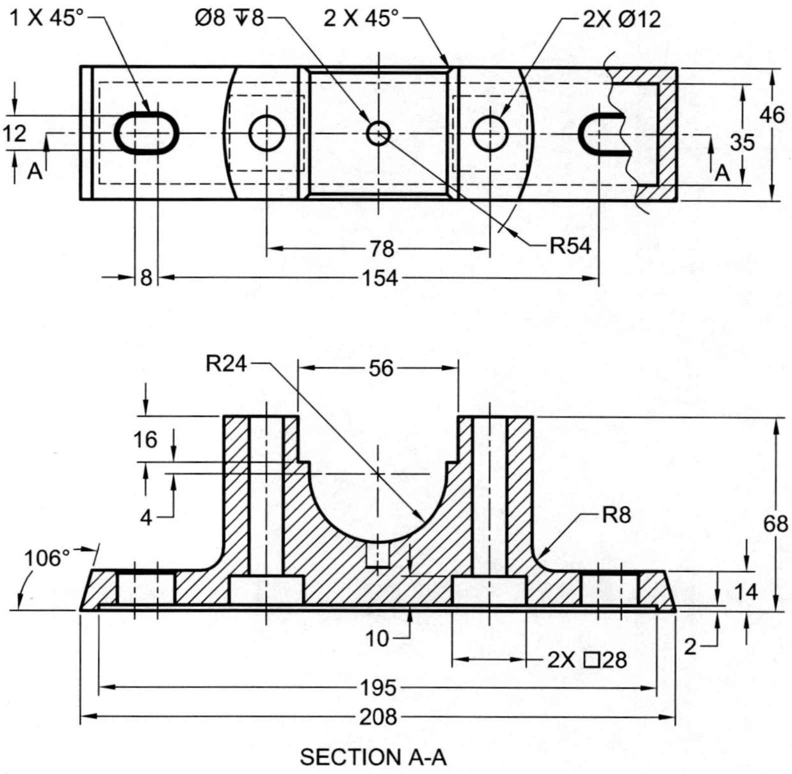

Figure 12-36 *Views and dimensions of the Casting*

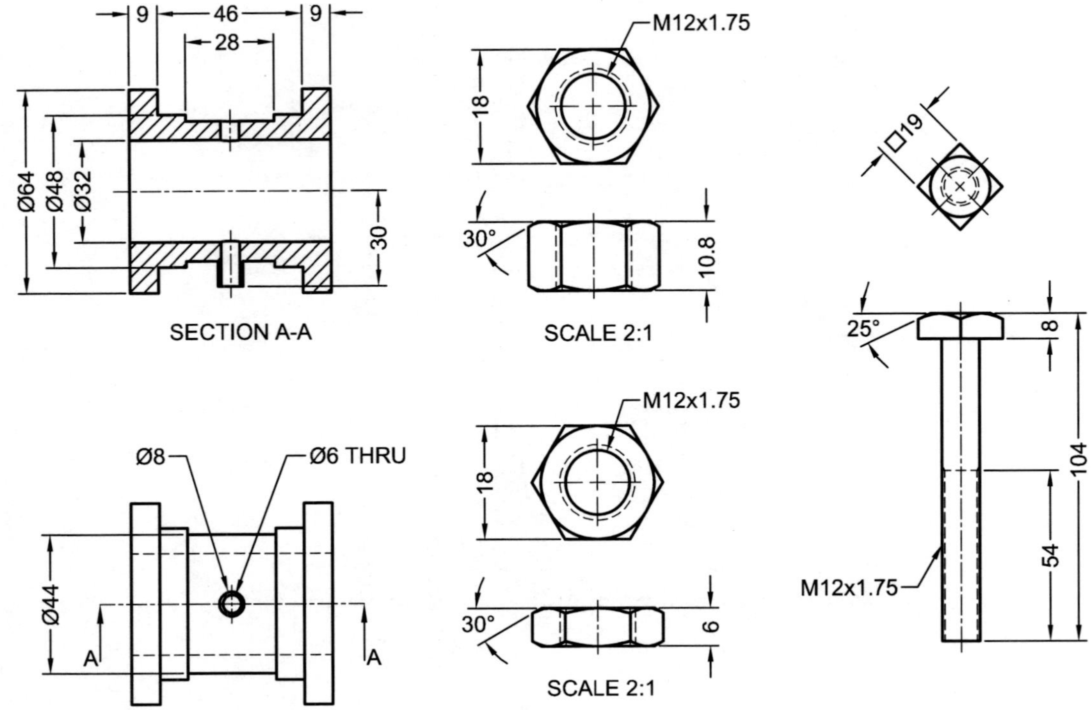

Figure 12-37 *Views and dimensions of the Brasses, Nut, Lock Nut, and Bolt*

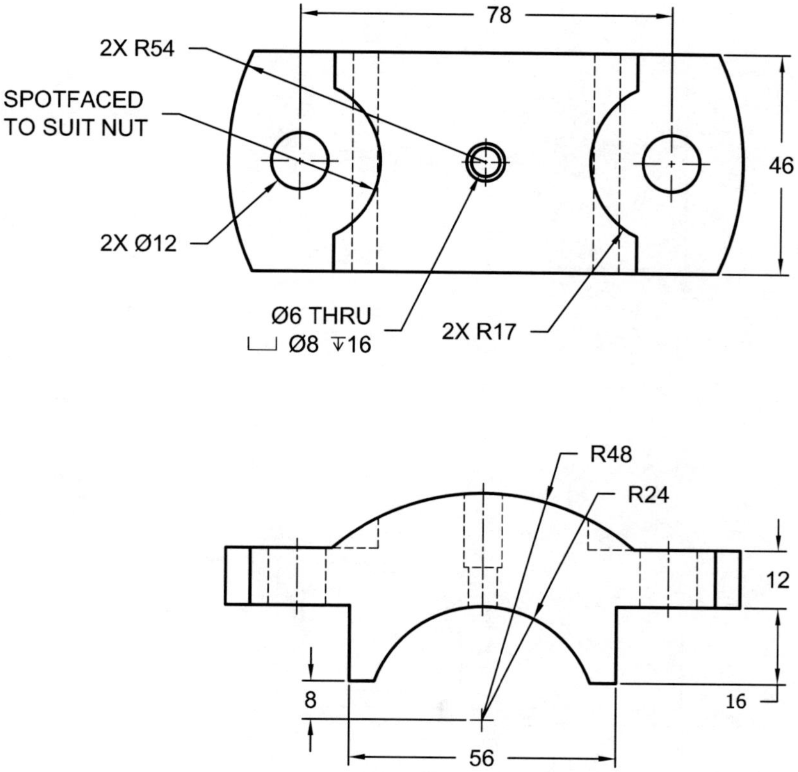

Figure 12-38 Views and dimensions of the Cap

Exercise 2

Create all the components of the Anti Vibration Mount assembly and assemble them. Its assembly is shown in Figure 12-39. The views and dimensions of various components of this assembly are given in Figures 12-40 through 12-45. Note that all dimensions are in mm.

(Expected time: 1 hr)

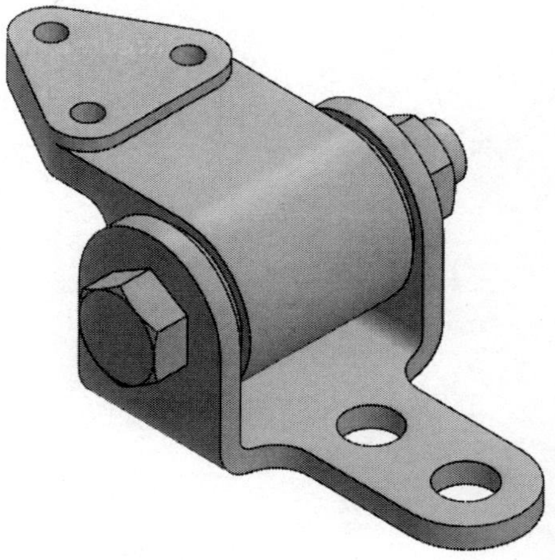

Figure 12-39 Anti Vibration Mount

ITEM NO.	PART NO.	DESCRIPTION	QTY.
1	Yoke Plate		1
2	Body		1
3	Bushing Rubber		2
4	Hex Bolt		1
5	Nut		1

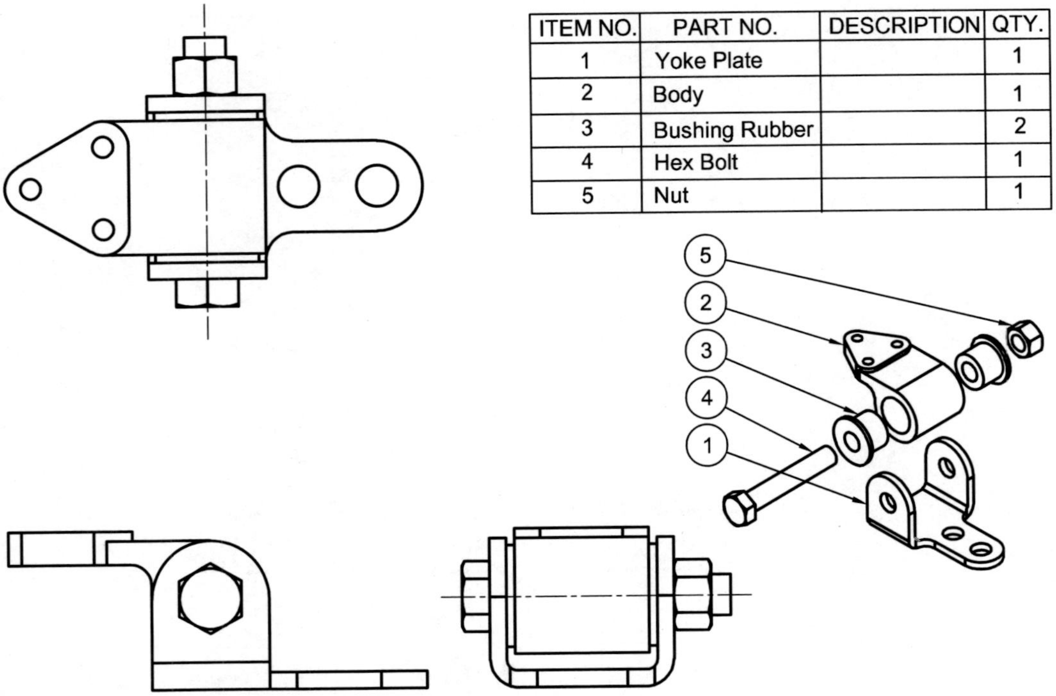

Figure 12-40 *Anti Vibration Mount*

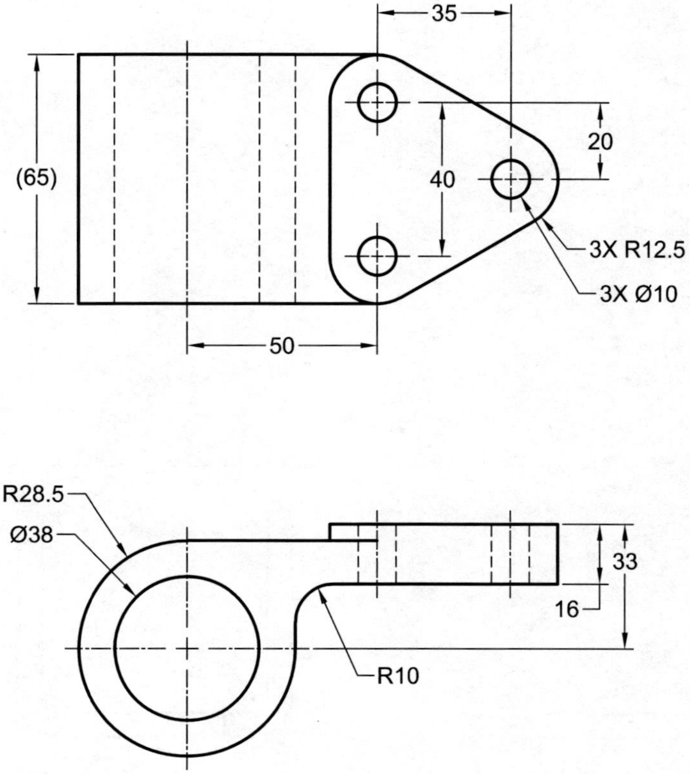

Figure 12-41 *Top and front views of Body*

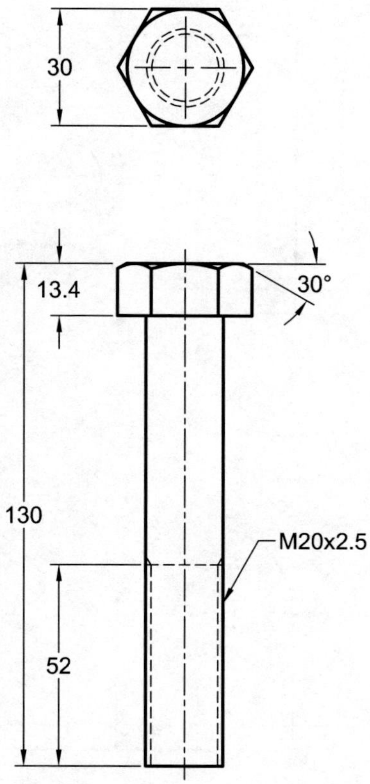

Figure 12-42 *Top and front views of Hex Bolt*

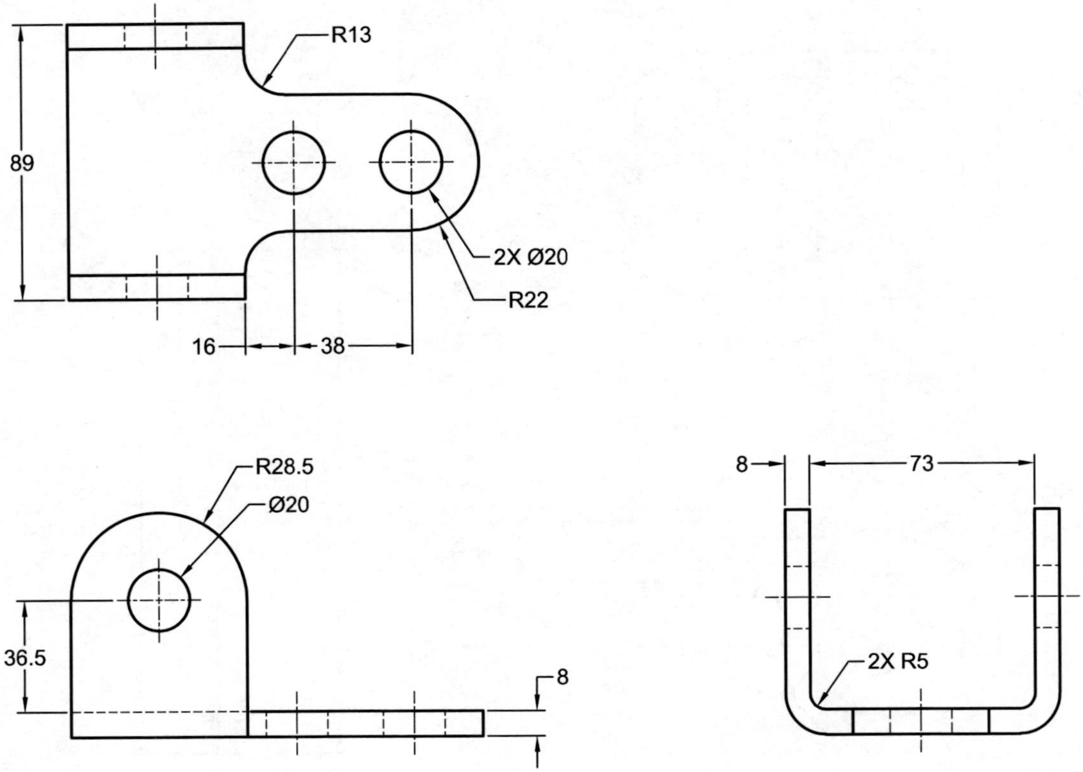

Figure 12-43 Top, front, and side views of Yoke plate

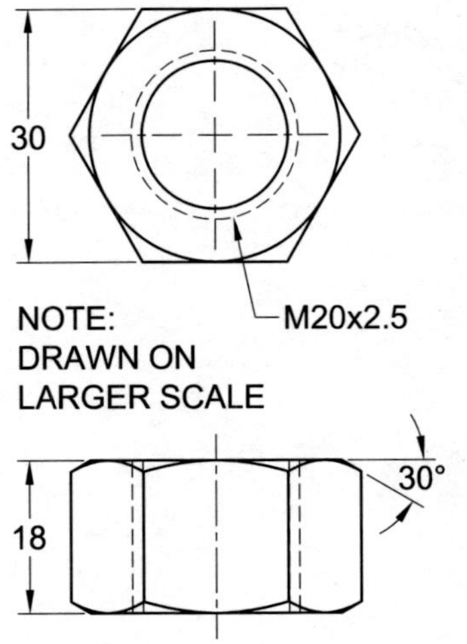

Figure 12-44 Top and front views of Nut

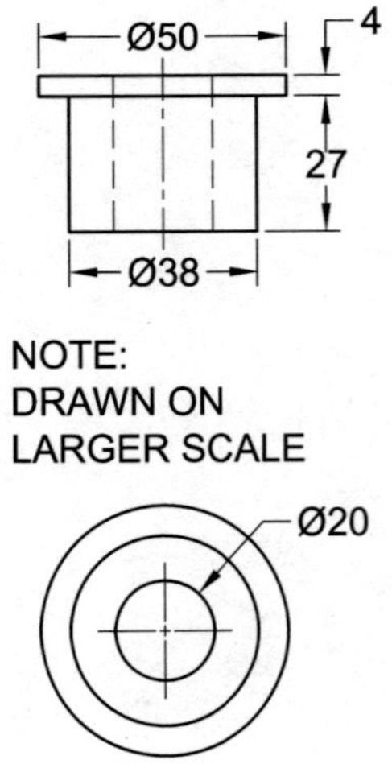

Figure 12-45 Top and front views of Bushing Rubber

Chapter 13

Assembly Modeling-II

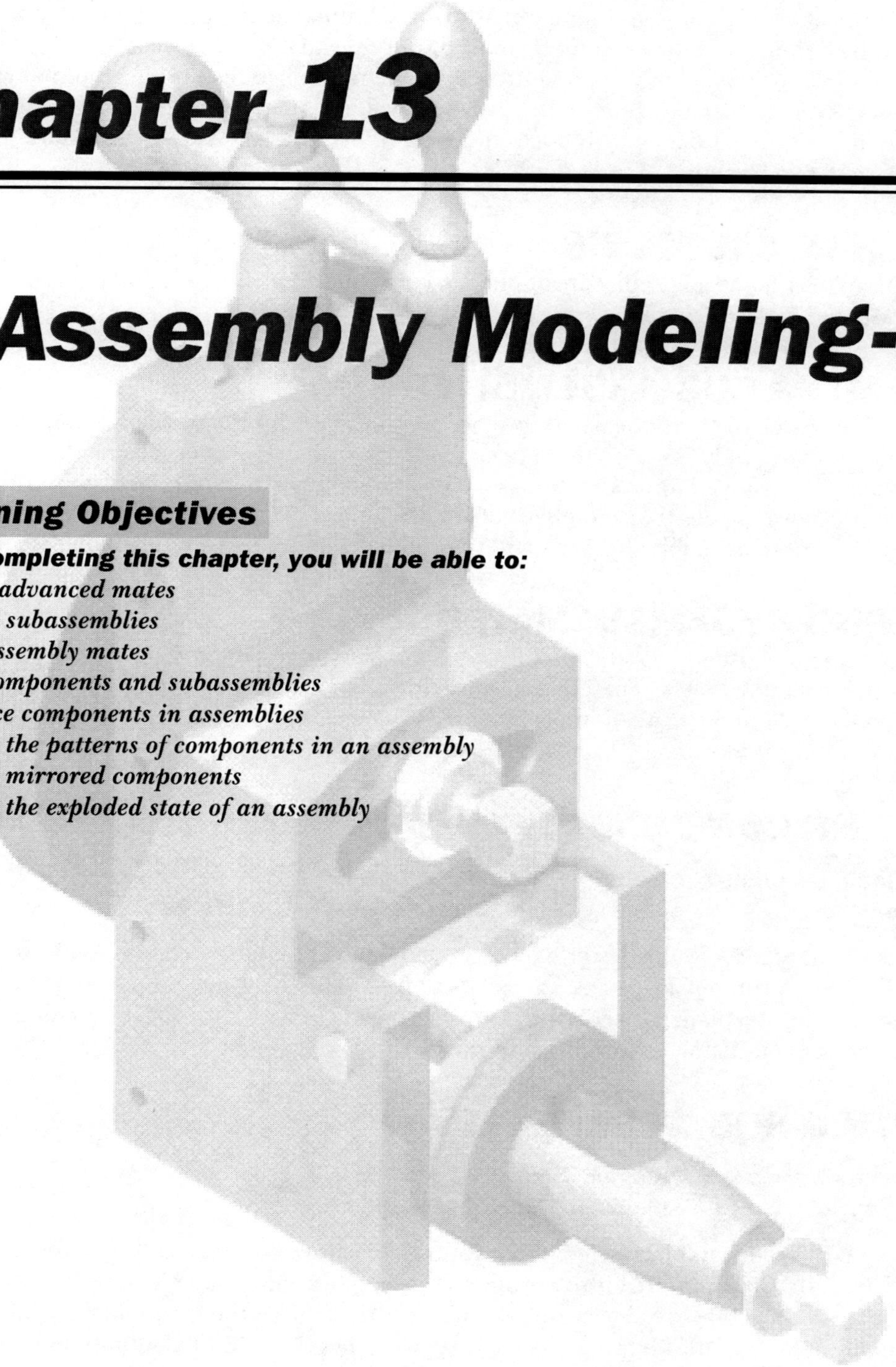

Learning Objectives

After completing this chapter, you will be able to:

- *Apply advanced mates*
- *Create subassemblies*
- *Edit assembly mates*
- *Edit components and subassemblies*
- *Replace components in assemblies*
- *Create the patterns of components in an assembly*
- *Create mirrored components*
- *Create the exploded state of an assembly*

ADVANCED ASSEMBLY MATES

In the previous chapter, you learned to place the components in an assembly document and also to apply the assembly mates to them. On applying all possible mates to the components, they will be fully defined and they cannot be moved. But in actual practice, you may need to create assemblies in which some of the parts such as gears, pulleys, and so on need to be moved. In SOLIDWORKS, this can be achieved by applying advanced mates. In this chapter, you will learn to apply the advanced mates such as Profile Center, Symmetric, Width, Path Mate, Distance, and Angle to components.

MECHANICAL MATES

In SOLIDWORKS, the mates that are applied to create mechanical drives and joints are grouped in the **Mechanical Mates** rollout.

CREATING SUB-ASSEMBLIES

In the previous chapter, you learned to place the individual parts in the assembly document and apply the assembly mates to them to create an assembly. However, this method will not be effective if you create a large assembly like assembling a machine tool, car, and so on. In such cases, you need to create the assemblies of smaller units. These are known as sub-assemblies. Then, you need to assemble the parts and the subassemblies to create the main assembly.d.

EDITING ASSEMBLY MATES

After creating the assembly or during the process of assembling the components, you may need to edit the assembly mates. The editing operations that can be performed include modifying the type of the assembly mate or angle and offset values, changing the component to which the mate was applied, and so on.

EDITING COMPONENTS

CommandManager:	Assembly > Edit Component	
Toolbar:	Assembly > Edit Component	*(Customize to add)*

At some stage of your design cycle, you may need to edit the components after placing and mating them in the assembly document. This may include editing the features, sketches, and sketch planes. To edit the components, you first need to select the component and invoke the part modeling environment in the assembly document.

EDITING SUB-ASSEMBLIES

CommandManager:	Assembly > Edit Component	
Toolbar:	Assembly > Edit Component	*(Customize to add)*

To edit sub-assemblies, select them from the **FeatureManager Design Tree** and choose the **Edit Component** button from the **Assembly CommandManager**. You can add components to a sub-assembly, modify the mates, and replace the components while they are in the editing mode. After editing the sub-assembly, choose the **Edit Component** button again to exit the editing mode.

DISSOLVING SUB-ASSEMBLIES

Dissolving the sub-assembly means the components of the selected sub-assembly will become the components of the next higher level sub-assembly. When you dissolve a sub-assembly, the sub-assembly will disappear from the **FeatureManager Design Tree** and the components of that sub-assembly will become the components of the next higher level sub-assembly.

REPLACING COMPONENTS

Sometimes in the assembly design cycle, you may need to replace a component of the assembly with some other component.

CREATING PATTERNS OF COMPONENTS IN AN ASSEMBLY

CommandManager: Assembly > Linear Component Pattern flyout
SOLIDWORKS menus: Insert > Component Pattern

While working on a complex assembly, you may need to assemble more than one instance of the component about a specified arrangement. For example, consider the case of a flange coupling where you have to assemble eight instances of nuts and bolts to fasten the coupling. If you assemble eight instances of the nuts and bolts with the couplings individually, it will be very tedious and time-consuming process. Therefore, to save the time, SOLIDWORKS has provided various pattern tools to create the patterns of the components. SOLIDWORKS has different types of component patterns.

COPYING AND MIRRORING COMPONENTS

CommandManager: Assembly > Linear Component Pattern flyout > Mirror Components
SOLIDWORKS menus: Insert > Mirror Components

In the assembly mode of SOLIDWORKS, you can mirror a component and place a new instance in the assembly document.

COPYING A COMPONENT ALONG WITH MATES

CommandManager: Assembly > Insert Components flyout > Copy with Mates
SOLIDWORKS menus: Insert > Component > Copy with Mates

In SOLIDWORKS, you can copy a component along with the mates. This will reduce the time in assembling the components that have similar mates.

MAGNETIC MATES

On working with large assembly models, such as plant layouts, it is very tedious and time consuming process to assemble a large number of parts. It takes lot of time to place such large number of components in an assembly by defining various mates between each component to get the desired assembly model. In SOLIDWORKS, assembly of such models can now be done within short intervals of time using the **Magnetic Mates** tool. Using various tools and options, you can publish a model as an asset in part or an assembly mode by defining ground plane and connection points on the model. Next, you will define a ground plane of the assembly so that when you insert an asset into the assembly, the ground plane of the asset will snap to the ground plane of the assembly and also the connection point of the asset will snap corresponding connection points of the other assets. After snapping corresponding connection points and ground plane, model gets placed and a **Magnetic** mate is automatically added between the components.

CREATING EXPLODED STATE OF AN ASSEMBLY

CommandManager: Assembly > Exploded View flyout > Exploded View
SOLIDWORKS menus: Insert > Exploded View

In SOLIDWORKS, you can create exploded state of an assembly using the **Explode PropertyManager**.

Creating the Explode Line Sketch

CommandManager: Assembly > Exploded View flyout > Explode Line Sketch
SOLIDWORKS menus: Insert > Explode Line Sketch

The explode lines are the parametric axes that display the direction of explosion of the components in an exploded state.

Creating Smart Explode Lines

CommandManager: Assembly > Exploded View flyout > Insert/Edit Smart Explode Lines
Toolbar: Assembly > Insert/Edit Smart Explode Lines *(Customize to add)*

In SOLIDWORKS 2018, you can automatically add route lines sketch to an assembly to display the direction of explosion of the components. Also, when you change any explode step, these route lines are updated automatically.

TUTORIALS

Tutorial 1

In this tutorial, you will create the radial engine assembly shown in Figure 13-1. This assembly will be created in two parts: sub-assembly and main assembly. You will also create the exploded state of the assembly and then create the explode line sketch. The exploded state of the assembly is displayed in Figure 13-2. The views and dimensions of all the components of this assembly are displayed in Figures 13-3 through 13-6. **(Expected time: 3 hr)**

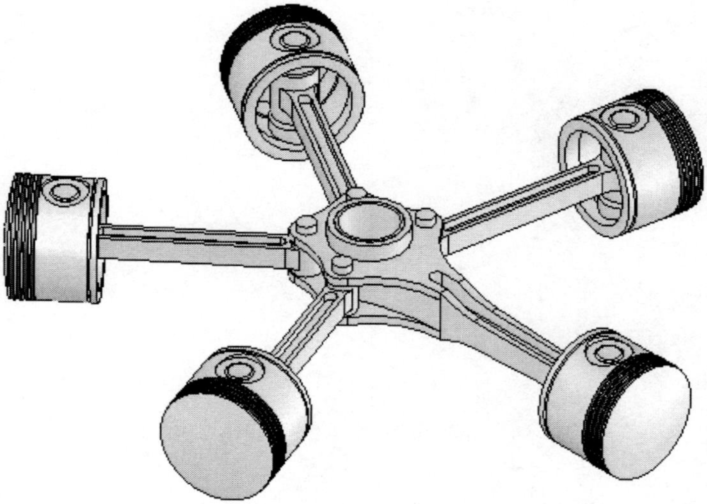

Figure 13-1 The radial engine assembly

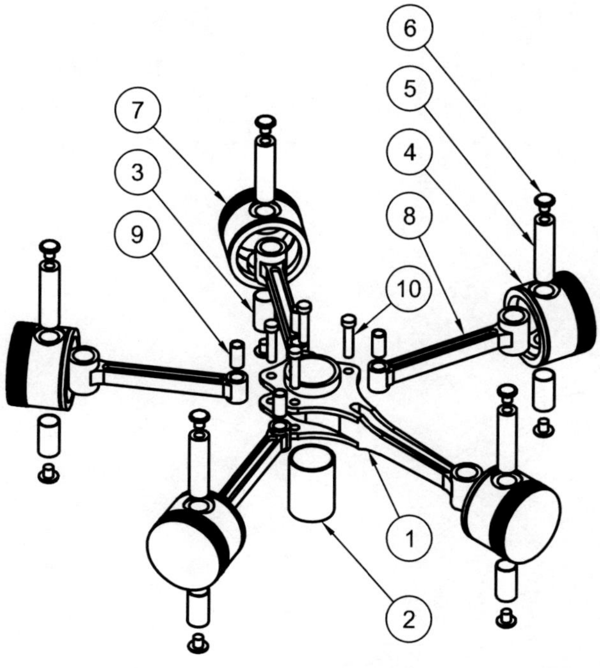

ITEM NO.	PART NO.	QTY.
1	Master Rod	1
2	Master Rod Bearing	1
3	Rod Bush Upper	5
4	Piston	5
5	Piston Pin	5
6	Piston Pin Plug	10
7	Piston Ring	20
8	Articulated Rod	4
9	Rod Bush Lower	4
10	Link Pin	4

Figure 13-2 Exploded view of the assembly

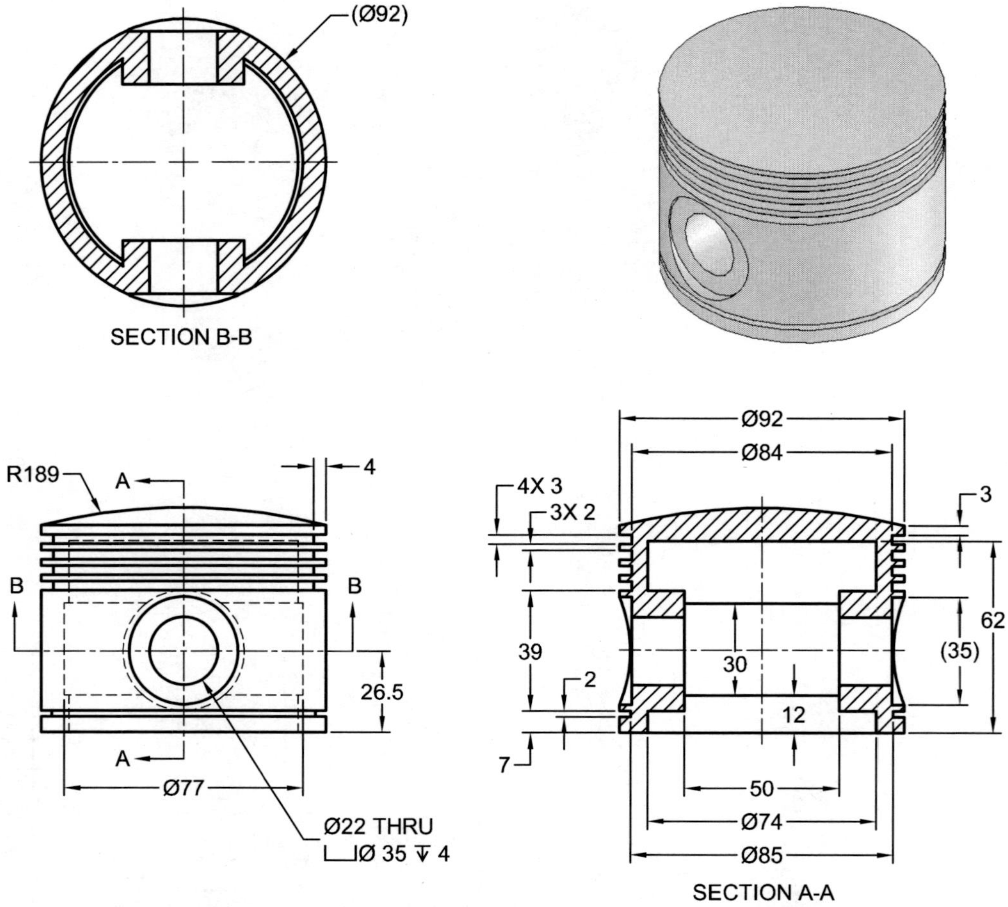

Figure 13-3 *Views and dimensions of the Piston*

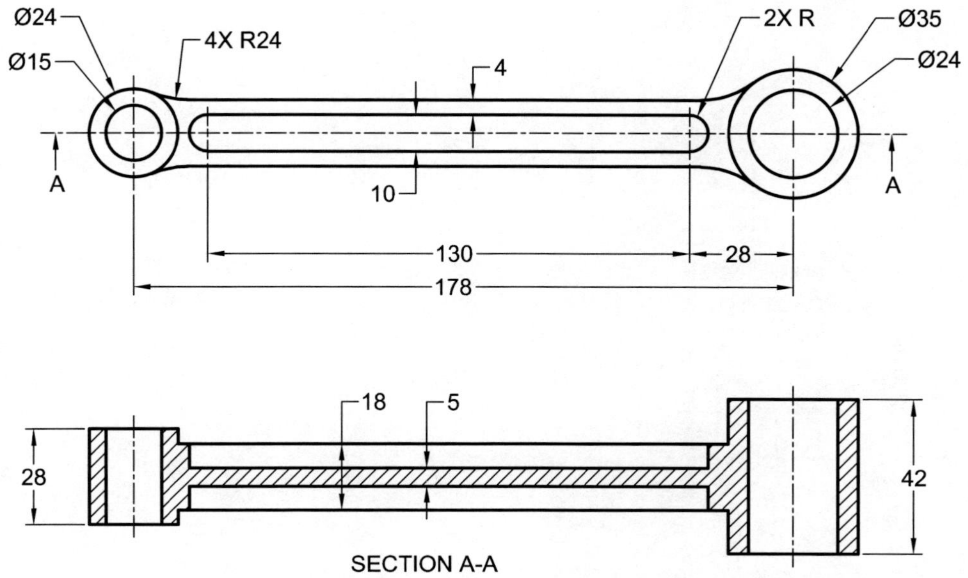

Figure 13-4 *Views and dimensions of the Articulated Rod*

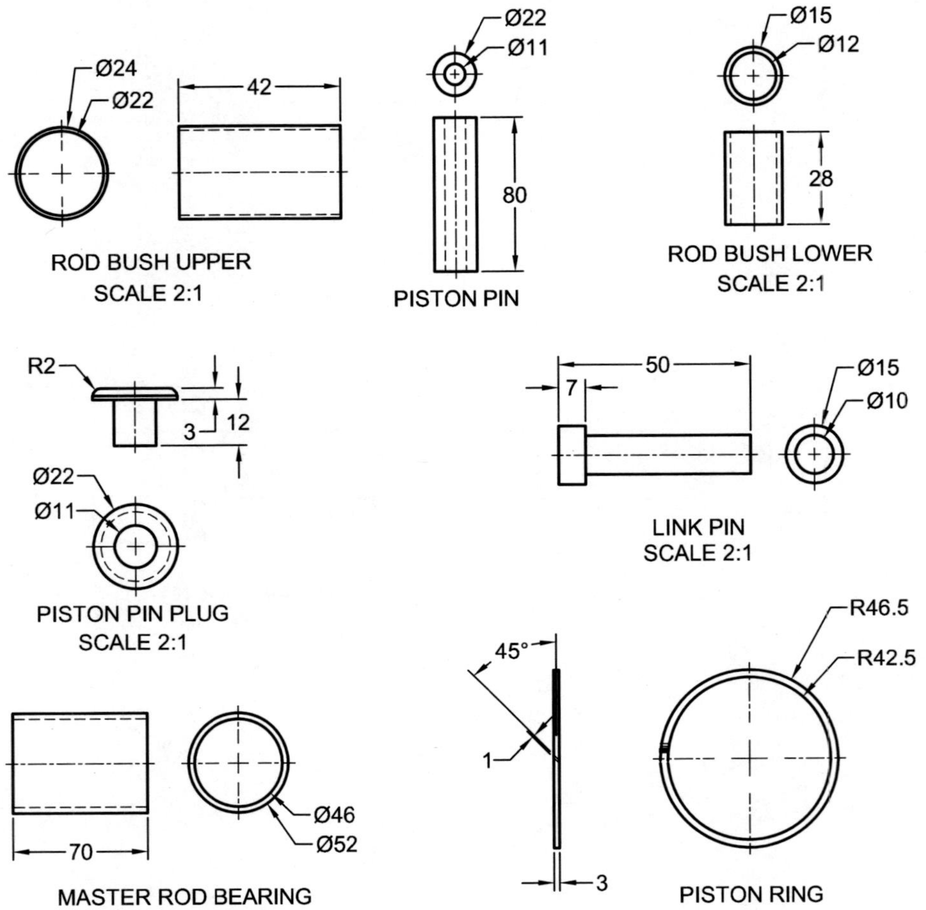

***Figure 13-5** Views and dimensions of other components*

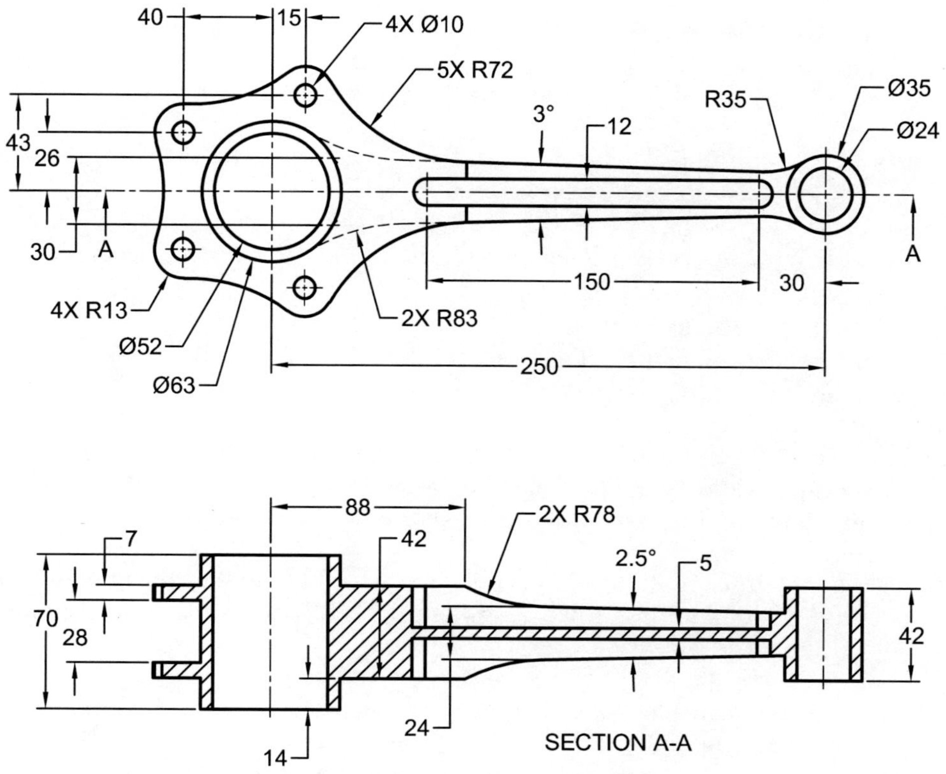

Figure 13-6 *Views and dimensions of the Master Rod*

You need to break this assembly in two steps because it is a large assembly. One will be the sub-assembly and the other will be the main assembly. First, you need to create the sub-assembly consisting of Articulated Rod, Piston, Piston Rings, Piston Pin, Rod Bush Upper, Rod Bush Lower, and Piston Pin Plug. Next, you need to create the main assembly by assembling the Master Rod with the Piston, Piston Rings, Piston Pin, Rod Bush Upper, and Piston Pin Plug. Finally, you will assemble the sub-assembly with the main assembly.

The following steps are required to complete this tutorial:

a. Create all components of the assembly in the **Part** mode and save them in the *Radial Engine Assembly* folder.
b. Start a new assembly document and assemble the components to complete the sub-assembly, refer to Figures 13-7 through 13-9.
c. Start a new assembly document and assemble the components of the main assembly, refer to Figure 13-10.
d. Assemble the sub-assembly in the main assembly, refer to Figures 13-11 through 13-15.
e. Create the exploded view of the assembly and then create the explode line sketch, refer to Figures 13-16 through 13-18.

Creating the Components

1. Create a folder with the name *Radial Engine Assembly* at the location *\Documents\SOLIDWORKS\c13*. Create all components in the individual part documents and save them in this folder.

Note

While creating the Master Rod, make sure that the holes on the left of the Master Rod are created using the sketch-driven pattern. This is done because while assembling the Link Pin, you need to create the derived pattern of the Link Pin using the sketch-driven pattern feature.

Creating the Sub-assembly

As discussed earlier, you first need to create the sub-assembly and then assemble it with the main assembly.

1. Start a new SOLIDWORKS assembly document and close the **Open** dialog box and **Begin Assembly PropertyManager**. Next, save the assembly with the name **Piston Articulated Rod Sub-assembly** in the same folder in which the parts are created.

2. First, place the Articulated Rod at the origin of the assembly and then the other components such as Piston, Piston Pin, Piston Pin Plug, Rod Bush Upper, and Rod Bush Lower in the assembly document.

3. Apply the required mates to assemble these components. Figure 13-7 shows the sequence of assembling the components. The exploded view and the explode line sketch are given only for your reference. The assembly after assembling the Articulated Rod, Piston, Piston Pin, Piston Pin Plug, Rod Bush Upper, and Rod Bush Lower is shown in Figure 13-8.

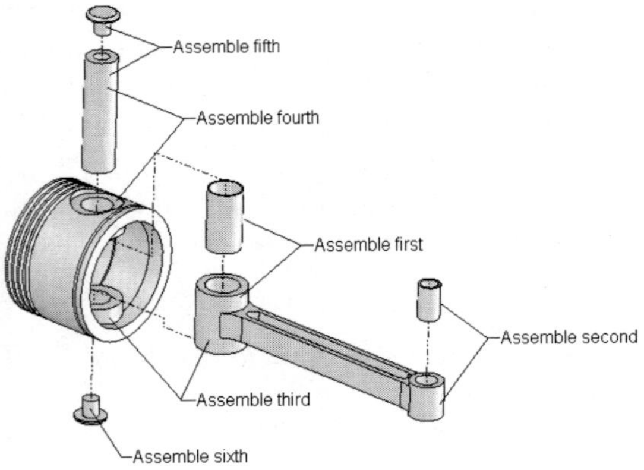

***Figure 13-7** Assembly of the Articulated Rod, Piston, Piston Pin, Piston Pin Plug, Rod Bush Upper, and Rod Bush Lower*

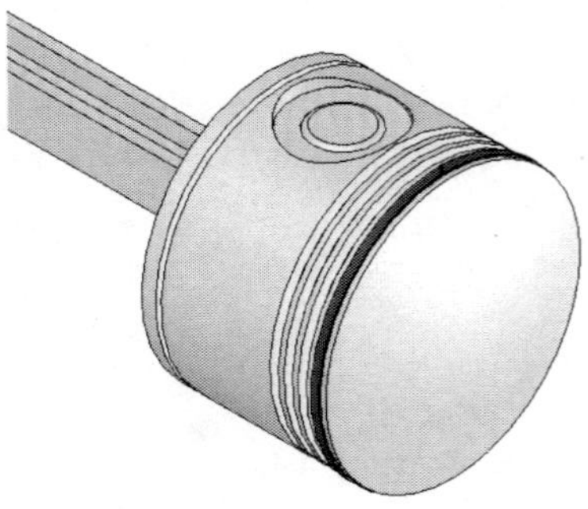

***Figure 13-8** First instance of the Piston Ring assembled with the Piston*

It is clear from the assembly that you need to assemble four instances of the Piston Ring. You will assemble only one instance of the Piston Ring at the uppermost groove of the ring and then create a local linear pattern.

4. Insert the Piston Ring in the assembly document and assemble the Piston Ring at the uppermost groove of the Piston using the assembly mates, refer to Figure 13-8. In this assembly, the color of the Piston Ring needs to be changed. To do so, select the Piston Ring; a pop-up toolbar is displayed. Choose the **Appearances** button from the pop-up toolbar; a flyout is displayed. Choose the name of the component; the **Color PropertyManager** is displayed. Set the color using the option available in this PropertyManager.

   Next, you need to create the local linear pattern of the Piston Ring.

5. Choose the **Linear Component Pattern** button from the **Assembly CommandManager**; the **Linear Pattern PropertyManager** is displayed.

6. Select any one of the horizontal edges of the Articulated Rod to define the direction of pattern creation.

7. Click once in the **Components to Pattern** selection box and then select the Piston Ring from the drawing area; preview of the linear pattern with default settings is displayed in the drawing area.

8. Choose the **Reverse Direction** button to reverse the direction of the pattern creation if required.

9. Set the value **5** in the **Spacing** spinner and the value **4** in the **Number of Instances** spinner.

10. Choose the **OK** button from the **Linear Pattern PropertyManager**; the sub-assembly after patterning the Piston Ring is shown in Figure 13-9.

11. Save and close the assembly document.

Creating the Main Assembly

Next, you need to create the main assembly and then assemble the sub-assembly with it.

1. Start a new SOLIDWORKS assembly document and close the **Open** dialog box and **Begin Assembly PropertyManager**. Now, save it with the name **Radial Engine assembly** in the same folder in which the parts are saved.

2. First, place the Master Rod at the origin of the assembly and then place the Piston, Piston Pin, Piston Pin Plug, Piston Ring, Rod Bush Upper, and Master Rod Bearing in the current assembly document.

3. Assemble all components of the main assembly using the assembly mates.

   The components after assembling them in the main assembly are displayed in Figure 13-10.

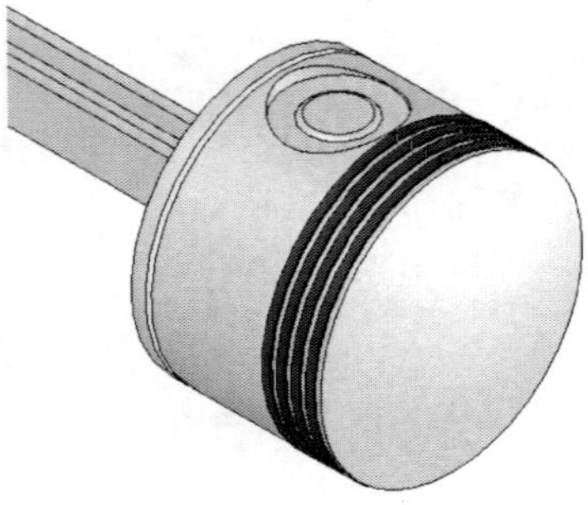

Figure 13-9 *Sub-assembly after patterning the Piston Ring*

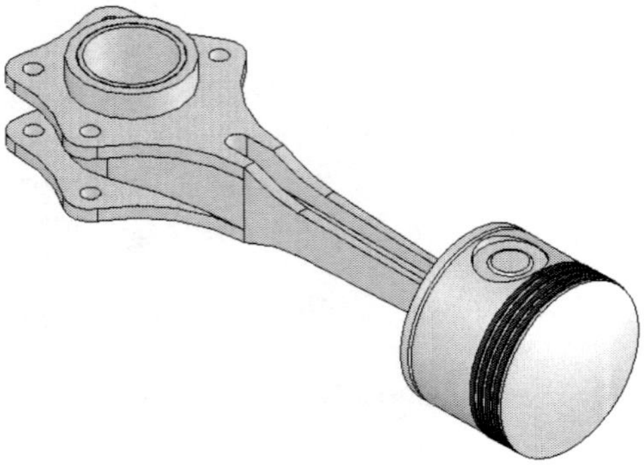

Figure 13-10 *Components assembled in the main assembly*

Assembling the Sub-assembly with the Main Assembly

Next, you need to place the sub-assembly in the main assembly and then assemble them together.

1. Choose the **Insert Components** button from the **Assembly CommandManager**; the **Insert Component PropertyManager** along with the **Open** dialog box is displayed.

2. If the sub-assembly document is not opened, browse the *Piston Articulated Rod Sub-assembly* file or else close the **Open** dialog box.

3. Select **Assembly** (**.asm, *.sldasm*) from the **Files of type** drop-down list for refining the search.

4. Double-click on *Piston Articulated Rod Sub-assembly* and place the sub-assembly in the main assembly. Figure 13-11 shows the sub-assembly and the main assembly placed together.

5. Assemble the sub-assembly with the main assembly using the assembly mates. Refer to Figure 13-12 which shows the assembly structure that will help you in assembling the instances of the sub-assembly.

   Figure 13-13 shows all instances of the sub-assembly assembled with the main assembly.

Assembling the Link Pin

After assembling the sub-assembly with the main assembly, you need to assemble the Link Pin with the main assembly.

1. Place the Link Pin in the current assembly document. Next, assemble the Link Pin with the main assembly using the assembly mates. Figure 13-14 shows the first instance of the Link Pin assembled with the main assembly.

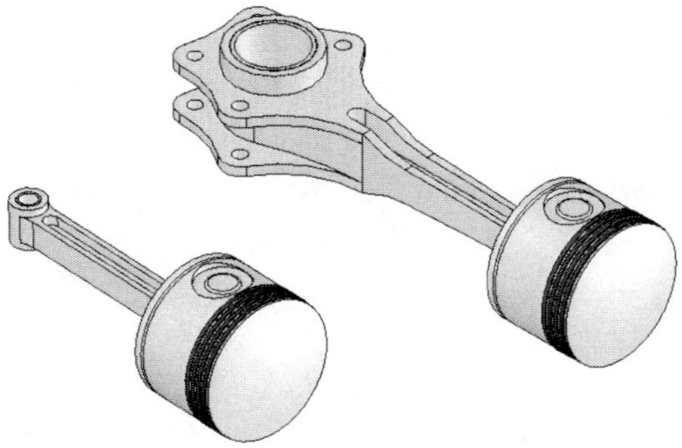

Figure 13-11 Sub-assembly and the main assembly placed together

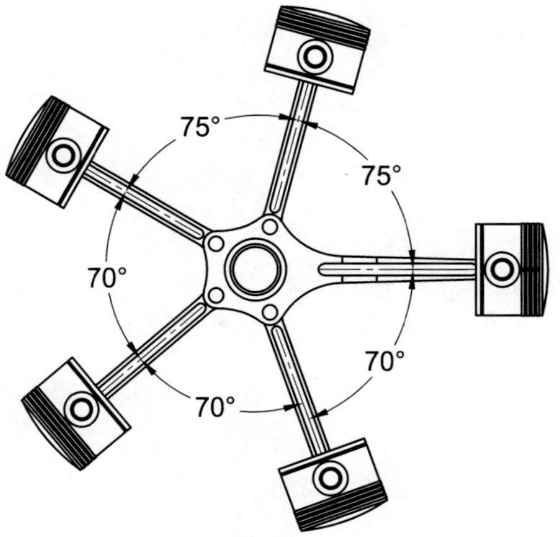

Figure 13-12 Assembly structure

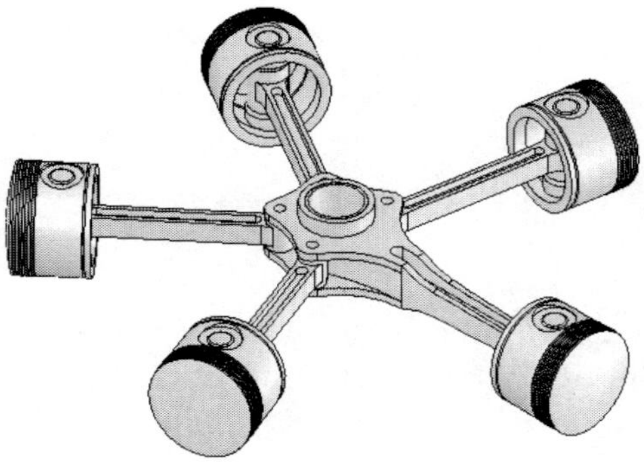

Figure 13-13 The sub-assembly assembled with the main assembly

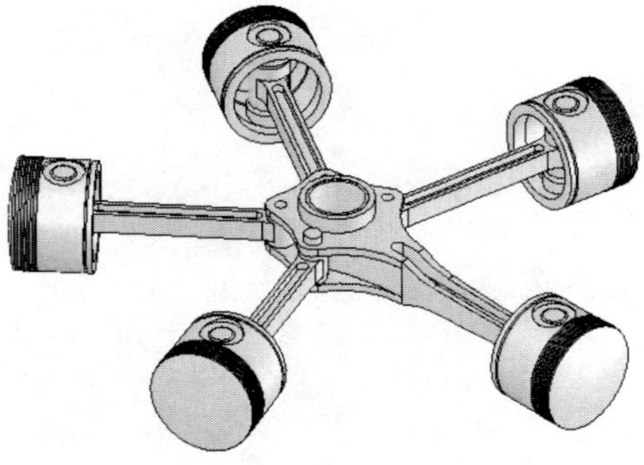

Figure 13-14 The first instance of the Link Pin assembled with the main assembly

As discussed earlier, the other instances of the Link Pin will be assembled using the sketch-driven pattern feature of the holes created on the left of the master rod.

2. Choose the **Pattern Driven Component Pattern** button from the **Linear Component Pattern** flyout; the **Pattern Driven PropertyManager** is displayed.

3. Select the Link Pin from the main assembly; its name is displayed in the **Components to Pattern** selection box.

4. Click once in the **Driving feature or component** selection box to activate the selection mode.

5. Select any one of the hole instances from the master rod; the name of the sketch pattern feature is displayed in the **Driving feature or component** selection box and the preview of the resulting pattern is also displayed.

6. If the instances are not placed properly, choose the **Select Seed Position** button and select the correct seed feature.

7. Choose the **OK** button from the **Pattern Driven PropertyManager**. Figure 13-15 shows the final assembly.

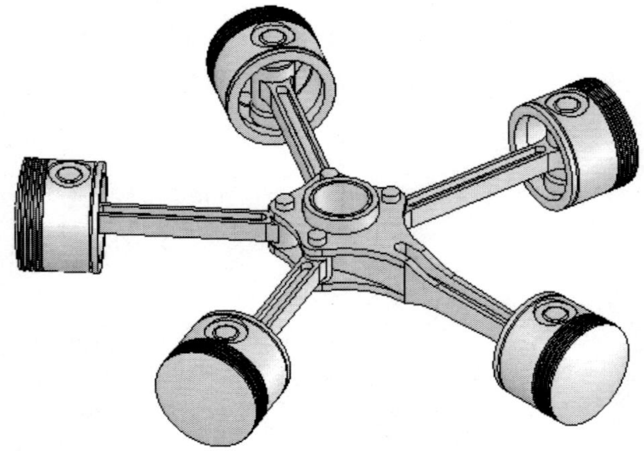

Figure 13-15 *The final assembly*

Exploding the Assembly

After creating the assembly, you need to explode it using the **Exploded View** tool. The current assembly contains a sub-assembly that needs to be exploded first to create the final exploded view. Therefore, you need to open the Piston Articulated Rod Sub-assembly.

1. Open the Piston Articulated Rod Sub-assembly document.

2. Choose the **Exploded View** button from the **Assembly CommandManager**; the **Explode PropertyManager** is displayed.

3. Choose the **Regular step (translate and rotate)** button from the **Explode Step Type** rollout of the **Explode PropertyManager**, if it is not selected by default.

4. Ensure that all check boxes are cleared.

5. Select the top face of the Piston Pin Plug; a triad is displayed.

6. Select the arrow of the triad that is normal to the selected face.

7. Set the value of the **Explode Distance** spinner to **170** and then choose the **Apply** button; the selected instance of the Piston Pin Plug is exploded and the component is removed

from the selection set. Also, the sequence of explosion is displayed as **Explode Step1** in the **Existing explode steps** list box of the **Explode Steps** rollout.

8. If the piston pin plug is moved downward, choose the **Reverse Direction** button on the left of the **Explode Direction** selection box in the **Settings** rollout and then choose **Done**.

   Similarly, explode the another instance of Piston Pin Plug in the opposite direction.

9. Select the flat face of the Piston Pin in the drawing area; a triad is displayed.

10. Select the arrow of the triad that is normal to the selected face.

11. Set the value of the **Explode Distance** spinner to **150**, and then choose the **Apply** button; the selected instance of the Piston Pin is exploded and the component is removed from the selection set. Also, the sequence of explosion is displayed as **Explode Step3** in the **Existing explode steps** list box of the **Explode Steps** rollout.

12. If the Piston Pin is moved downward, choose the **Reverse Direction** button on the left of the **Explode Direction** edit box in the **Settings** rollout and then choose **Done**.

    When the triad is displayed on selecting the face, you can select an arrow to specify the direction and drag the arrow to relocate the component.

13. Explode all components of the sub-assembly.

14. Save the sub-assembly and then open the main assembly document; the **SOLIDWORKS** message box is displayed stating that the models in the assembly have changed and would you like to rebuild them. Choose **OK** to save the models and rebuild the sub-assembly.

15. Choose the **Exploded View** button from the **Assembly CommandManager**; the **Explode PropertyManager** is displayed. Ensure that all check boxes are cleared.

16. As discussed earlier, explode the components of the main assembly only.

    Next, you need to explode the parts of the remaining sub-assemblies.

17. Select all sub-assemblies from the drawing area and choose the **Reuse Subassembly Explode** button in the **Explode PropertyManager**; all sub-assemblies are exploded.

18. Choose **OK** to exit the **Explode PropertyManager**. The assembly after exploding the components is shown in Figure 13-16.

Creating the Explode Line Sketch

After exploding the assembly, you need to create the explode line sketch of the exploded state of the assembly.

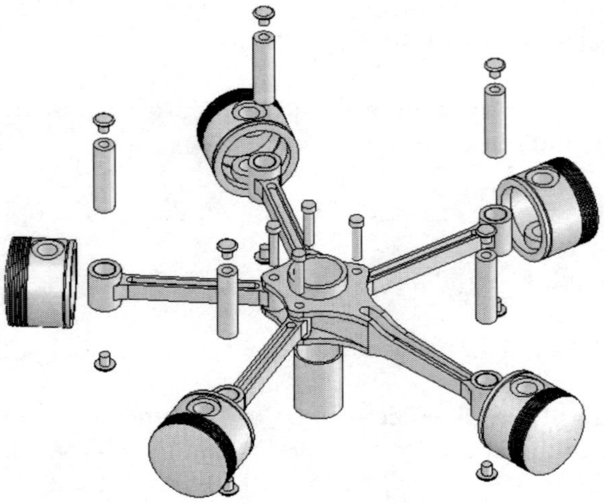

Figure 13-16 Final exploded assembly

1. Choose the **Explode Line Sketch** button from the **Assembly CommandManager**; the **Route Line PropertyManager** is displayed and you are prompted to select a cylindrical face, planar face, vertex, point, arc, or a line entity.

2. Select the cylindrical face of piston pin plug, refer to Figure 13-17, as the first selection; the name of the selected face is displayed in the **Items To Connect** selection box. Also, the preview of the explode line sketch is displayed at the center of the selected face.

3. Refer to Figure 13-17 and select the other cylindrical faces to create the explode line sketch.

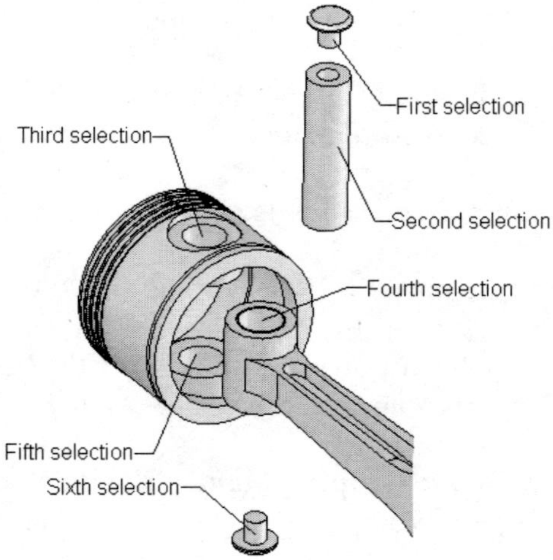

Figure 13-17 Faces to be selected to create the explode line sketch

4. Next, choose the **OK** button; an exploded line is created.

5. Similarly, create explode lines between the other parts of the exploded assembly. Figure 13-18 shows the assembly after creating the explode line sketch.

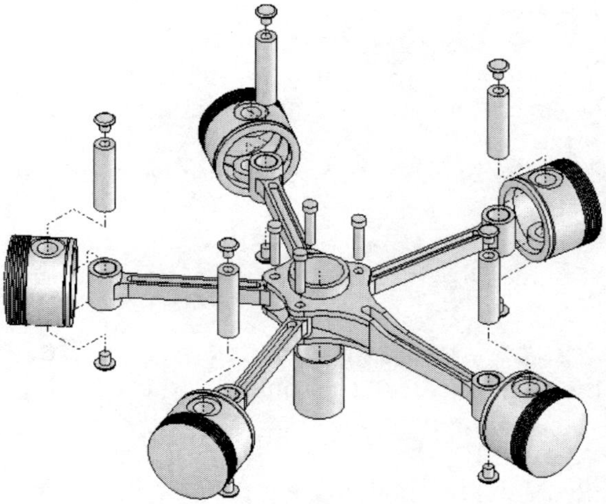

Figure 13-18 Explode line sketch created for the exploded state of the assembly

6. Invoke the **ConfigurationManager** and expand the **Default** node. Next, select **ExplView1** node and right-click; a shortcut menu is displayed. 

7. Choose the **Animate collapse** option from the menu to view the animation of the exploded view.

8. Save and close the assembly document.

Tutorial 2

In this tutorial, you will modify the assembly created in Tutorial 1 (Bench Vice) of Chapter 12. You will modify the design of the components of the assembly and then suppress some mates that enable it to have a particular degree of freedom. Next, you will check the assembly for collision detection when the assembly is in motion. Then, you will modify the assembly and check the interference. **(Expected time: 1 hr)**

The following steps are required to complete this tutorial:

a. Copy and save the Bench Vice assembly folder in the *c13* directory and then open the Bench Vice assembly.
b. Modify the design of the components within the context of the assembly, refer to Figures 13-19 through 13-23.
c. Suppress the mate to enable the Vice Jaw to move along the slide ways of the Vice Body.
d. Check the new assembly design for the collision detection when the assembly is in motion. Modify the design, if there is any collision between the components, refer to Figures 13-24 and 13-25.
e. Check the interference in the modified assembly.

Opening the Bench Vice Assembly

The assembly created in Tutorial 1 of Chapter 12 is the Bench Vice assembly. You need to copy and save it in the current folder of Chapter 13.

1. Copy the folder in which the Bench Vice assembly is saved and paste it in the *c13* folder.

2. Start SOLIDWORKS, invoke the **Open** dialog box, and then browse to the Bench Vice assembly document. Double-click on it to open the assembly document.

Modifying the Design of the Components of the Bench Vice Assembly

You need to modify the components in the context of the assembly because of alteration in the design of some of the components.

Before you start modifying the components, it is recommended that you hide some of them. This will simplify the assembly and facilitate the selection of components while editing and modifying them.

1. Press and hold the CTRL key and select the Clamping Plate, Base Plate, four Set Screw 1, two Set Screw 2, Oval Fillister, Screw Bar, Bar Globes, and Jaw Screw from the **FeatureManager Design Tree**.

2. When you release the CTRL key, a pop-up toolbar is displayed. Choose the **Hide Components** option from the pop-up toolbar; the visibility of the selected components is turned off.

The design alteration includes creating a through slot on the right face of the Vice Jaw. To modify its design, you first need to enable the part editing environment.

3. Select the Vice Jaw from the assembly and choose the **Edit Component** button from the **Assembly CommandManager**; the part modeling environment is invoked.

4. Make sure that the Vice Body is transparent. If not, choose the **Assembly Transparency** button from the **Features CommandManager**; a flyout is displayed. Choose the **Force Transparency** option from the flyout. The Vice Body becomes transparent, as shown in Figure 13-19.

5. Select the right face of the Vice Jaw and invoke the sketching environment.

6. Create the sketch of the slot, as shown in Figure 13-20.

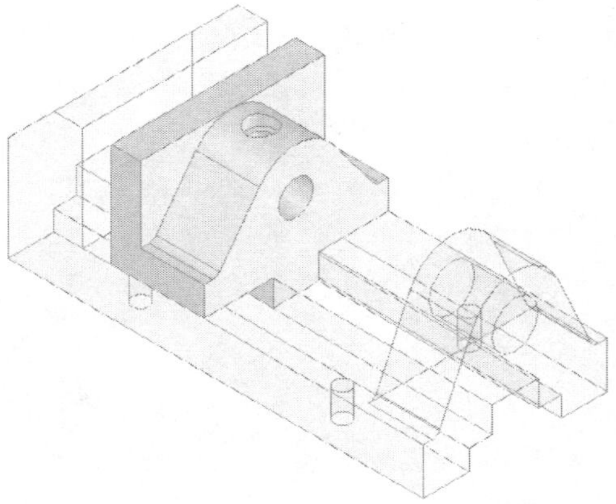

Figure 13-19 Vice Jaw in the part edit mode in the assembly document

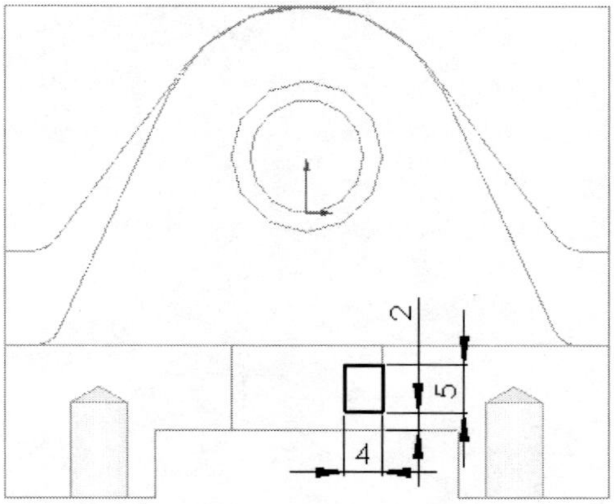

Figure 13-20 Sketch of the slot

7. Invoke the **Cut-Extrude PropertyManager**.

8. Create the cut feature using the **Through All** option and exit the **PropertyManager**.

9. Choose the **Edit Component** button to exit the part editing environment.

   Figure 13-21 shows the assembly after modifying the design of the Vice Jaw.

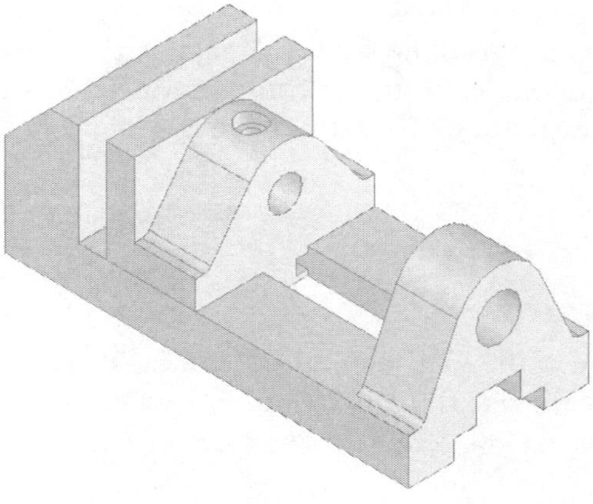

Figure 13-21 Modified Vice Jaw

10. Similarly, modify the design of the Vice Body by creating a blind extruded boss feature up to 60 mm depth on the right face of the Vice Body. Note that you need to reverse the direction of feature creation.

    The sketch of the extruded boss feature is shown in Figure 13-22. Figure 13-23 shows the assembly after exiting the part editing environment.

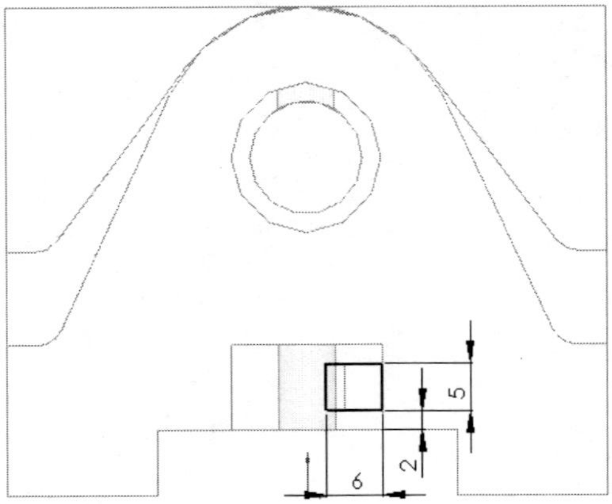

Figure 13-22 *Sketch of the extruded boss feature*

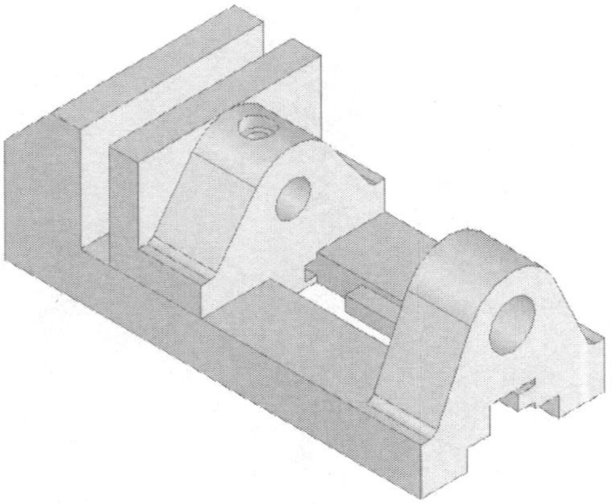

Figure 13-23 *Modified assembly*

11. Exit the **Edit Component** mode and choose the **Save** button from the Menu Bar; the **Save Modified Documents** dialog box is displayed. Choose **Save All** to save the referenced models also.

Suppressing the Mate Constraint to Move the Vice Jaw Freely

To analyze the movement of the Bench Vice assembly, you need to make the movement of the Vice Jaw free in the X direction. On doing so, the Vice Jaw will slide on the sideways of the Vice Body.

1. Expand the **Vice Jaw** node from the **FeatureManager Design Tree** and then expand the **Mates in Bench Vice** sub node. Next, select the **LimitDistance1** mate; the planar faces of the Vice Jaw and the Vice Body, to which this mate is applied, get highlighted and a pop-up toolbar is displayed.

2. Choose **Suppress** from the pop-up toolbar; the degree of freedom along the X direction becomes free.

3. Select a horizontal edge of the Vice Jaw and choose the **Move Component** button from the **Assembly CommandManager**; the **Move Component PropertyManager** is displayed. On dragging the cursor, you will observe that you can move the Vice Jaw in the X direction.

4. Drag the Vice Jaw back to its original position and choose the **OK** button from the **Move Component PropertyManager**.

Analyzing the Collision between the Components when the Assembly is in Motion

Next, you need to analyze the collision between the components of the assembly when the assembly is in motion.

1. Choose the **Move Component** button from the **Assembly CommandManager**. Next, select the **Collision Detection** radio button.

2. Select the Vice Jaw and drag the cursor to move in the direction shown in Figure 13-24.

   On moving the Vice Jaw in the specified direction, you will observe that the right face of the Vice Jaw and the newly created extrusion feature of the Vice Body are highlighted, refer to Figure 13-25. This indicates that the Vice Jaw has collided with the Vice Body. Leave the assembly at this location.

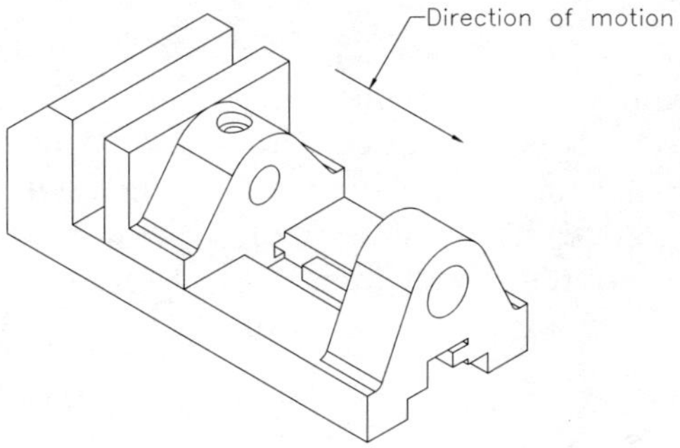

Figure 13-24 *Direction in which the Vice Jaw will move*

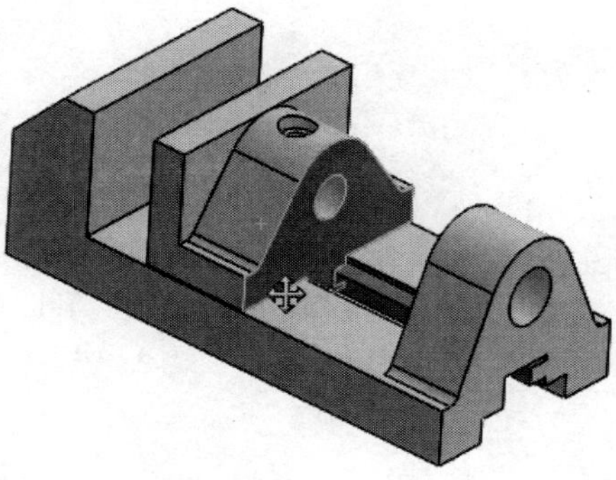

Figure 13-25 *Faces of the Vice Jaw and the Vice Body highlighted in different colors*

3. Choose the **OK** button from the **Move Component PropertyManager**.

   The collision has been detected in the assembly, and therefore you need to modify the design of one of the components. In this case, you will modify the dimensions of the extruded boss feature.

4. Double-click on the newly created extrusion feature of the Vice Body; the dimensions of the newly created feature are displayed.

5. Double-click on the dimension having the value **6**; the **Modify** dialog box is displayed. Set the value of the dimension to **4** and then press the ENTER key.

6. Press CTRL+B on the keyboard to rebuild the entire assembly.

7. Choose the **Interference Detection** button from the **Evaluate CommandManager**; the **Interference Detection PropertyManager** is displayed.

8. Choose the **Calculate** button. You will observe that **No Interferences** is displayed in the **Interference Results** selection box in the **Results** rollout of this PropertyManager.

9. Choose the **Cancel** button from the **Interference Detection PropertyManager**.

   Next, you need to show all the components of this assembly.

10. Press and hold the CTRL key, select the hidden components from the **FeatureManager Design Tree**, and then choose **Show Components** from the pop-up toolbar.

11. Expand the **Mates** node from the **FeatureManager Design Tree**. Select the **LimitDistance1** mate that is suppressed, and then choose **Unsuppress** from the pop-up toolbar.

12. Save the assembly document and all the referenced part documents.

EXERCISE

Exercise 1

Create the assembly shown in Figure 13-26. Ensure that its back plate is fixed and the entire assembly can move in the Y direction with respect to the back plate. Keep the rotational degree of freedom of the screw rod free, so that it can also rotate on its axis. After creating the assembly, explode it and create the explode line sketch. The exploded view of the assembly with the explode line sketch is shown in Figure 13-27. The dimensions of the model are given in Figures 13-28 through 13-33. Assume the missing dimensions. **(Expected time: 4 hr)**

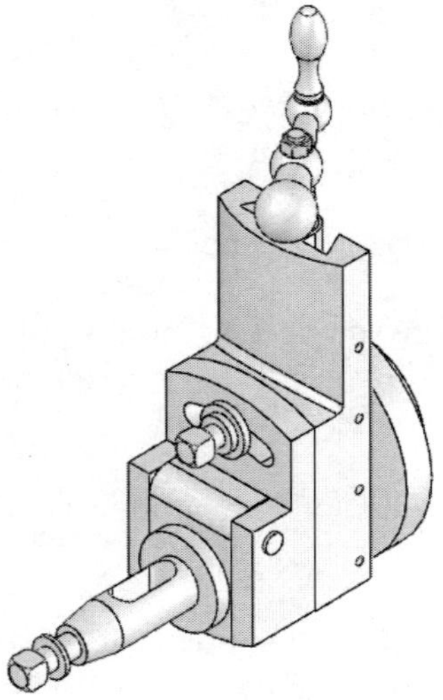

Figure 13-26 *Shaper tool holder assembly*

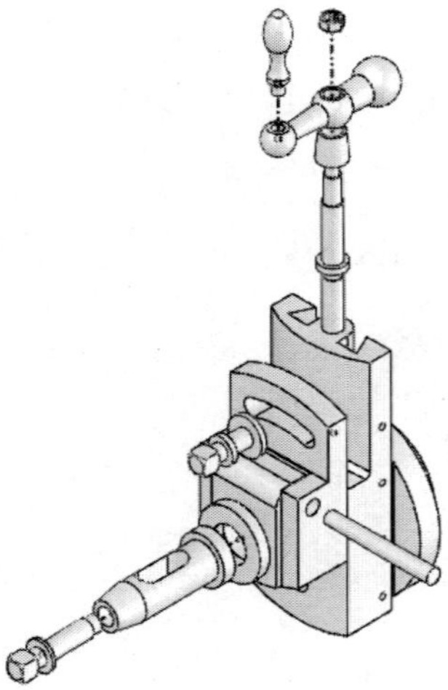

Figure 13-27 *Exploded view of the Shaper tool holder assembly with explode lines*

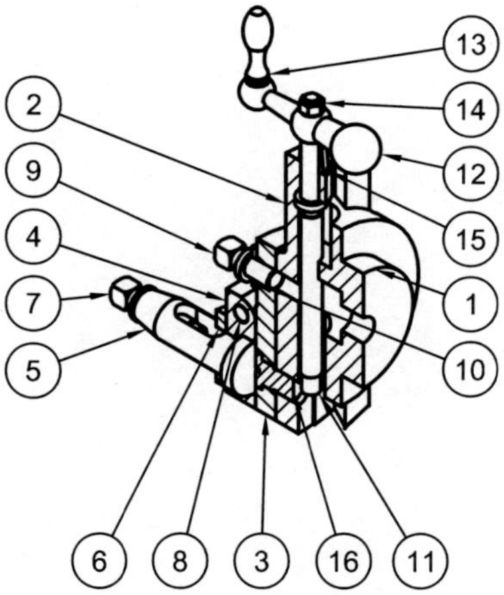

ITEM NO.	PART NO.	DESCRIPTION	QTY.
1	Back Plate		1
2	Vertical Slide		1
3	Swivel Plate		1
4	Drag Plate		1
5	Tool Holder		1
6	Washer		1
7	Tool Fixing Screw		1
8	Pivot Pin		1
9	Clamping Screw		1
10	Small Washer		1
11	Screw Bar		1
12	Handle Bar		1
13	Handle		1
14	Nut M10		1
15	Spacer Bush		1
16	Swivel Screw Pin		1

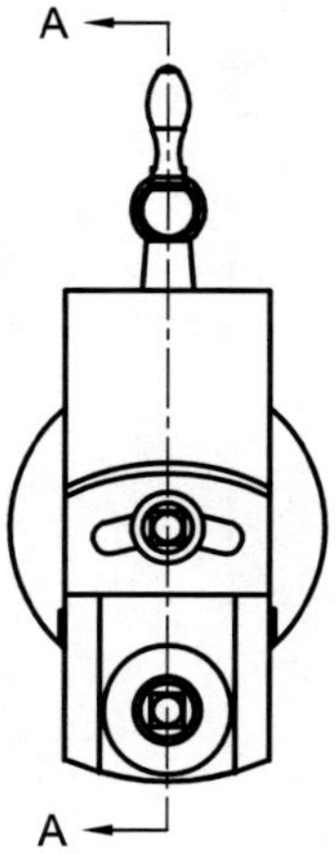

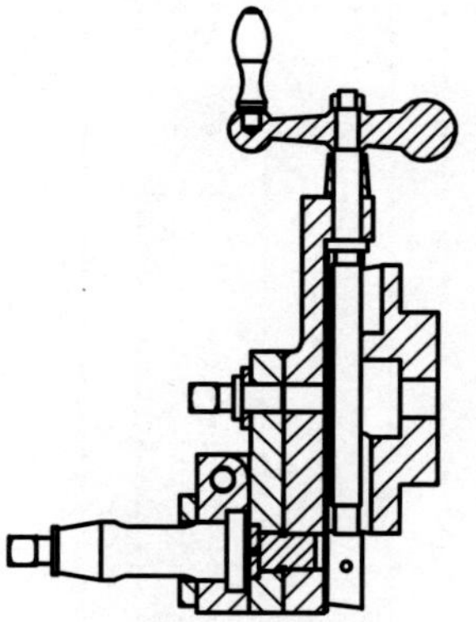

SECTION A-A

Figure 13-28 *Shaper tool holder assembly*

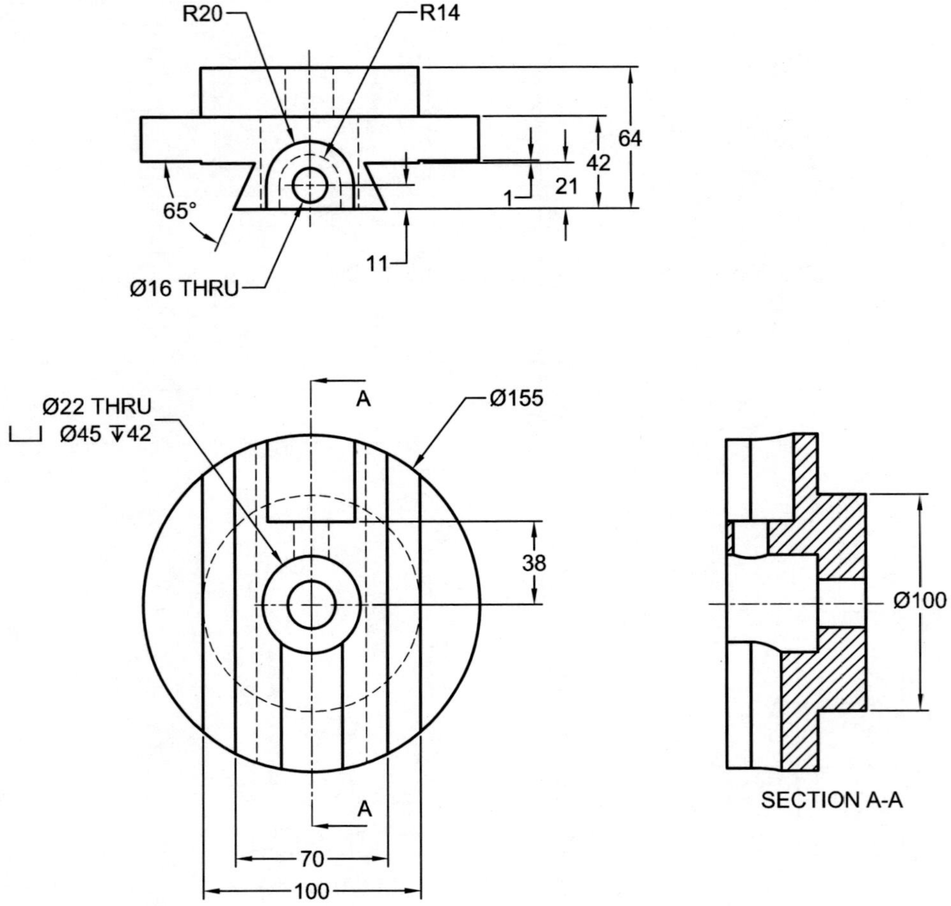

Figure 13-29 *Views and dimensions of the Back Plate*

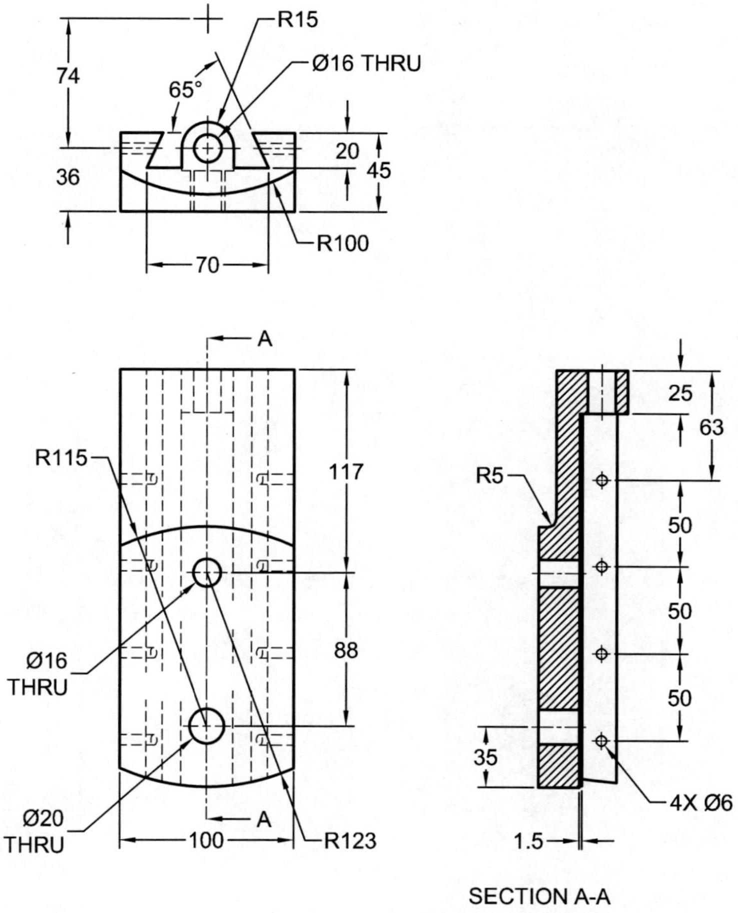

Figure 13-30 *Views and dimensions of the Vertical Slide*

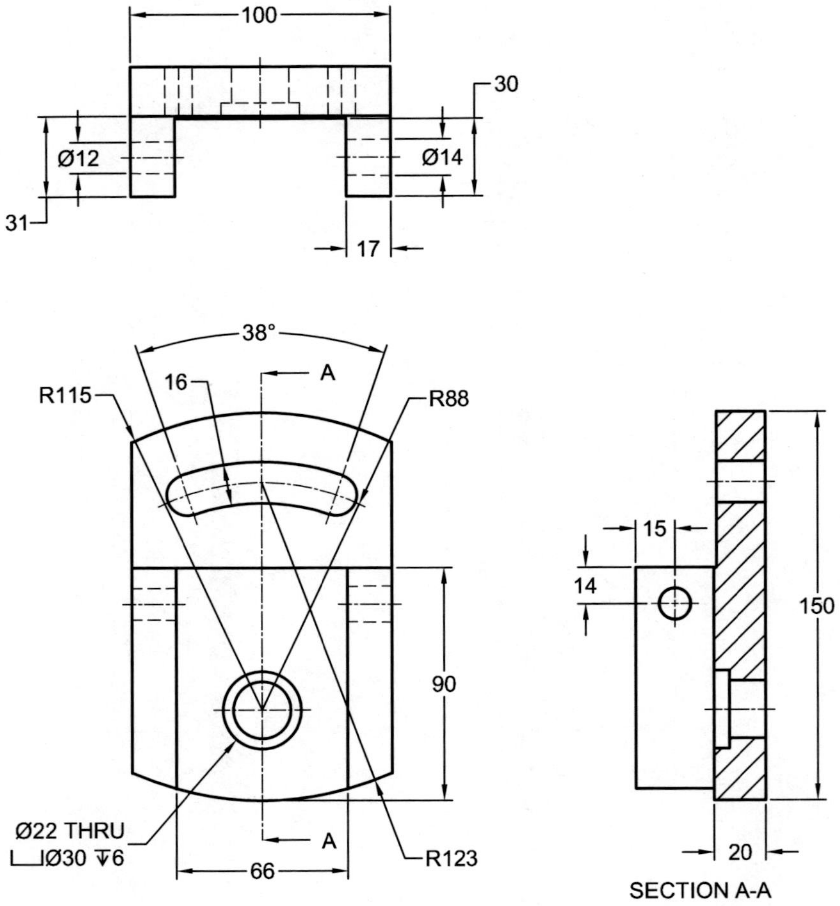

Figure 13-31 *Views and dimensions of the Swivel Plate*

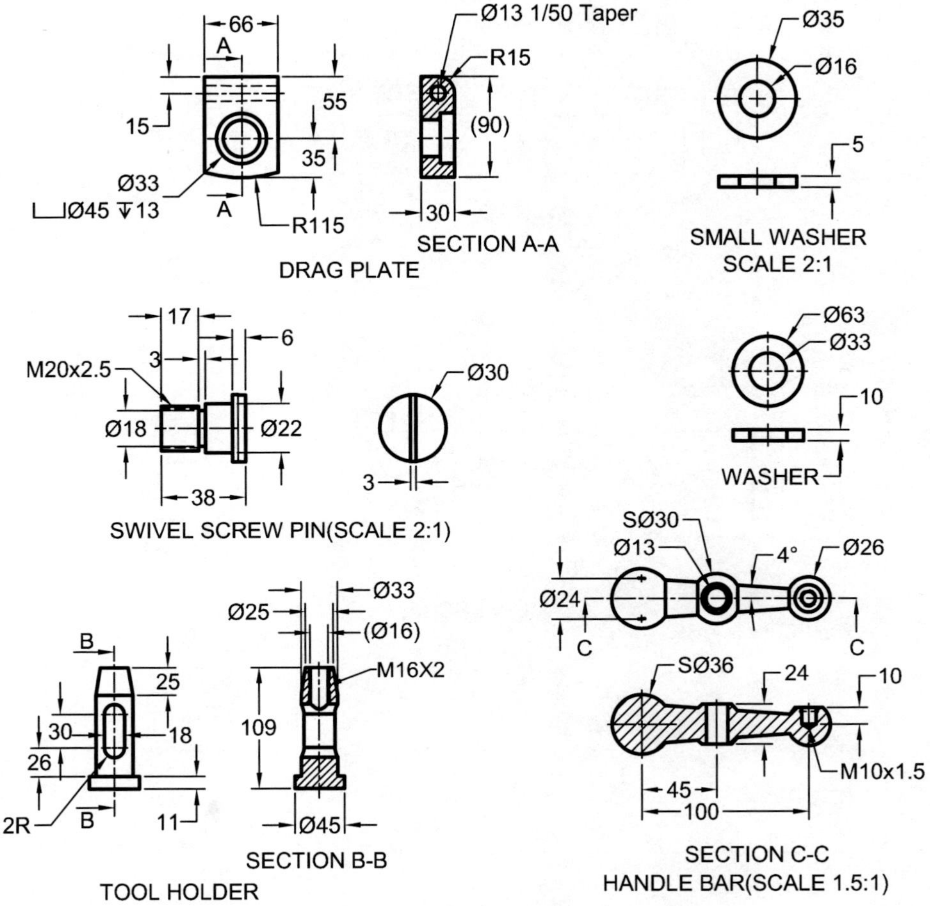

Figure 13-32 *Views and dimensions of other components*

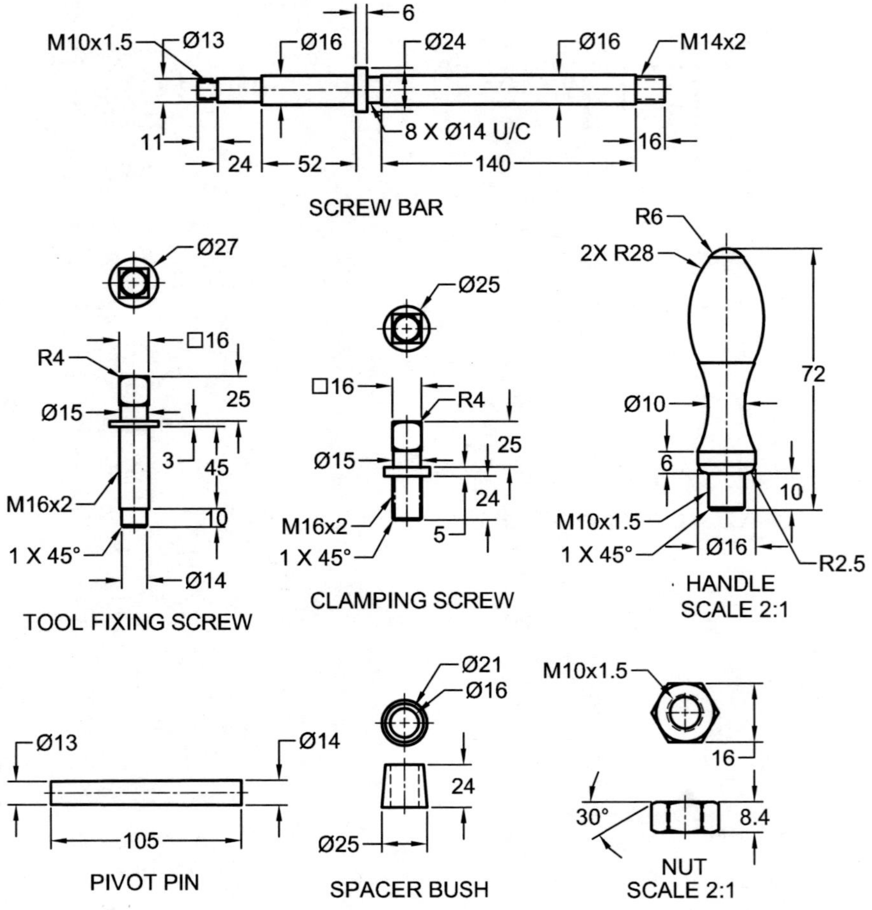

Figure 13-33 *Views and dimensions of the components*

Chapter 14

Working with Drawing Views-I

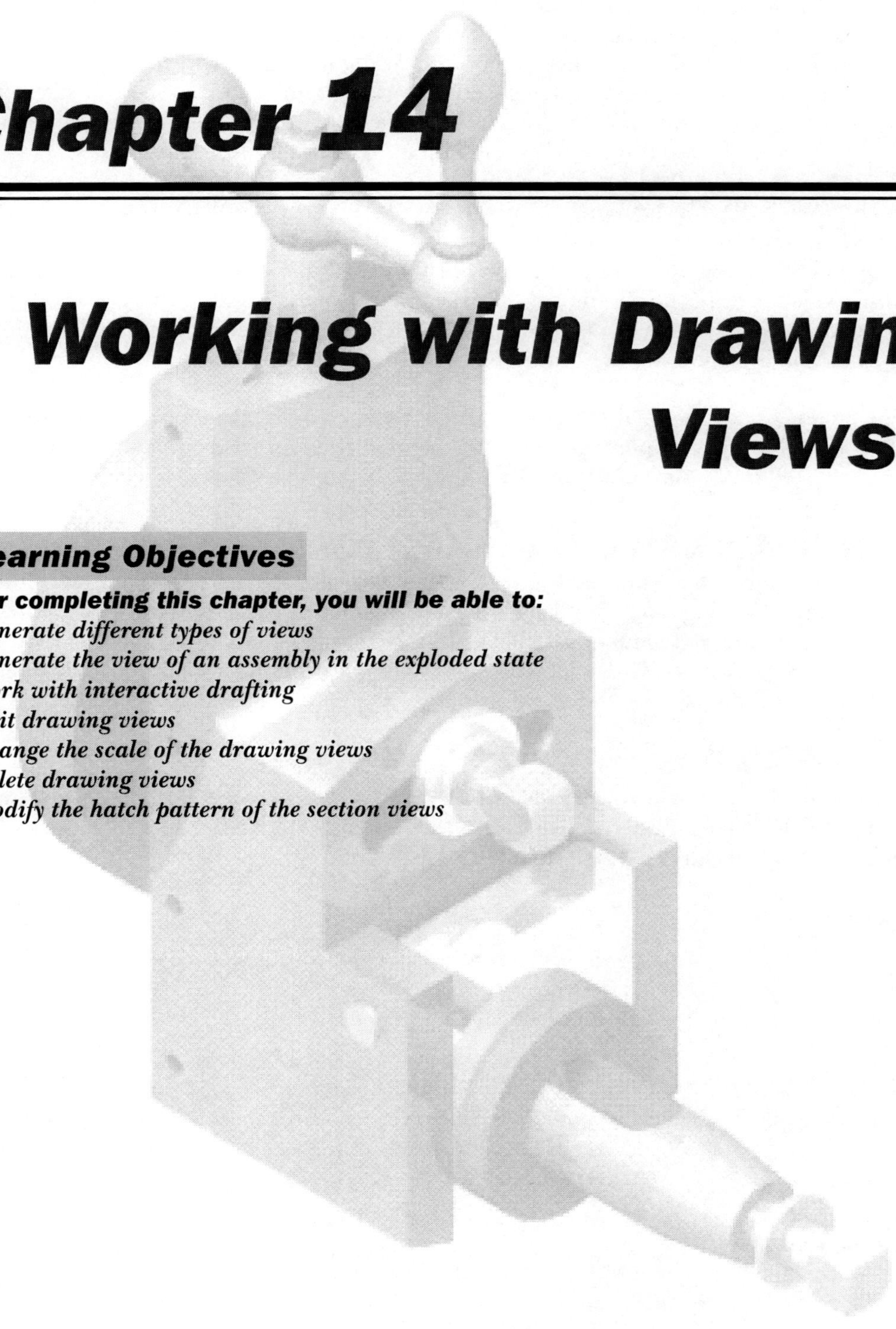

Learning Objectives

After completing this chapter, you will be able to:

- *Generate different types of views*
- *Generate the view of an assembly in the exploded state*
- *Work with interactive drafting*
- *Edit drawing views*
- *Change the scale of the drawing views*
- *Delete drawing views*
- *Modify the hatch pattern of the section views*

THE DRAWING MODE

After creating solid models or assemblies, you need to generate their two-dimensional (2D) drawing views. These views are the lifeline of all manufacturing systems because at the shop floor or the machine floor, the machinist needs the 2D drawing for manufacturing. SOLIDWORKS provides a specialized environment known as the **Drawing** mode. This mode provides all the tools required to generate and modify the drawing views and add dimensions and annotations to them. In other words, you can get the final shop floor drawing using this mode.

You can also sketch 2D drawings in the **Drawing** mode of SOLIDWORKS using the sketching tools provided in this mode. In other words, there are two types of drafting methods available in SOLIDWORKS: Generative and Interactive. Generative drafting is a technique of generating the drawing views by using a solid model or an assembly. Interactive drafting is a technique of sketching the drawing views in the **Drawing** mode by using the sketching tools. In this chapter, you will learn about generating the drawing views of parts or assemblies.

One of the major advantages of working in SOLIDWORKS is that this software has bidirectional associative property. This property ensures that the modifications made in a model in the **Part** mode are reflected in the **Assembly** and **Drawing** modes, and vice versa.

STARTING A DRAWING DOCUMENT

To generate drawing views, you need to start a new drawing document. There are different methods to start a drawing document in SOLIDWORKS. The first method is to start a drawing document by using the **Welcome - SOLIDWORKS 2018** dialog box, the second method is by using the **New SOLIDWORKS Document** dialog box and the third method is by using the options available in the part or assembly document.

TYPES OF VIEWS

You can generate different types of views in SOLIDWORKS. You first need to generate a standard view such as the top view or the front view, and then use it to derive the remaining views from the standard view. You can generate the following types of drawing views:

Model View

A model view is used to create the base view in the drawing sheet. You can generate orthogonal views such as the front, top, left, and so on as the model view. You can also generate an isometric, a trimetric, or a dimetric view as the model view.

Projected View

A projected view is generated by using an existing view as the parent view. It is generated by projecting the lines normal to the parent view or at an angle. The resulting view will be an orthographic view or an isometric view.

Section View

A section view is generated by chopping a part of an existing view using a section plane and then viewing the parent view from a direction normal to that plane. In SOLIDWORKS, the section plane is defined using one or more sketched line segments.

Aligned Section View

An aligned section view is used to section the features that are created at a certain angle to the main section planes. The aligned sections straighten these features by revolving them about an axis that is normal to the view plane. Remember that the axis about which the feature is straightened should lie on the cutting planes.

Auxiliary View

An auxiliary view is generated by projecting the lines normal to the specified edge of an existing view.

Detail View

A detail view is used to display the details of a portion of an existing view. You can select a portion whose detailing has to be shown in the parent view. On doing so, the selected portion will be magnified and placed as a separate view. You can control the magnification of the detail view.

Break View

A break view is the one in which a portion of the drawing view is removed from the existing view keeping the ends of the drawing view intact. This type of view is used to display the components whose length to width ratio is very high. This means that either the length is very large as compared to the width or the width is very large as compared to the length. The broken view will break the view along the horizontal or vertical direction such that the drawing view fits the required area.

Broken-out Section View

A broken-out section view is used to remove a part of an existing view and display the area of a model or an assembly that lies behind the removed portion. This type of view is generated by using a closed sketch associated with the parent view.

Crop View

A crop view is used to crop an existing view enclosed in a closed sketch associated to that view. The portion of the view that lies inside the associated sketch is retained and the remaining portion is removed.

Alternate Position View

The alternate position view is used to create a view in which you can show two different positions of an assembly while it is in motion. The main position is displayed in the drawing view in continuous lines and the alternate position of the assembly is displayed in the same view in dashed lines (phantom lines).

WORKING WITH INTERACTIVE DRAFTING IN SOLIDWORKS

As mentioned earlier, you can also sketch the 2D drawings in the drawing document of SOLIDWORKS. In technical terms, sketching 2D drawings is known as interactive drafting. Before starting the drawing, it is recommended that you insert an empty view. To create an empty view, choose **Insert > Drawing View > Empty** from the SOLIDWORKS menus; an empty view will be attached to the cursor. Select a point at the desired location to place an empty view. Now, select the empty view to activate and use the tools in the **Sketch CommandManager** to sketch the view. If you move the empty view by selecting and dragging it, the sketched entities will also move. This is because the sketch that you have drawn is associated to the empty view.

TUTORIALS

Tutorial 1

In this tutorial, you will generate the front view, top view, right view, aligned section view, detail view, and isometric view of the model created in Tutorial 2 of Chapter 8. Use the Standard A4 Landscape sheet format for generating these views. **(Expected time: 30 min)**

The following steps are required to complete this tutorial:

a. Copy the part document of Tutorial 2 of Chapter 8 in the folder of the current chapter.
b. Open the copied part document and start a new drawing document from within the part document.
c. Select the standard A4 landscape sheet format and generate the parent view, refer to Figure 14-1.
d. Generate the projected views using the **Projected View** tool, as shown in Figure 14-1.
e. Generate the aligned section view using the **Section View** tool, refer to Figures 14-2 and 14-3.
f. Generate the detail view, refer to Figure 14-4.
g. Save and close the drawing document.

Copying and Opening the Part Document

1. Create a folder with the name *c14* in the *SOLIDWORKS* directory and copy *c08_tut02.sldprt* from the location *\Documents\SOLIDWORKS\c08*.

2. Start SOLIDWORKS and open the part document that you copied in the *c14* folder.

Starting a New Drawing Document

As mentioned earlier in this chapter, you can start a new drawing document from the part document. This way, the model in the part document is automatically selected and you can generate its drawing views.

1. Choose **New > Make Drawing from Part/Assembly** from the Menu Bar; the **Sheet Format/Size** dialog box is displayed.

Note

*If you are using the advanced form of the **New SOLIDWORKS Document** dialog box, the **New SOLIDWORKS Document** dialog box will be displayed every time you choose the **Make Drawing from Part/Assembly**. In the **New SOLIDWORKS Document** dialog box, select a drawing template from the **Templates** tab and choose **OK**. Remember that if you are not using the advanced form of the **New SOLIDWORKS Document** dialog box, then a new drawing document will start directly and the **Sheet Format/Size** dialog box will be displayed.*

2. Clear the **Only show standard formats** check box and select the **A4 (ANSI) Landscape** sheet from the list box in this dialog box and choose the **OK** button; a new drawing document is

started with the standard A4 sheet and the **View Palette** task pane is displayed automatically. By default, the model of Tutorial 2 of Chapter 8 is selected for generating the drawing views.

Generating the Parent View and the Projected Views

Before you proceed to generate the drawing views, you need to confirm whether the projection type for the current sheet is set to the third angle.

1. Click anywhere on the sheet to close the **View Palette** task pane. Select **Sheet1** from the **FeatureManager Design Tree** and then right-click to display shortcut menu. Choose the **Properties** option from the shortcut menu; the **Sheet Properties** dialog box is displayed.

2. Select the **Third angle** radio button from the **Type of projection** area if it is not selected by default and then, choose the **Apply Changes** button.

3. Choose the **View Palette** tab to view the **View Palette** task pane and make sure the **Auto-start projected view** check box is selected.

4. Select the **Front** view from the **View Palette** task pane and drag it to the middle left of the drawing sheet above the title block. Drop the view at this location to place the front view, refer to Figure 14-1. The **Projected View PropertyManager** is invoked automatically and the preview of the projected view is attached to the cursor.

5. Next, move the cursor vertically upward from the front view; a preview of the top view is displayed. Specify a point to place the top view, refer to Figure 14-1. Preview of another projected view with the front view as the parent view is attached to the cursor.

6. Next, move the cursor horizontally toward the right side from the front view, a preview of the side view is displayed. Specify a point to place the side view, refer to Figure 14-1.

7. Move the cursor horizontally toward the right and then move it upward; a preview of the isometric view is displayed. Specify a point to place the isometric view. Next, choose the **OK** button in the PropertyManager.

   The current location of the isometric view of the model is such that it will interfere with the aligned section view that you need to place next. Therefore, you need to move the isometric view close to the top right corner of the drawing sheet.

8. Move the cursor over the isometric view; the bounding box of the view is displayed in orange.

9. Click to select the view; the border of the view is highlighted.

10. Move the cursor on one of the borderlines of the view; the cursor changes into a move cursor.

11. Press and hold the left mouse button and drag the view close to the upper right corner of the drawing sheet. The drawing sheet after generating and moving the isometric view is shown in Figure 14-1.

Tip

*You can turn on/off the origins displayed in the drawing views by using the **View (Heads-Up)** toolbar.*

The center marks are automatically created in the drawing views of the circular features in a model.

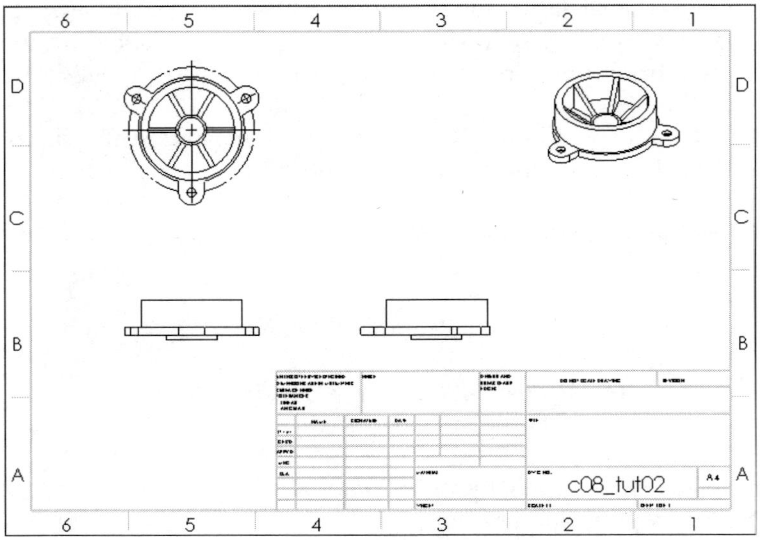

***Figure 14-1** Drawing sheet after generating the front, top, right, and isometric views*

Generating the Aligned Section View

Next, you need to generate the aligned section view on the top circular feature of the model. Before doing so, you need to activate the view from which you will generate the section view.

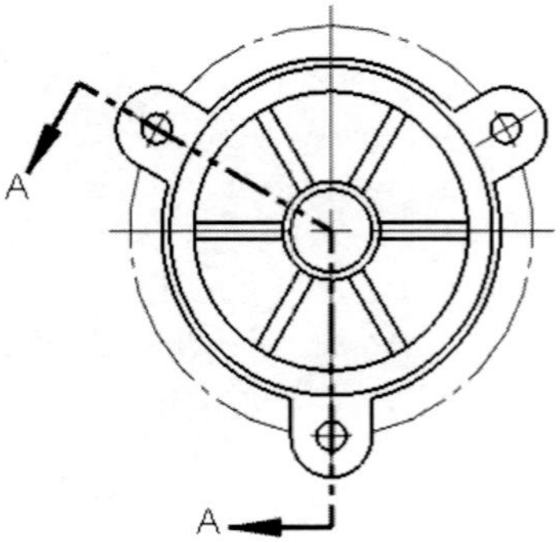

***Figure 14-2** Sketch of planes created for the aligned section view*

1. Click on the top view to activate it.

2. Invoke the **View Layout CommandManager** and choose the **Section View** button; the **Section View Assist PropertyManager** is displayed.

3. Choose the **Aligned** button in the **Cutting Line** rollout and create the cutting plane line, as shown in Figure 14-2; the **Section View** pop-up toolbar is displayed. Choose the **OK** button from the pop-up toolbar; the **SOLIDWORKS** information box is displayed with the message that the selected model has a rib feature that will vary the hatching of the section view. It also asks you whether to make an aligned section view or a foreshortened section view.

4. Choose the **Make aligned section view** option from the **SOLIDWORKS** information box; the aligned section view is attached to the cursor.

The view generated is normal to the vertical line. If the direction of viewing the aligned section view is in opposite direction, you need to flip it after placing the view.

5. Move the cursor to the right of the top view and place the aligned section view; the **Section View PropertyManager** is displayed. If the direction of viewing is not the required one, select the **Flip Direction** button. Click anywhere on the sheet to exit the PropertyManager.

The sheet after generating the aligned section view is shown in Figure 14-3.

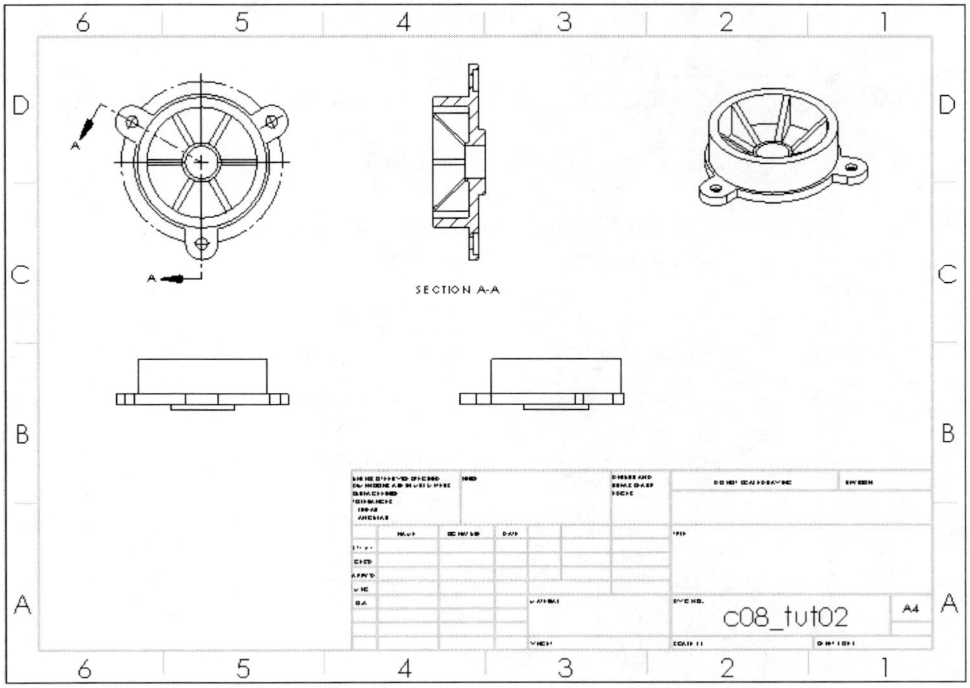

***Figure 14-3** Sheet after generating the aligned section view*

Modifying the Hatch Pattern of the Aligned Section View

The gap between the hatching lines in the aligned section view is large. Therefore, you need to modify the spacing.

1. Select the hatch pattern from one of the sections in the aligned section view; the **Area Hatch/Fill PropertyManager** is displayed.

2. Clear the **Material crosshatch** check box; the **Hatch Pattern**, **Hatch Pattern Scale**, and **Hatch Pattern Angle** options become available. Set the value **2** in the **Hatch Pattern Scale** spinner and exit the PropertyManager by choosing the **OK** button.

Generating the Detail View

Next, you need to generate the detail view of the right circular feature of the model. Before doing so, you need to activate the view from which you will derive the detail view.

1. Activate the top view and choose the **Detail View** button from the **View Layout CommandManager**; the **Detail View PropertyManager** is displayed and you are prompted to sketch a circle to continue viewing the creation. Also, the cursor is replaced by the circle cursor.

2. Draw a small circle on the right circular feature of the model in the top view, refer to Figure 14-4. As you draw the circle, the detail view is attached to the cursor.

3. Place the view on the right side of the drawing sheet above the title block, refer to Figure 14-4.

4. Select the **Use custom scale** radio button in the **Scale** rollout. Then, select the **User Defined** option from the drop-down list below the **Use custom scale** radio button.

5. Set the value of the scale factor of the detail view to **3:1**.

6. Choose the **Hidden Lines Visible** button from the **Display Style** rollout and choose the **OK** button from the **Detail View PropertyManager**.

   You may need to move the drawing view and its label so that the view does not overlap the title block. Figure 14-4 shows the final drawing sheet displayed after generating the detail view from the top view.

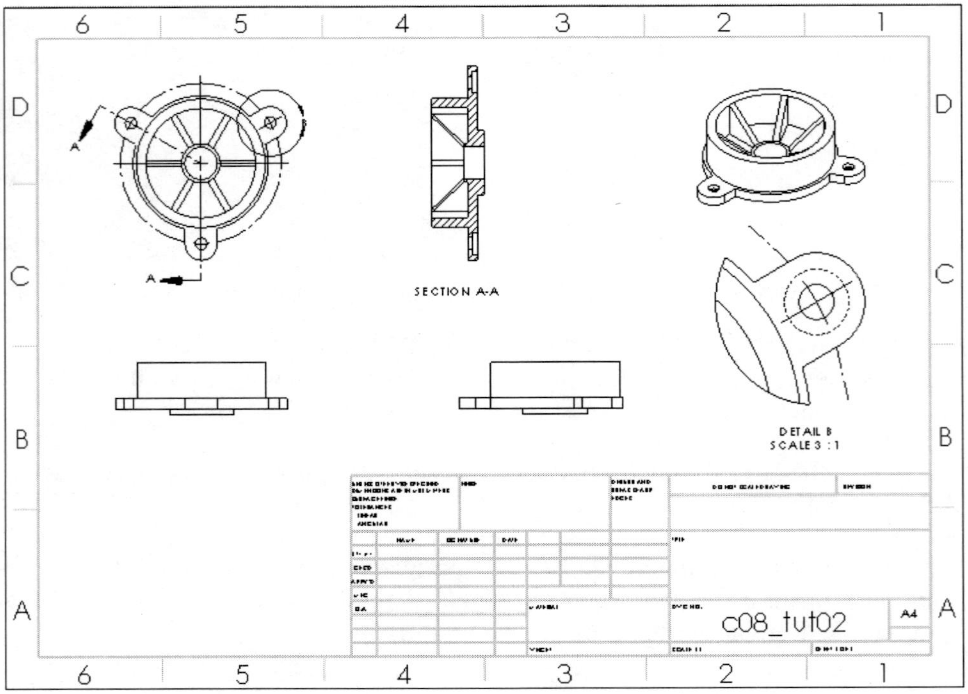

Figure 14-4 *The detail view derived from the top view*

Saving the Drawing

1. Choose the **Save** button from the Menu Bar and save the drawing file with the name *c14_tut01* at the following location:

   \Documents\SOLIDWORKS\c14

2. Choose **File > Close** to close this document. Also, close the part file of Tutorial 2 of Chapter 8.

Tutorial 2

In this tutorial, you will generate the drawing view of the Bench Vice assembly created in Chapter 12. You will generate the top view, sectioned front view, right view, and isometric view of the assembly in the exploded state. **(Expected time: 45 min)**

The following steps are required to complete this tutorial:

a. Copy the folder of the Bench Vice assembly from *c12* to the *c14* folder.
b. Create the exploded view of the Bench Vice assembly, refer to Figure 14-5.
c. Start a new drawing document from the assembly document using A4 landscape sheet format and generate the top view, refer to Figure 14-6.
d. Generate the section view using the **Section View** tool, refer to Figure 14-7.
e. Generate the right view using the **Projected View** tool, refer to Figure 14-8.
f. Generate the isometric view and change the state of the isometric view to the exploded state, refer to Figure 14-9.
g. Save and close the drawing and assembly documents.

Copying the Folder of the Bench Vice Assembly

First, you need to copy the folder of the Bench Vice assembly to the *c14* folder.

1. Copy the folder of the Bench Vice assembly from *Documents\SOLIDWORKS\c12* to the *c14* folder.

Creating the Exploded View of the Assembly

Before proceeding further to generate the drawing views of the assembly, you need to create the exploded state of the assembly in the **Assembly** mode.

1. Open the Bench Vice assembly and create the exploded state as well as the explode lines, as shown in Figure 14-5.

   It is recommended that whenever you create an exploded state of an assembly, you must revert to the collapsed state. If you save the assembly in the exploded state and then generate the drawing views of the assembly, the view will be generated in the exploded state.

2. Right-click on the assembly name in the **FeatureManager Design Tree** and choose **Collapse** to unexplode the assembly.

3. Save the assembly.

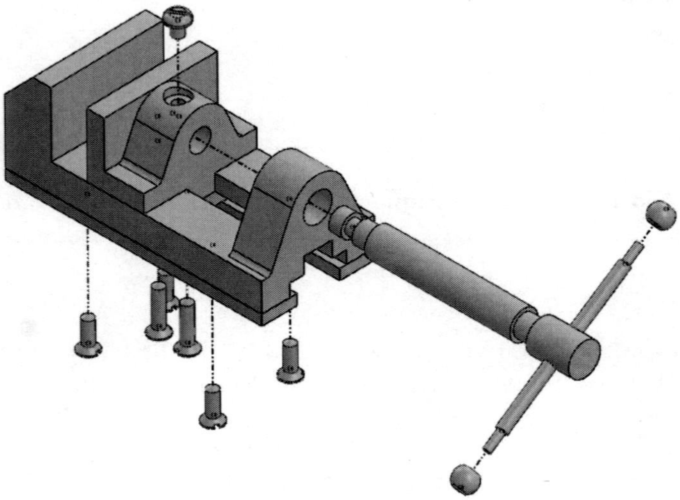

Figure 14-5 Exploded view of the assembly with explode lines

Starting a New Drawing Document from the Assembly Document

As mentioned earlier, you can also start a drawing document from the assembly document.

1. Choose **New > Make Drawing from Assembly** from the Menu Bar; the **Sheet Format/Size** dialog box is displayed.

2. Select the **A4 (ANSI) Landscape** sheet and choose the **OK** button. Right-click on **Sheet1** and choose **Properties** from the shortcut menu. Set the current projection type to third angle using the **Sheet Properties** dialog box if not set by default.

Generating the Top View

1. Invoke the **Model View** tool and double-click on **Bench Vice** in the **Part/Assembly to Insert** rollout; the **Model View PropertyManager** gets modified.

2. Choose the **Top** button from the **Standard views** area in the **Orientation** rollout of the **Model View PropertyManager**.

3. Select the **Use custom scale** radio button from the **Scale** rollout.

4. Set the value of the scale factor to **1:2** and then select the **Preview** check box from the **Orientation** rollout; a preview of the top view of the assembly is displayed.

5. Make sure that the **Auto-start projected view** check box is cleared in the **Options** rollout.

6. Place the view close to the top left corner of the drawing sheet, refer to Figure 14-6. Click anywhere on the sheet to exit the PropertyManager.

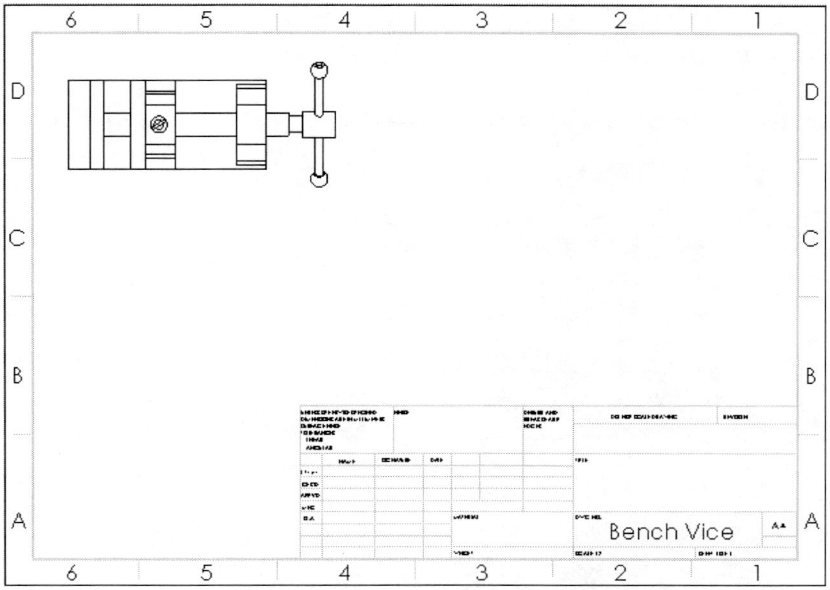

*Figure 14-6 Top view generated using the **Model View** tool*

Creating the Sectioned Front View

Next, you need to generate the sectioned front view that is derived from the top view.

1. Activate the top view and choose the **Section View** button from the **View Layout CommandManager**; the **Section View Assist PropertyManager** is displayed.

2. Choose the **Horizontal** button from the **Cutting Line** rollout of the **Section View Assist PropertyManager** and create the cutting plane line, refer to Figure 14-7; the **Section View** pop-up toolbar is displayed. Choose the **OK** button from the pop-up toolbar; the **Section View** dialog box is displayed.

   This dialog box is used to exclude the components from the section cut.

3. Invoke the **FeatureManager Design Tree** and click on the ▸ sign located on the left of the **Drawing View1** to display the name of the assembly. Next, expand the assembly.

4. Select Screw Bar, Bar Globes, Jaw Screw, Oval Fillister, Set Screw1, and Set Screw2. Next, click on the **Section View** dialog box in the drawing area to activate it. Select one of the components displayed in the **Exclude components/rib features** selection box and then select the **Don't cut all instances** check box. Similarly, select all the components one by one from this selection box and then select the **Don't cut all instances** check box.

5. Next, select the **Auto hatching** check box and choose the **OK** button from the **Section View** dialog box; the preview of the section view gets attached to the cursor. Also, the **Section View PropertyManager** is displayed.

6. If the direction of viewing of the section view is not as required, flip the direction by selecting the **Flip Direction** button from the **Cutting Line** rollout.

7. Place the section view below the top view. Click anywhere on the sheet to exit the PropertyManager.

8. Modify the hatch scale for the components. Figure 14-7 shows the section view generated using the **Section View** tool after modifying the hatch scale.

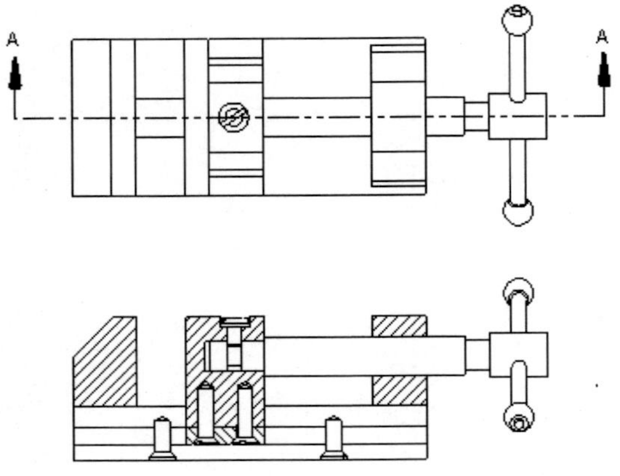

Figure 14-7 *Section view generated using the* ***Section View*** *tool*

Generating the Right Side View

The next view that you need to generate is the right side view derived from the sectioned front view and it will be generated using the **Projected View** tool.

1. Select the sectioned front view and invoke the **Projected View** tool; the **Projected View PropertyManager** is displayed and a projected view is attached to the cursor.

2. Move the cursor to the right of the sectioned front view and place the view on the right of the sectioned front view. Next, exit the PropertyManager.

3. The sheet after generating the projected view is shown in Figure 14-8.

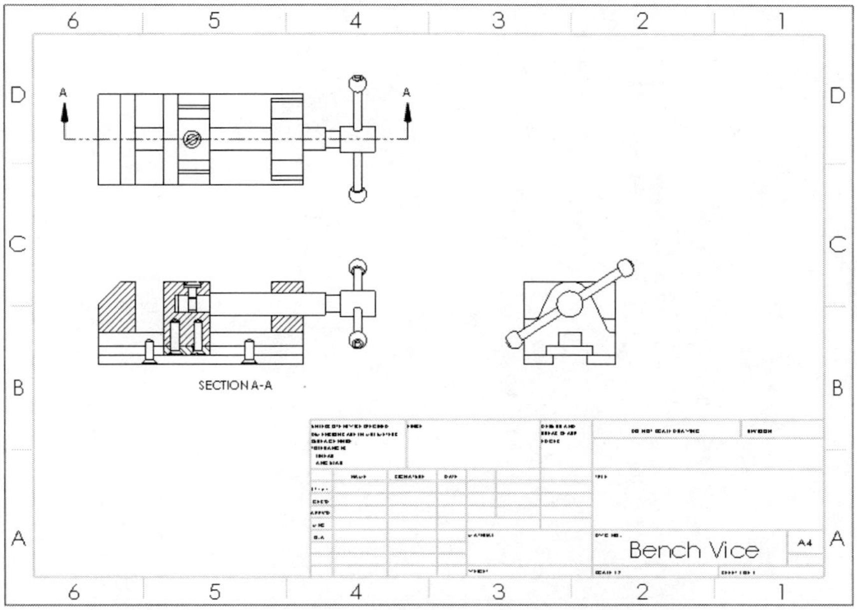

*Figure 14-8 Right side view generated using the **Projected View** tool*

Creating the Isometric View of the Assembly in the Exploded State

The last view to be generated is the isometric view of the assembly in the exploded state.

1. Invoke the **Model View** tool and generate the isometric view, and then place it close to the upper right corner of the drawing sheet.

2. Set the scale factor of the drawing view to **1:2**.

3. Right-click on the view to invoke the shortcut menu. Choose the **Properties** option from the shortcut menu; the **Drawing View Properties** dialog box is displayed.

4. Select the **Show in exploded or model break state** check box in the **View Properties** tab of the **Drawing View Properties** dialog box and choose the **OK** button.

5. Move the views to place them in the drawing sheet. Figure 14-9 shows the final drawing sheet after generating all the drawing views.

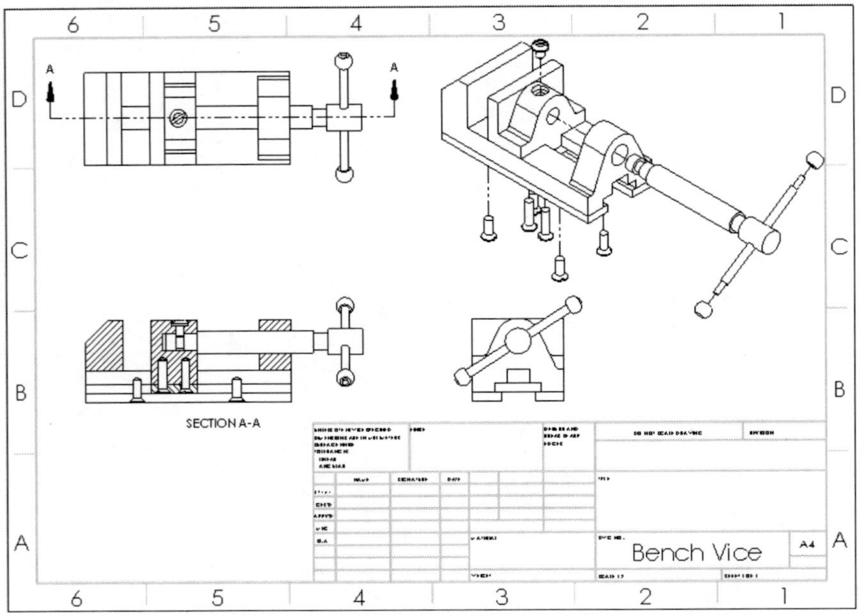

Figure 14-9 Drawing sheet after generating all views

Saving the Drawing

Next, you need to save the drawing file.

1. Choose the **Save** button from the Menu Bar and save the drawing document with the name *c14_tut02* at the following location:

   \Documents\SOLIDWORKS\c14

2. Close the drawing and assembly files.

EXERCISE

Exercise 1

In this exercise, you will generate the front view, section right view, isometric view, and the alternate position view on the isometric view of Exercise 1 of Chapter 13. You need to scale the parent view to the scale factor of 1:3. The views that you need to generate are shown in Figure 14-10. **(Expected time: 30 min)**

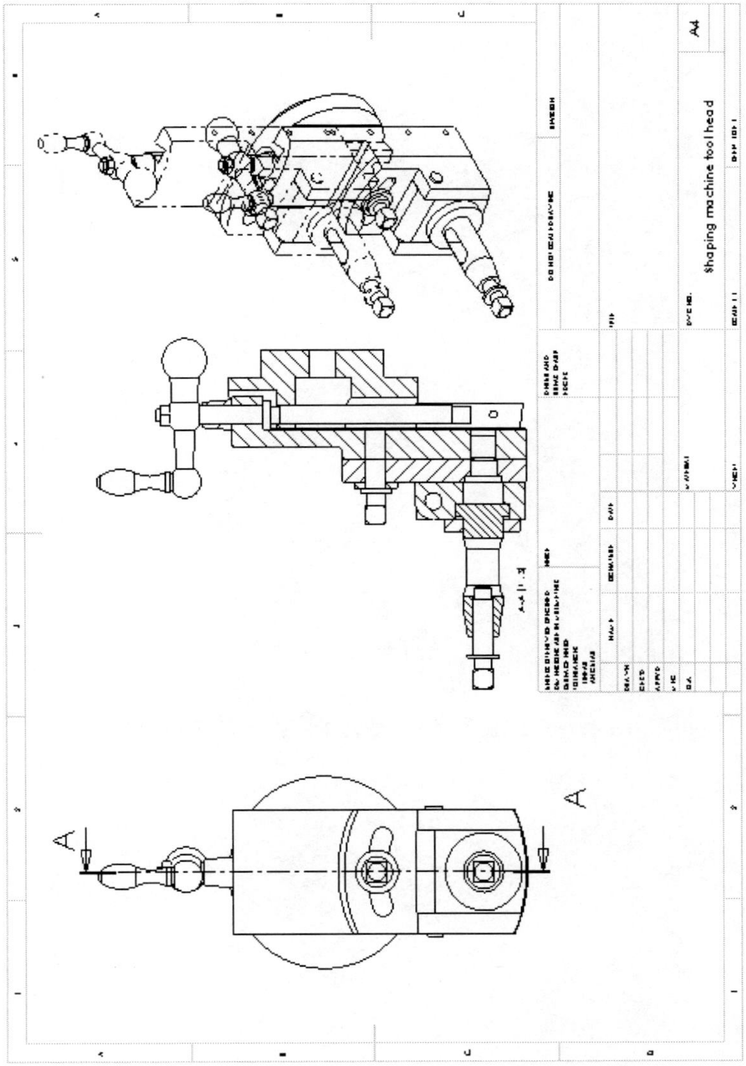

Figure 14-10 *Views to be generated for Exercise 1*

This page is intentionally left blank

Chapter 15

Working with Drawing Views-II

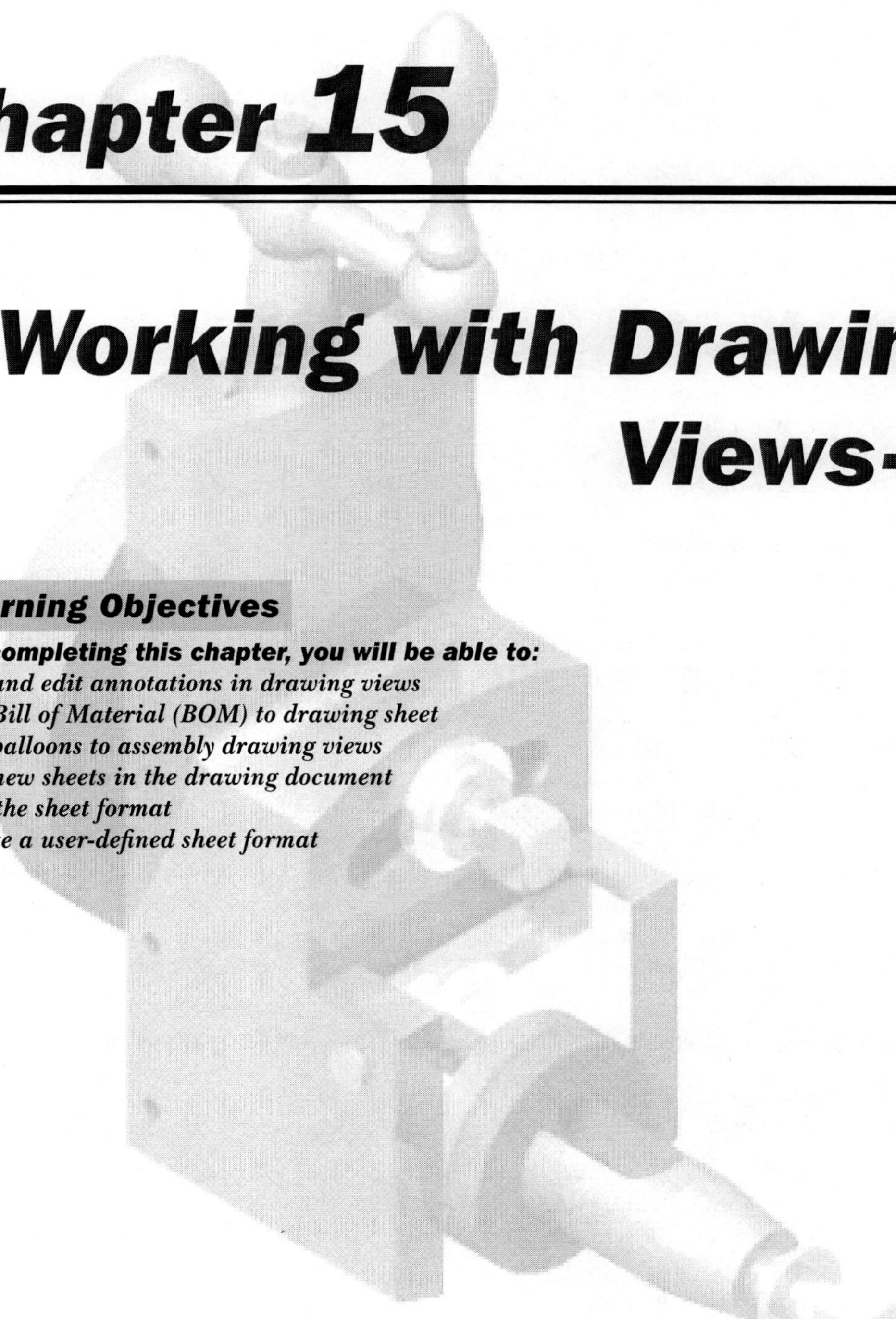

Learning Objectives

After completing this chapter, you will be able to:

- *Add and edit annotations in drawing views*
- *Add Bill of Material (BOM) to drawing sheet*
- *Add balloons to assembly drawing views*
- *Add new sheets in the drawing document*
- *Edit the sheet format*
- *Create a user-defined sheet format*

ADDING ANNOTATIONS TO DRAWING VIEWS

After generating drawing views, you need to generate the dimensions and add annotations such as notes, surface finish symbols, geometric tolerance, and so on to them. Two types of annotations can be generated in the drawing views. The first type of annotations are the generative annotations that are added while creating the part in the Part mode. For example, the dimensions that you add to the sketch and features of the part are generative annotations. The second type of annotations are added manually to the geometry of drawing views such as reference dimensions, notes, surface finish symbols, and so on.

ADDING THE BILL OF MATERIALS (BOM) TO A DRAWING

CommandManager: Annotation > Tables > Bill of Materials
SOLIDWORKS menus: Insert > Tables > Bill of Materials

The Bill of Materials (BOM) is a table that displays the list of components used in an assembly or a part. This table can also be used to provide information related to the number of components in an assembly or a part, their names, quantity, or any other information required to assemble the components. Remember that the sequence of parts in the BOM depends on the sequence in which they were inserted in the assembly document. The BOM, placed in the drawing document, is parametric. Therefore, if you add or delete a part from the assembly in the assembly document, changes will be reflected in the corresponding BOM in the drawing document.

LINKING BILL OF MATERIALS

In SOLIDWORKS, you can add Bill of Materials both in the Part as well as in the Assembly environment. To do so, open the part or assembly files and then choose **Insert > Tables > Bill of Materials** from the SOLIDWORKS menus; the BOM will be added to the part or assembly. Now, if you add BOM in the drawing view by choosing the **Bill of Materials** button, the **Bill of Materials PropertyManager** will be displayed with different options as compared to those discussed earlier. In this case, the **BOM options** rollout will have two options. To add a new BOM as discussed earlier, select the **Create new table** radio button and choose the **OK** button. To copy the BOM created in part or assembly, select the **Copy existing table** radio button. If you need to link it with the copied BOM then select the **Linked** check box and choose the **OK** button.

ADDING BALLOONS TO THE DRAWING VIEWS

CommandManager: Annotation > Balloon
SOLIDWORKS menus: Insert > Annotations > Balloon

After adding the BOM, you need to add the balloons to the components in the drawing views. Figure 15-1 shows the drawing sheet in which the balloons have been added to the assembly drawing. The balloons can be added manually using the **Balloon** tool.

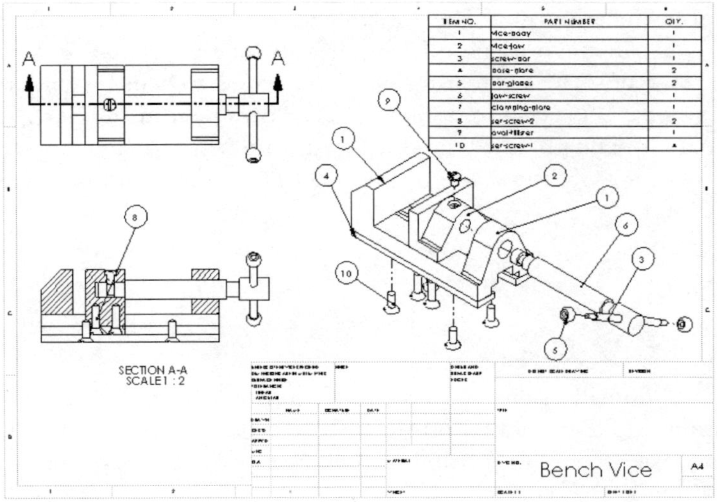

Figure 15-1 *Balloons added to the assembly drawing view*

The naming of balloons depends on the sequence of the parts in the BOM. The method of adding balloons to the components in the drawing views is discussed next.

ADDING NEW SHEETS TO THE DRAWING DOCUMENT

You can also add new sheets to a drawing document. A multisheet drawing document can be used to generate the drawing views of all the components and the drawing views of the assembly in the same document. You can switch between the sheets easily to refer to the drawing views of different parts of an assembly within the same document, without opening separate drawing documents.

EDITING THE SHEET FORMAT

You can edit the standard sheet format according to your requirement. But before editing a sheet ensure that it is active. To make a sheet active, select it from the **FeatureManager Design Tree** and choose the **Activate** option from the shortcut menu displayed. Now, to edit the sheet format, again right-click on the sheet to invoke the shortcut menu and choose the **Edit Sheet Format** option from it. On doing so, all entities, annotations, and views will disappear from the drawing sheet. You can edit the sheet format by using the sketching tools in the **Sketch CommandManager**. After editing the sheet format, select the active sheet again and invoke the shortcut menu. Choose the **Edit Sheet** option from the shortcut menu to switch back to the edit sheet environment.

CREATING USER-DEFINED SHEET FORMATS

In SOLIDWORKS, you can also create a user-defined sheet format. To do so, when you start a new drawing document, select the **Custom sheet size** radio button from the **Sheet Format/Size** dialog box. On the basis of your design requirements, set the size of the sheet in the **Width** and **Height** edit boxes in this dialog box and choose **OK**; the new customized drawing sheet will be displayed. Select **Sheet1** from the **FeatureManager design tree** and invoke the shortcut menu. Choose the **Edit Sheet Format** option; the edit sheet format environment will be invoked. Note that the **Edit Sheet Format** option will be available in the shortcut menu only when the selected sheet is activated. You can create or modify the sheet format by using the sketching tools in the **Sketch CommandManager**. After creating or modifying the sheet format, switch back to the edit sheet environment. Choose **File > Save Sheet Format** from the SOLIDWORKS menus; the **Save Sheet Format** dialog box will be displayed. Browse to the location where you need to save the sheet format. Specify the name of the sheet format and choose the **Save** button from the **Save Sheet Format** dialog box.

TUTORIALS

Tutorial 1

In this tutorial, first you will open the drawing created in Tutorial 1 of Chapter 14 and then generate dimensions and add annotations to it. Next, you will change the display of the front and right views to make hidden lines visible. Finally, you will change the display of the isometric view to the shaded mode. **(Expected time: 45 min)**

The following steps are required to complete this tutorial:

a. Copy the part document from Chapter 8 and the drawing document from Chapter 14 to the folder of the current chapter.
b. Configure the font settings and generate the dimensions using the **Model Items** tool.
c. Arrange the dimensions and delete the unwanted ones, refer to Figures 15-3 and 15-4.
d. Add the datum symbol and geometric tolerance to the drawing views, refer to Figures 15-5 and 15-6.
e. Change the model display state of the drawing views, refer to Figure 15-7.

Copying the Documents in the Folder of the Current Chapter

Before proceeding, you need to copy the model and the drawing document in the folder of the current chapter.

1. Create a folder with the name *c15* in the *SOLIDWORKS* folder. Next, copy *c08_tut02.sldprt* and *c14_tut01.slddrw* from *\Documents\SOLIDWORKS\c08* and *\Documents\SOLIDWORKS\c14*, respectively and paste them in the *c15* folder.

Opening the Drawing Document

Next, you need to open the drawing document in the SOLIDWORKS window.

1. Invoke the **Open** dialog box and open the *c14_tut01.slddrw* document from the folder of the current chapter.

   The drawing document in which you need to add the dimensions is displayed in the drawing area, as shown in Figure 15-2.

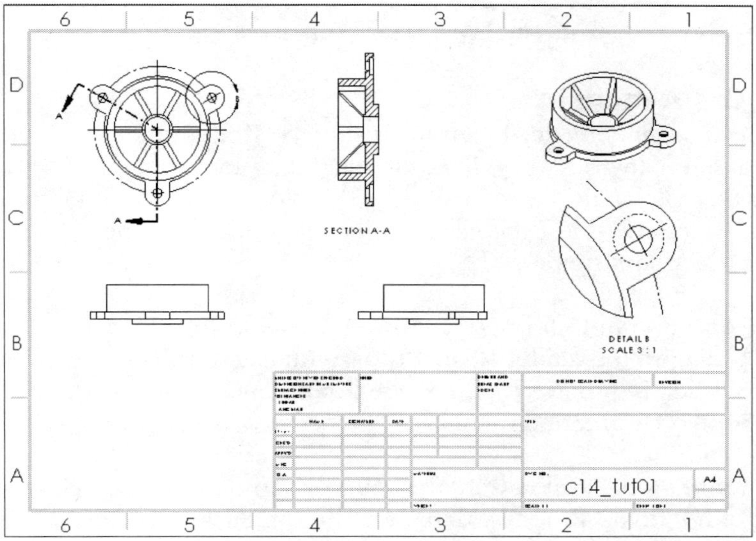

Figure 15-2 *Drawing views generated in Chapter 14*

Applying the Document Settings

Before generating the model dimensions, you need to configure the document settings. These settings will allow the dimensions and other annotations in the current sheet to be viewed properly.

1. Invoke the **System Options - General** dialog box by choosing the **Options** button from the Menu Bar.

2. Next, choose the **Document Properties** tab in the **System Options - General** dialog box to invoke the **Document Properties - Drafting Standard** dialog box and then choose **Annotations > Notes** from the area on the left in the dialog box.

3. Choose the **Font** button from the **Text** area of the dialog box; the **Choose Font** dialog box is invoked.

4. Select the **Points** radio button from the **Height** area and set the value of the font size as **9** from the list box.

5. Choose the **OK** button from the **Choose Font** dialog box.

6. Choose the **Dimensions** option from the area on the left in the dialog box; the related options are displayed on the right.

7. Set the values of height, width, and length of the arrows as **1**, **3**, and **6**, respectively using their respective edit boxes in the **Arrows** area.

8. Now, choose **Views > Section** from the left in the dialog box; the related options are displayed on the right. Set the values of height, width, and length of the section arrows as **2**, **4**, and **8** using their respective edit boxes in the **Section/view size** area.

9. Choose the **OK** button from the **Document Properties - Section** dialog box.

Generating the Dimensions

Next, you need to generate the dimensions using the **Model Items** tool. As discussed earlier, if you do not select any view on generating the dimensions using the **Model Items** tool, all the dimensions will be displayed in all the views. Sometimes, the dimensions may overlap each other. Therefore, select the view in which you need to generate the dimension and then invoke the **Model Items** tool.

1. Select the top view and choose the **Model Items** button from the **Annotation CommandManager**; the **Model Items PropertyManager** is displayed and the name of the selected view is displayed in the **Place annotations in these views** selection box under the **Source/Destination** rollout.

2. Select the **Entire model** option from the **Source** drop-down list and choose the **OK** button from the **Model Items PropertyManager**; the dimensions of the model which can be displayed in the selected view are generated.

   Note that the generated dimensions are scattered arbitrarily on the drawing sheet. Therefore, you need to arrange the dimensions by moving them to the required locations.

3. Select all the dimensions by dragging a window around them; the **Dimension Palette** button is displayed.

4. Move the cursor over this button; the **Dimension Palette** is displayed.

5. Choose the **Auto Arrange Dimensions** button from this palette; the dimensions are arranged at equal distance.

6. Select the dimensions one by one and drag them to the desired location. You can reverse the direction of arrowheads by clicking on the control point that is displayed on them. Any radial dimension attached to the counterbore hole in the top view needs to be deleted because you will add a hole callout to this counterbore hole later.

7. Select the radial dimension and press the DELETE key to delete the dimension. Similarly, delete the diameter dimension value 20. The drawing view after arranging the dimensions and deleting the diameter dimension is shown in Figure 15-3.

> **Note**
> *In Figure 15-3, the display of hidden lines is turned on. To hide them, first select the top view from the drawing area and then choose the **Hidden Lines Removed** button from the **Display Style** rollout of the **Drawing View PropertyManager**.*

8. Select the aligned section view and generate the dimensions using the **Model Items** tool. After placing the dimensions, move them to appropriate places, as shown Figure 15-4. You may need to change the arrowheads to closed filled arrowheads if they are not closed filled already.

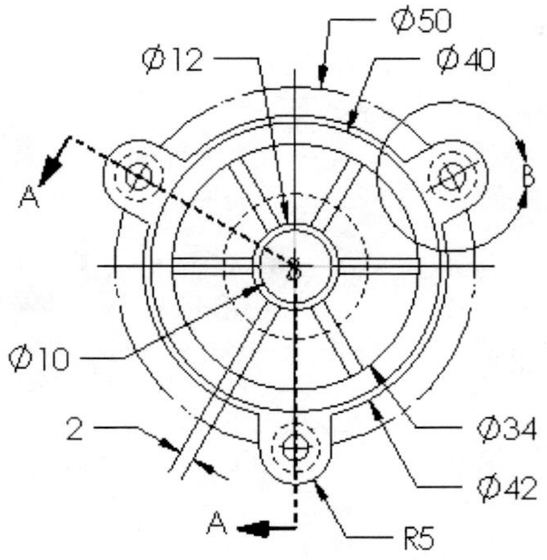

Figure 15-3 Top view after generating and arranging the dimensions

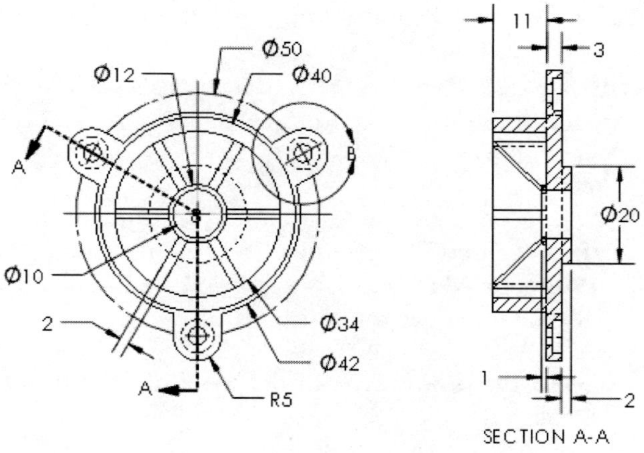

Figure 15-4 Partial view of the sheet after generating and arranging the dimensions in the aligned section view

9. Choose the **Hole Callout** button from the **Annotation CommandManager** and select the outer circle of one of the counterbore features in the top view; the hole callout is attached to the cursor. Pick a point on the drawing sheet to place the hole callout, as shown in Figure 15-5.

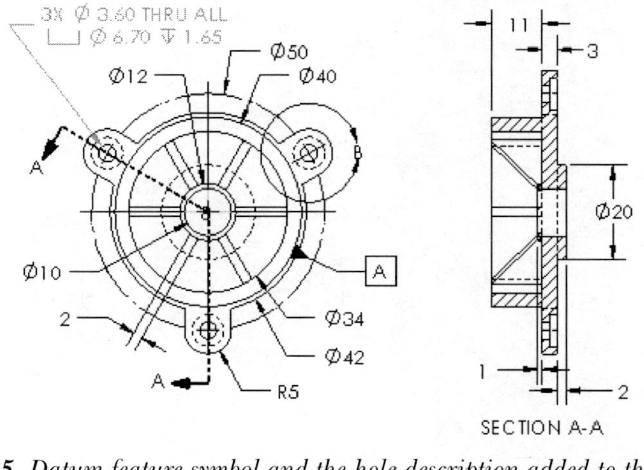

Figure 15-5 Datum feature symbol and the hole description added to the top view

Adding the Datum Feature Symbol to the Drawing Views

After generating the dimensions, you need to add the datum feature symbol to the drawing view. The datum feature symbols are used as datum reference for adding the geometric tolerance to the drawing views.

1. Select the edge of the outer cylindrical feature from the top view and then choose the **Datum Feature** button from the **Annotation CommandManager**; the **Datum Feature PropertyManager** is displayed and a datum callout is attached to the cursor.

2. Place the datum symbol at an appropriate location, refer to Figure 15-5.

3. Choose the **OK** button from the **Datum Feature PropertyManager**.

Adding the Geometric Tolerance to the Drawing Views

After defining the datum feature symbol, you need to add the geometric tolerance to the drawing view.

1. Select the circular edge that has a diameter of 12 mm from the top view and choose the **Geometric Tolerance** button from the **Annotation CommandManager**; the **Properties** dialog box is displayed. This dialog box is used to specify the parameters of the geometric tolerance.

2. Select the down arrow in the **Symbol** drop-down list in the first row; the **Symbols** flyout is displayed.

3. Choose the **Concentricity** option from this flyout and then enter **0.002** in the **Tolerance 1** edit box.

4. Enter **A** in the **Primary** edit box to define the primary datum reference.

5. Choose the **OK** button from the **Properties** dialog box; the geometric tolerance is attached to the selected circular edge. You may need to move the geometric tolerance, if it overlaps the dimensions. Figure 15-6 shows the drawing view after adding and rearranging the geometric tolerance.

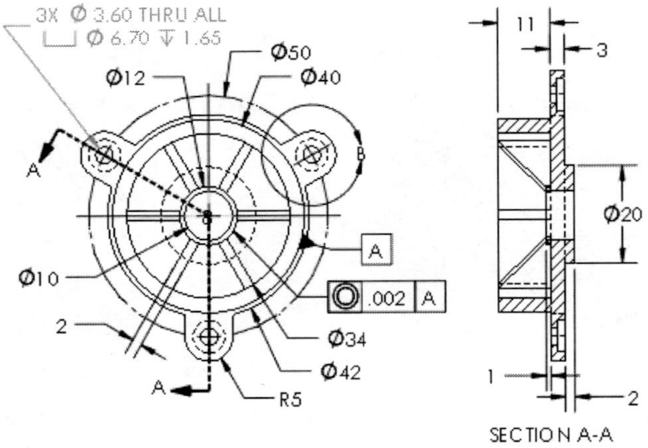

Figure 15-6 *Geometric tolerance added and arranged in the drawing view*

Changing the View Display Options

After adding all annotations to the drawing views, you need to change the display setting of the drawing views.

1. Press and hold the CTRL key, select the front view and then the right-side view from the drawing sheet.

2. Choose the **Hidden Lines Visible** button from the **View (Heads-Up)** toolbar; the hidden lines are displayed in the selected drawing views. You can also choose this button from the **Display Style** rollout of the **Multiple Views PropertyManager** that is displayed when you select the front and right-side views.

3. Now, select the isometric view from the drawing sheet.

4. Choose the **Shaded With Edges** button from the **View (Heads-Up)** toolbar or the **Drawing View PropertyManager**. Figure 15-7 shows the final drawing sheet after changing the display view settings.

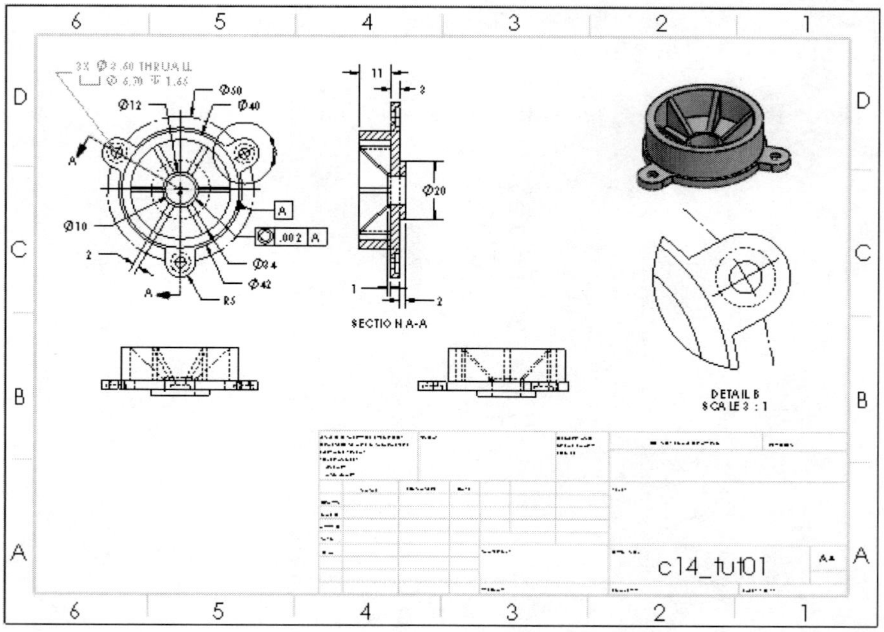

Figure 15-7 *Final drawing sheet*

5. Save and close the drawing document.

Tutorial 2

In this tutorial, you will generate the Bill of Materials (BOM) of the Bench Vice assembly and then add balloons to the isometric view in the exploded state. **(Expected time: 45 min)**

The following steps are required to complete this tutorial:

a. Copy the Bench Vice folder, which contains parts, assembly, and the drawing document, from Chapter 14 to the folder of the current chapter.
b. Delete the views that are not required in the drawing sheet, refer to Figure 15-9.
c. Move the views and arrange them in the drawing sheet, refer to Figure 15-9.
d. Set the anchor on the drawing sheet where the BOM will be attached.
e. Generate the BOM, refer to Figure 15-10.
f. Add balloons to the isometric view, refer to Figure 15-11.

Copying the Bench Vice Assembly Folder to the Current Folder

Before proceeding, you need to copy the model and the drawing document in the folder of the current chapter.

1. Copy the *Bench Vice* folder from the *\Documents\SOLIDWORKS\c14* folder to the folder of the current chapter.

Opening the Drawing Document

After copying the folder, you need to open the drawing document in the SOLIDWORKS window.

1. Open the *c14_tut02.slddrw* document.

The drawing document in which you need to generate the BOM and balloons is displayed in Figure 15-8.

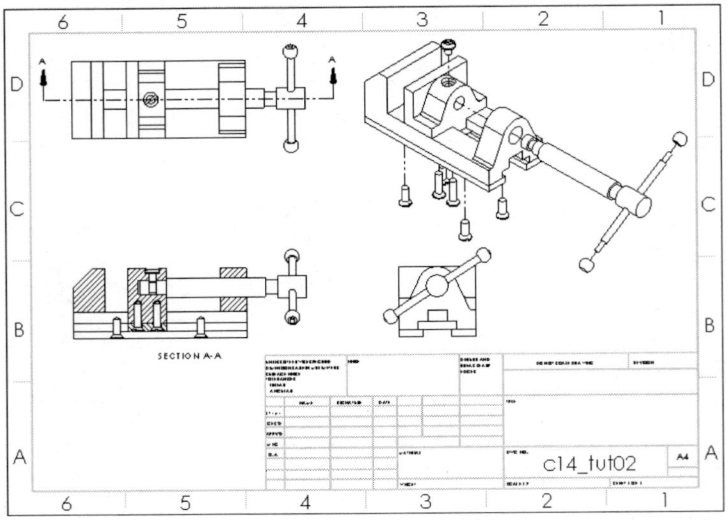

Figure 15-8 *Drawing views*

Deleting the Unwanted View

You need to delete the right view because it is not required in this tutorial.

1. Select the right-side view and press the DELETE key; the **Confirm Delete** dialog box is displayed.

2. Choose the **Yes** button from this dialog box; the view is deleted from the current drawing sheet.

Moving the Isometric View

You need to move the exploded isometric view because the BOM will be generated and placed at the top right corner of the drawing sheet.

1. Select the isometric view; the border of the view is highlighted.

2. Move the cursor to the border; the cursor is replaced by the move cursor.

3. Select the border and drag the cursor to move the drawing view and then place it at the required location, as shown in Figure 15-9.

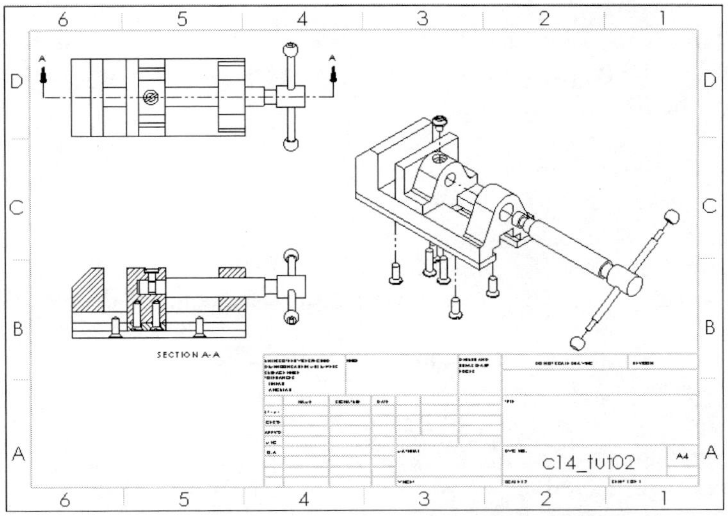

***Figure 15-9** Drawing sheet after deleting and moving the drawing views*

Setting the Anchor for the BOM

Before generating the BOM, you need to set its anchor. The anchor is a point on the drawing sheet to which one of the corners of the BOM coincides. By default, the anchor is defined at the top left corner of the drawing sheet. But in this tutorial, you need to add the BOM at the top right corner of the drawing sheet. Therefore, you need to set the anchor before generating the BOM.

1. Expand **Sheet1** from the **FeatureManager Design Tree** and then expand **Sheet Format1**.

2. Select the **Bill of Materials Anchor1** option, right-click on it to invoke a shortcut menu and then choose the **Set Anchor** option from it; the drawing views disappear from the sheet.

3. Specify the anchor point on the inner top right corner of the drawing sheet; a point is placed at the selected location.

After you specify the anchor point, the drawing views are displayed automatically in the sheet because the sheet editing environment is invoked automatically.

Generating the BOM

Next, you need to generate the BOM. As discussed earlier, the BOM generated in SOLIDWORKS is parametric. If a component is deleted or added in the assembly, the change is reflected automatically in the BOM. But before generating the BOM, you need to set its text parameters.

1. Invoke the **System Options - General** dialog box by choosing the **Options** button from the Menu Bar.

2. Next, choose the **Document Properties** tab in the **System Options - General** dialog box to invoke the **Document Properties - Drafting Standard** dialog box and then choose **Annotations > Notes** from the area on the left.

3. Choose the **Font** button from the **Text** area of the dialog box; the **Choose Font** dialog box is invoked. Select the **Points** radio button from the **Height** area and set the value of the font size to **9** from the list box.

4. Choose the **OK** button from the **Choose Font** dialog box. Similarly, change the text height of balloons to **14** and close the dialog box.

5. Select the isometric view and choose **Tables > Bill of Materials** from the **Annotation CommandManager**; the **Bill of Materials PropertyManager** is displayed.

6. Select the **Attach to anchor point** check box in the **Table Position** rollout.

7. Choose the **OK** button from the **Bill of Materials PropertyManager**; the BOM is generated. If the BOM is displayed outside the drawing sheet, move the cursor over the BOM; an anchor symbol is displayed. Click on the symbol; the **Bill of Materials PropertyManager** is displayed. Select an appropriate position for placing the BOM from the **Table Position** rollout and then choose the **OK** button; you will notice that the **Description** column is also displayed in the BOM. But this column is not required, so you need to delete it.

8. Move the cursor over the **Description** heading and right-click to display a shortcut menu. Choose **Delete > Column** from the shortcut menu; the column is deleted. The drawing sheet after generating the BOM and deleting the **Description** column is displayed, as shown in Figure 15-10.

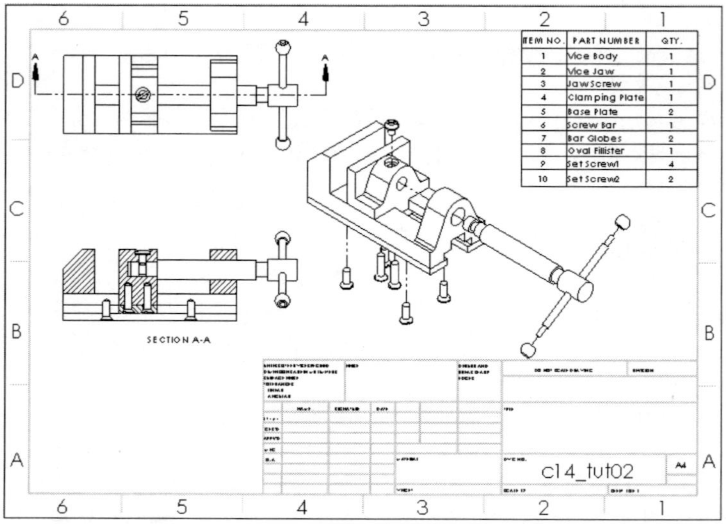

***Figure 15-10** Drawing sheet after generating the BOM*

Adding Balloons to the Components

After generating the BOM, you need to add balloons to the components. Before proceeding, make sure that you have changed the font height of balloons to **14**, as discussed in the previous section.

1. Select the isometric view and choose the **Auto Balloon** button from the **Annotation CommandManager**; the balloons are automatically added to all the components in the isometric view and the **Auto Balloon PropertyManager** is also displayed.

   The multiple instances of any component are ignored because the **Ignore multiple instances** check box is already selected in the **Balloon Layout** rollout. Also, make sure that the **Insert magnetic line(s)** check box is cleared.

2. Select **1 Character** from the **Size** drop-down list in the **Balloon Settings** rollout. Next, choose **OK** to close this PropertyManager.

   The balloons are added to all the components. You will notice that the balloons are not properly arranged on the sheet and are placed arbitrarily. Therefore, you need to drag each balloon manually to place it properly.

3. Move the cursor over any one of the balloons and when it is highlighted, drag it to place it at another location, refer to Figure 15-11.

4. Similarly, drag and place the remaining balloons at proper locations. The final drawing sheet after adding and rearranging balloons is shown in Figure 15-11.

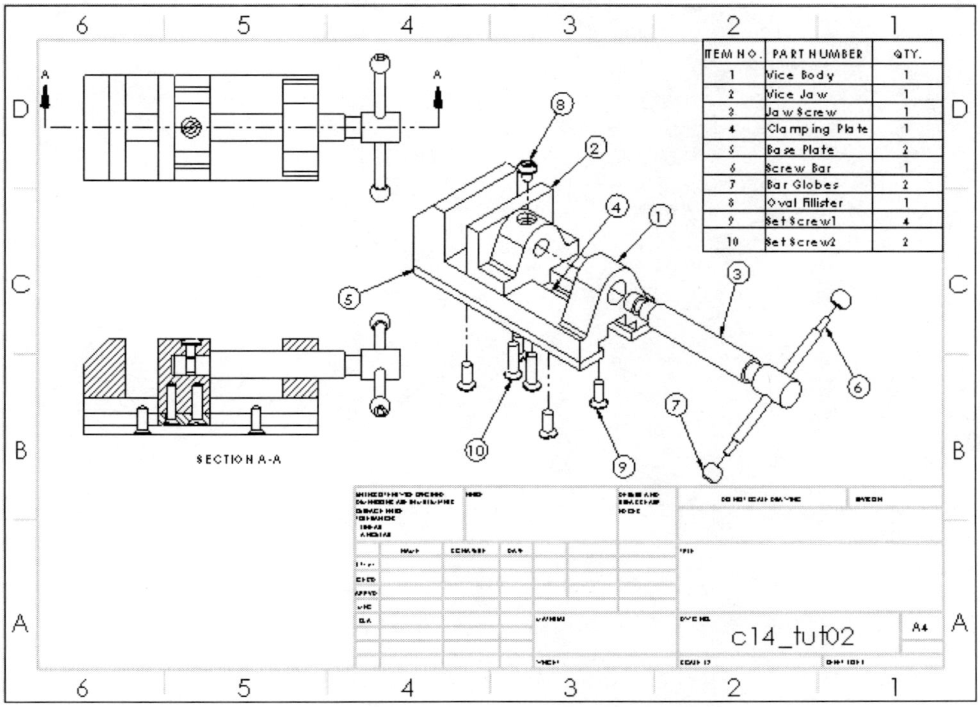

Figure 15-11 Final drawing sheet after adding balloons

5. Save and close the drawing document.

EXERCISE

Exercise 1

Generate the isometric view of the exploded view of the assembly created in Tutorial 1 of Chapter 13 on the standard A4 sheet format. The scale of the view will be 1:5. After generating the view, generate the BOM and add balloons to the assembly view, as shown in Figure 15-12.

(Expected time: 30 min)

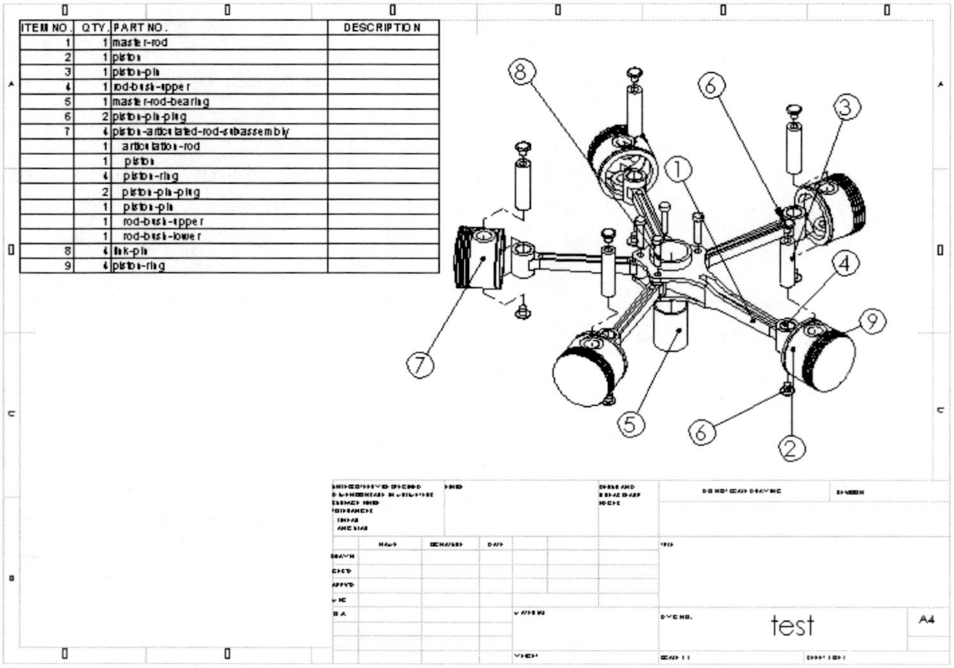

Figure 15-12 Drawing view for Exercise 1

Chapter 16

Surface Modeling

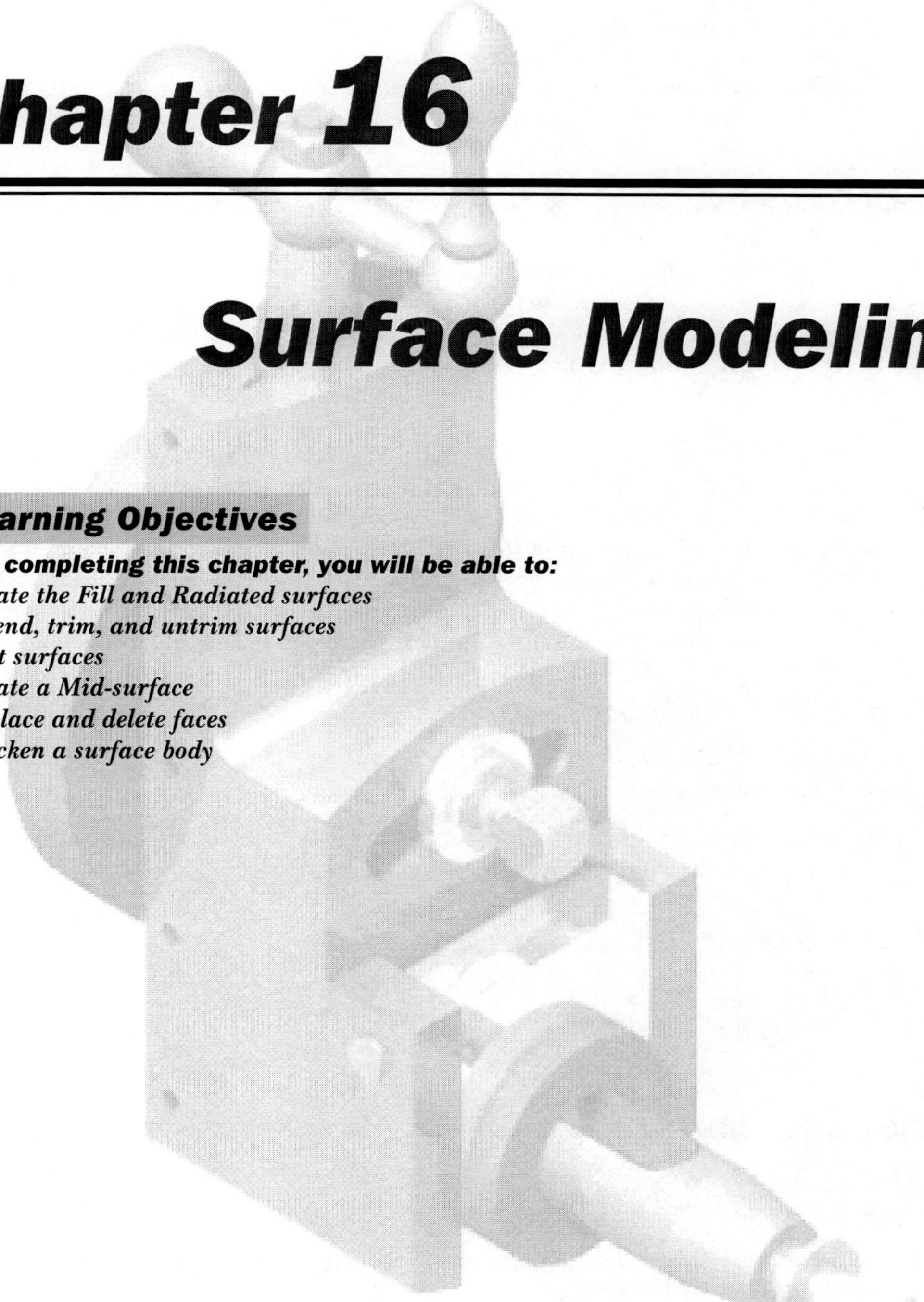

Learning Objectives

After completing this chapter, you will be able to:

- *Create the Fill and Radiated surfaces*
- *Extend, trim, and untrim surfaces*
- *Knit surfaces*
- *Create a Mid-surface*
- *Replace and delete faces*
- *Thicken a surface body*

SURFACE MODELING

Surface modeling is a technique of creating planar or non-planar geometry of zero thickness. This zero thickness geometry is known as surface. Surfaces are generally used to create models of complex shapes. You can easily convert surface models into solid models. You can also extract a surface from a solid model using the surface modeling tools. This chapter deals with the surface modeling tools in SOLIDWORKS. Using these tools, you can create complex shapes as surfaces and then convert them into solid models, if required.

Most of the real world components are created using solid modeling. But sometimes, you may need to create some complex features that can only be created by surface manipulation. Surface manipulation is done by using surface modeling tools. After creating the required complex surface, you can convert it into a solid model. The reasons to convert a surface model into a solid model are that a surface is a zero-thickness geometry and it has no mass and mass properties. But, while designing real world models, you may need mass and mass properties. The other reason is that you can generate a section view only if the model is a solid.

In SOLIDWORKS, surface modeling is done in the **Part** mode and the tools used for surface modeling are available in the **Surfaces CommandManager**. The **Surfaces CommandManager** will not be available, by default. Therefore, you need to right-click on any one of the **CommandManager** tabs and then choose the **Surfaces** option from the shortcut menu. The surface modeling tools can also be invoked by choosing **Insert > Surface** from the SOLIDWORKS menus. You will notice that some of the tools available in the **Surfaces CommandManager**, such as extrude, revolve, sweep, and loft are similar to those discussed in the solid modeling.

Creating a Boundary Surface

CommandManager:	Surfaces > Boundary Surface
SOLIDWORKS menus:	Insert > Surface > Boundary Surface

The **Boundary Surface** tool is used to create complex models with high accuracy as well as high surface quality, while maintaining the curvature continuity.

Creating a Planar Surface

CommandManager:	Surfaces > Planar Surface
SOLIDWORKS menus:	Insert > Surface > Planar

A planar surface is generally used to fill gaps between surfaces using a planar patch.

Creating a Fill Surface

CommandManager:	Surfaces > Filled Surface
SOLIDWORKS menus:	Insert > Surface > Fill

The **Filled Surface** tool is used to create a surface patch along N number of sides. The sides to be selected for creating a filled surface can be the edges of the existing model, 2D or 3D sketch entities, or 2D or 3D curves. The difference between a planar surface and a fill surface is that you cannot create a planar surface using 3D curves or edges.

Creating a Radiated Surface

CommandManager:	Surfaces > Radiate Surface (Customize to Add)
SOLIDWORKS menus:	Insert > Surface > Radiate

In SOLIDWORKS, you can also create a surface by radiating a surface along an edge or a split line. The radiated surface is always created parallel to the plane or the face selected as the radiate direction reference. This type of surface is generally used in mold design as parting surface for extracting the core and cavity.

Trimming Surfaces

CommandManager:	Surfaces > Trim Surface
SOLIDWORKS menus:	Insert > Surface > Trim

The **Trim Surface** tool is used to trim surfaces using an entity as the trim tool. A surface, a sketched entity, or an edge can be used as a trim tool.

Untrimming Surfaces

CommandManager:	Surfaces > Untrim Surface
SOLIDWORKS menus:	Insert > Surface > Untrim

The **Untrim Surface** tool is used to create a surface patch by extending the existing surfaces. Using this tool, you can fill the trimmed portion of a surface with a surface patch.

Knitting Surfaces

CommandManager:	Surfaces > Knit Surface
SOLIDWORKS menus:	Insert > Surface > Knit

The **Knit Surface** tool is used to knit multiple surfaces together to create a single surface. You can also knit a surface with the faces of a solid body. The surfaces to be knitted together must be in contact with each other. This means that you cannot knit disjointed surfaces or faces. The **Knit Surface** tool is widely used for extracting core and cavity while designing a mold.

Creating a Mid-Surface

CommandManager:	Surfaces > Mid-Surface (Customize to add)
SOLIDWORKS menus:	Insert > Surface > Mid Surface

The **Mid-Surface** tool is used to create a surface between the two parallel faces of a solid model. You can define the placement of a surface in terms of percentage value with respect to the face selected first. Note that the faces to be selected to create a mid-surface should be two parallel faces or two concentric curved faces.

Replacing Faces

CommandManager:	Surfaces > Replace Face
SOLIDWORKS menus:	Insert > Face > Replace

In SOLIDWORKS, you can replace the selected faces of a solid body with one or more surfaces. When you replace the selected faces with another surface or surfaces, the resultant solid body retains the shape of the replaced surface by adding or subtracting material from the solid body.

Deleting Faces

CommandManager:	Surfaces > Delete Face
SOLIDWORKS menus:	Insert > Face > Delete

The **Delete Face** tool is used to delete the faces of the selected surface or the solid body. When you delete a face of a solid body, the solid body is converted into a surface body. If you patch the deleted face, the solid body will not be converted into a surface body.

Adding Thickness to Surface Bodies

CommandManager:	Surfaces > Thicken
SOLIDWORKS menus:	Insert > Boss/Base > Thicken

In SOLIDWORKS, you can also add thickness to the surface bodies. There are two methods of adding thickness to surface bodies. In the first method, you need to add wall thickness to the surface body. In the second method, you need to solidify the closed, stitched surface body to create a solid body.

TUTORIALS

Tutorial 1

In this tutorial, you will create the model shown in Figure 16-1. You will create the model using the surface modeling tools available in the **Surfaces CommandManager** and then add wall thickness to the surface model. The views and dimensions of the model are shown in Figure 16-2.

(Expected time: 1hr)

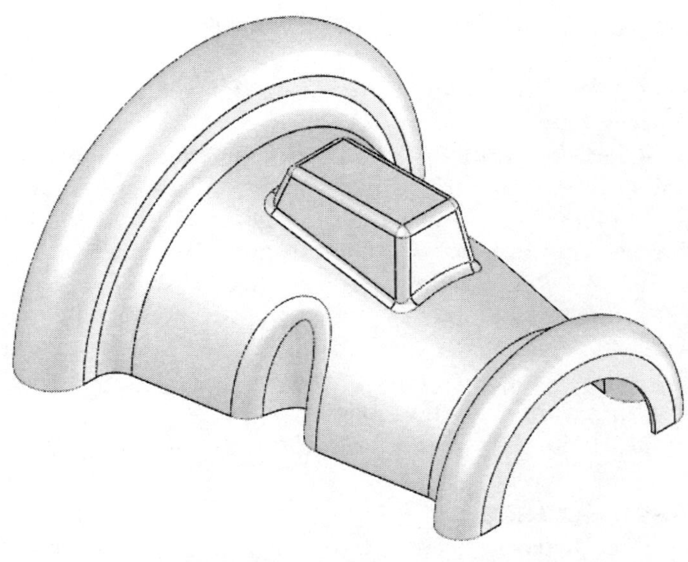

Figure 16-1 *Model for Tutorial 1*

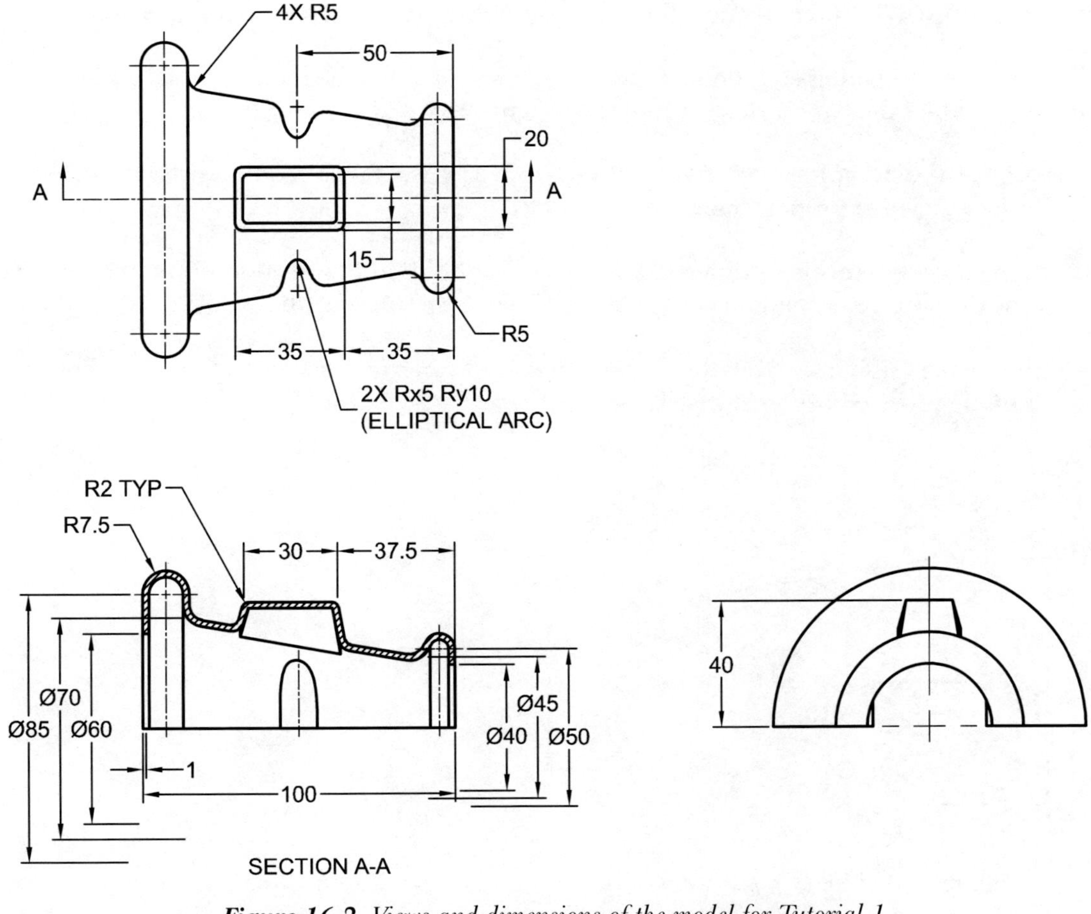

Figure 16-2 *Views and dimensions of the model for Tutorial 1*

The following steps are required to complete this tutorial:

a. Create the base surface of the model by revolving the sketch using the **Mid Plane** option through 180 degrees, refer to Figures 16-3 and 16-4.
b. Create the second surface feature, refer to Figures 16-5 and 16-6.
c. Trim the extruded surface using the trim tool, refer to Figures 16-7 and 16-8.
d. Add fillet to the trimmed base surface, refer to Figures 16-9 and 16-10.
e. Create a plane at an offset distance of 40 mm from the Top plane.
f. Create a lofted surface, refer to Figures 16-11 through 16-15.
g. Create a planar surface on the top of the lofted feature and trim the base feature using the lofted feature, refer to Figures 16-16 and 16-17.
h. Trim and knit all surfaces together, refer to Figure 16-18.
i. Add fillets to the surface model, refer to Figure 16-19.
j. Add thickness to the knitted surface, refer to Figure 16-20.

Creating the Base Surface

To create this model, first you need to create the base surface. The base surface will be created by revolving the sketch created on the Front plane.

1. Start SOLIDWORKS part document using the **New SOLIDWORKS Document** dialog box.

2. Invoke the sketching environment using the Front plane as the sketching plane and create the sketch of the base surface, as shown in Figure 16-3.

3. Choose the **Revolved Surface** button from the **Surfaces CommandManager**; the **Surface-Revolve PropertyManager** is displayed.

4. In the **Surface-Revolve PropertyManager**, select the **Mid Plane** option from the **Revolve Type** drop-down list and set the value of the **Angle** spinner to **180**.

5. Choose the **OK** button from the **Surface-Revolve PropertyManager**. Figure 16-4 shows the resulting revolved base feature.

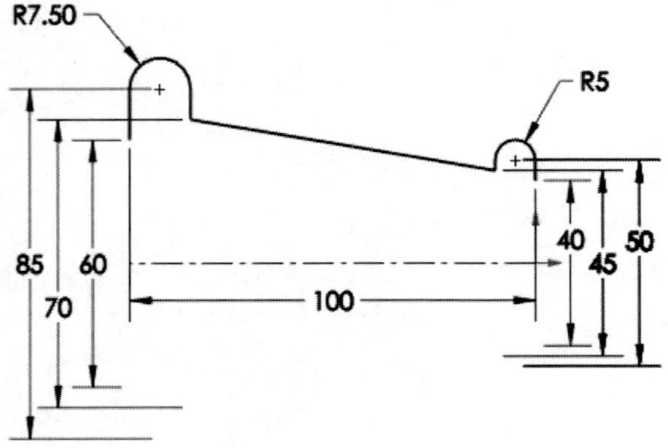

Figure 16-3 *Sketch of the base surface*

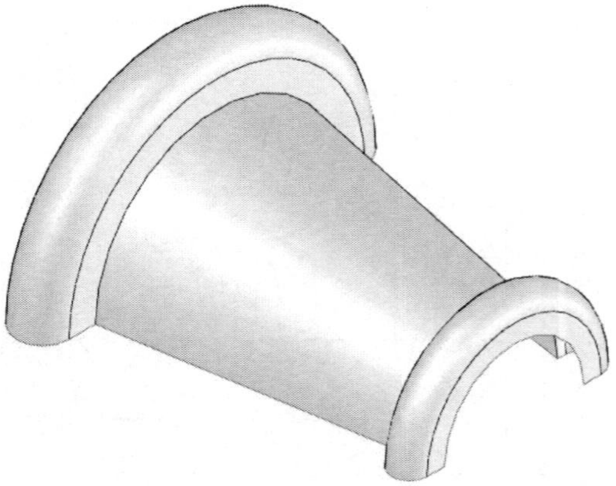

Figure 16-4 *Revolved base surface*

Creating the Second Surface Feature

The second surface feature is an extruded surface. This surface will be created by extruding a sketch created on the Top plane.

1. Select the Top plane as the sketching plane and invoke the sketching environment.

2. Create the sketch of the second surface feature, as shown in Figure 16-5. Note that you may have to apply the **Horizontal** or **Vertical** relation between the points of the ellipse to constrain them.

3. Next, choose the **Extruded Surface** button from the **Surfaces CommandManager**; the **Surface-Extrude PropertyManager** is displayed.

4. Set **40** in the **Depth** spinner and choose the **OK** button from the **Surface-Extrude PropertyManager**.

   The surface created after extruding the sketch is shown in Figure 16-6.

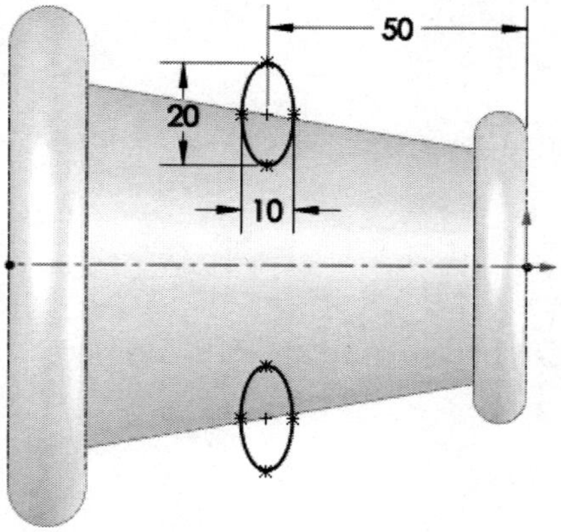

Figure 16-5 *Sketch of the second surface feature*

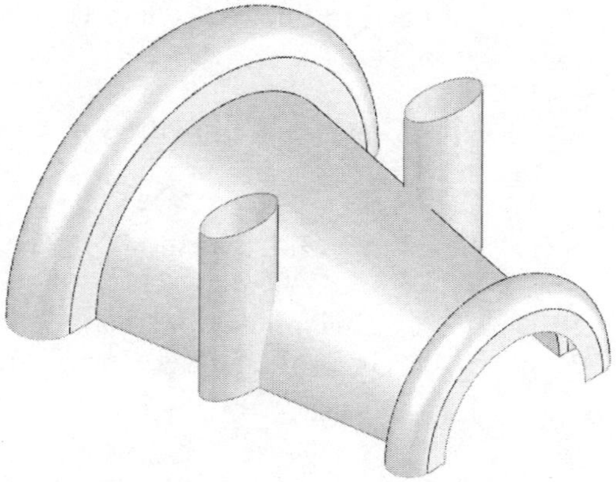

Figure 16-6 *Extruded surface*

Trimming the Base Surface Using the Extruded Surface

Next, you need to trim the unwanted portions of the base surface by using the extruded surface.

1. Choose the **Trim Surface** button from the **Surfaces CommandManager**; the **Trim Surface PropertyManager** is displayed.

2. Select the **Mutual** radio button from the **Trim Type** rollout and then select the surfaces, refer to Figure 16-7.

3. Next, click in the **Pieces to Keep** selection box to invoke the selection mode. Next, select the pieces to be kept, refer to Figure 16-7.

4. Choose the **OK** button from the **Trim Surface PropertyManager**. The model after trimming the base surface is displayed in Figure 16-8.

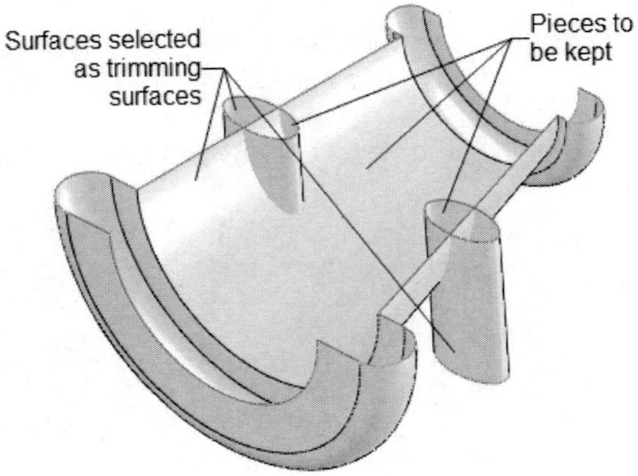

Figure 16-7 Surfaces selected for trimming

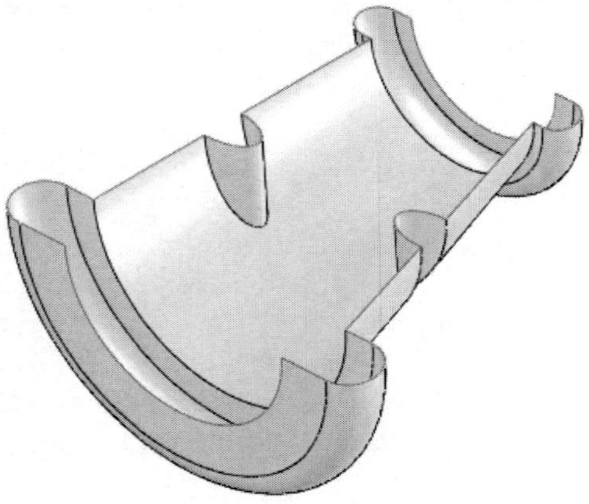

Figure 16-8 Model after trimming the base surface

Filleting the Edges of the Trimmed Surface

Next, you need to fillet the edges created at the intersection of the base surface with the extruded surfaces.

1. Choose the **Fillet** button from the **Surfaces CommandManager**; the **Fillet PropertyManager** is displayed.

2. Select the edges to fillet, as shown in Figure 16-9.

3. Set **5** in the **Radius** spinner and then choose the **OK** button from the **Fillet PropertyManager**. Figure 16-10 shows the model after adding the fillet.

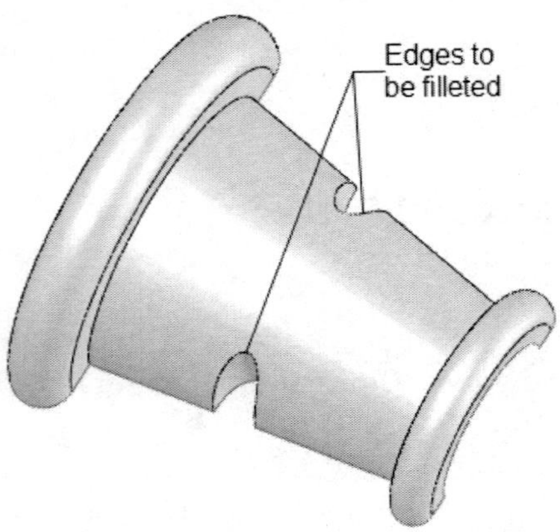

Figure 16-9 Edges to be filleted

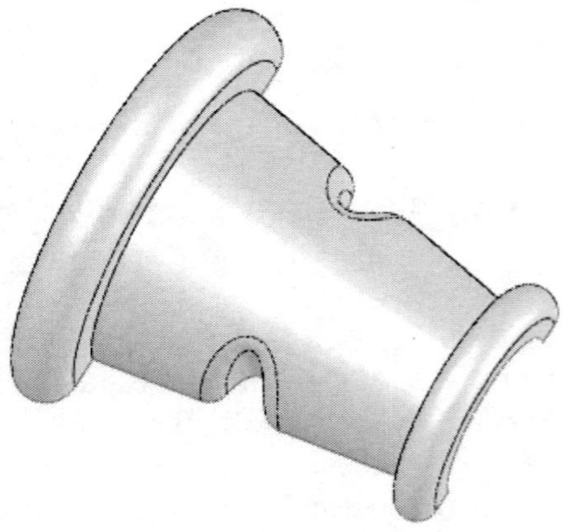

Figure 16-10 Resultant fillet feature

Creating the Lofted Surface

Next, you need to create a lofted surface. The lofted surface will be created between two curves. The first curve is to be created on the Top plane and projected on the surface. The next curve is to be created on an offset plane.

1. Create a plane at an offset distance of 40 mm from the Top plane and then invoke the sketching environment using the top plane as the sketching plane.

2. Create the sketch, as shown in Figure 16-11, and exit the sketching environment.

   Next, you need to project this sketch on the base surface.

3. Choose **Curves > Project Curve** from the **Surfaces CommandManager**; the **Projected Curve PropertyManager** is invoked.

4. Select the **Sketch on faces** radio button from the **Projection type** area.

5. Next, select the sketch. As you select the sketch the **Projection Faces** selection box gets activated.

6. Select the middle portion of the base surface. Choose the **OK** button from the **Projected Curve PropertyManager**.

   Next, you need to create second curve for creating the loft surface.

7. Invoke the sketching environment with the newly created plane as the sketching plane and then create the sketch, as shown in Figure 16-12. Then, exit the sketching environment. Figure 16-13 shows the model after creating the sketch and the projected curve.

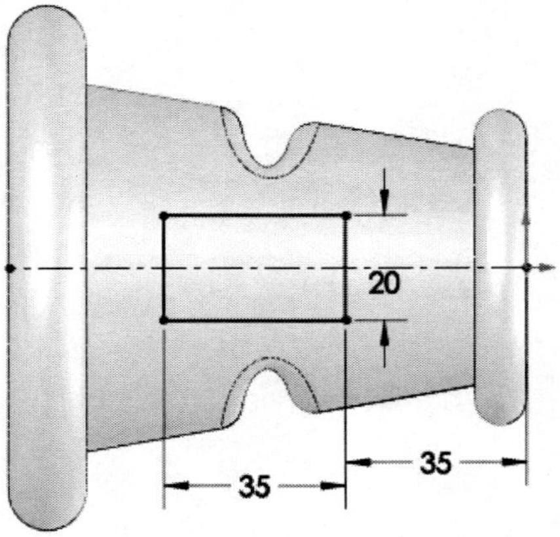

Figure 16-11 *Sketch to create the projected curve*

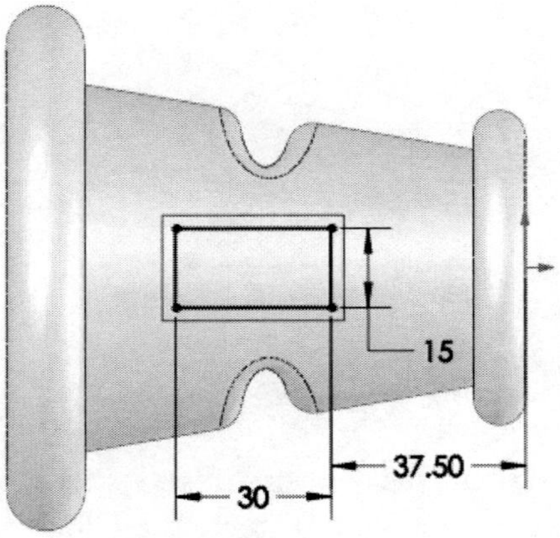

Figure 16-12 *Second sketch for the loft surface*

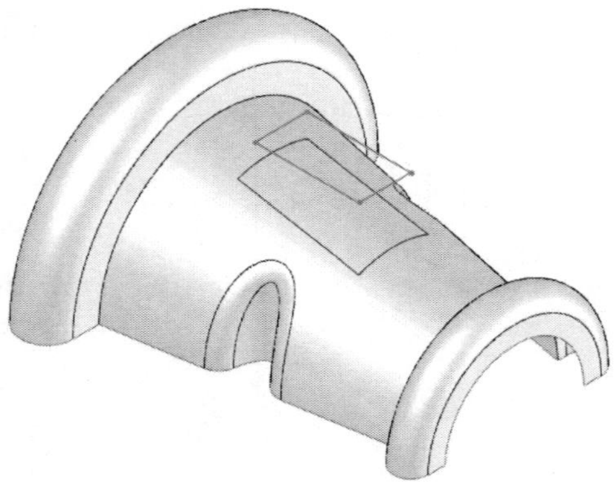

Figure 16-13 *Model after creating the sketch and the projected curve*

8. Choose the **Lofted Surface** button from the **Surfaces CommandManager**; the **Surface-Loft PropertyManager** is displayed.

9. Select the loft section, as shown in Figure 16-14, and choose the **OK** button from the **Surface-Loft PropertyManager**. Figure 16-15 shows the model after creating the lofted surface.

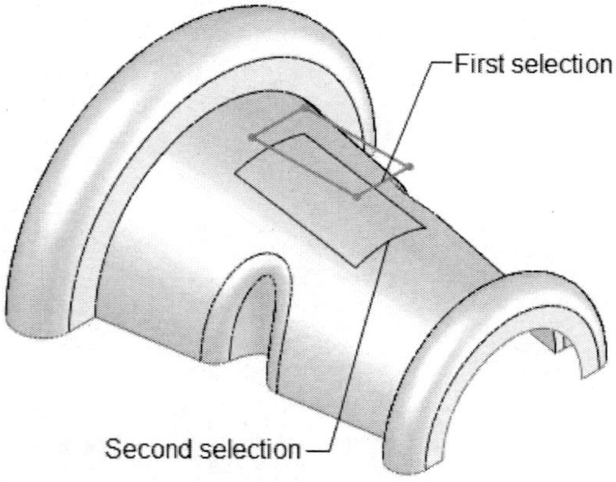

Figure 16-14 *Section selected for the lofted surface*

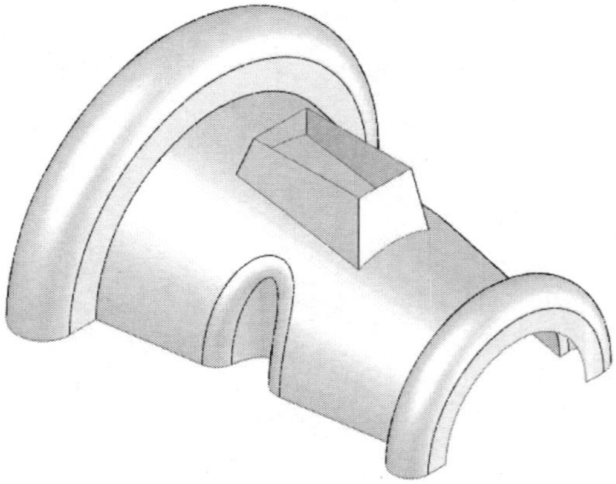

Figure 16-15 *Resultant lofted surface*

Creating the Planar Surface

Next, you need to create the planar surface by using the top edges of the lofted surface.

1. Choose the **Planar Surface** button from the **Surfaces CommandManager**; the **Planar Surface PropertyManager** is displayed.

2. Select the edges, as shown in Figure 16-16, to create the planar surface; the name of the selected edges is displayed in the **Bounding Entities** selection box of the **Planar Surface PropertyManager**.

3. Choose the **OK** button from the **Planar Surface PropertyManager**; the planar surface is created. Figure 16-17 shows the resultant planar surface.

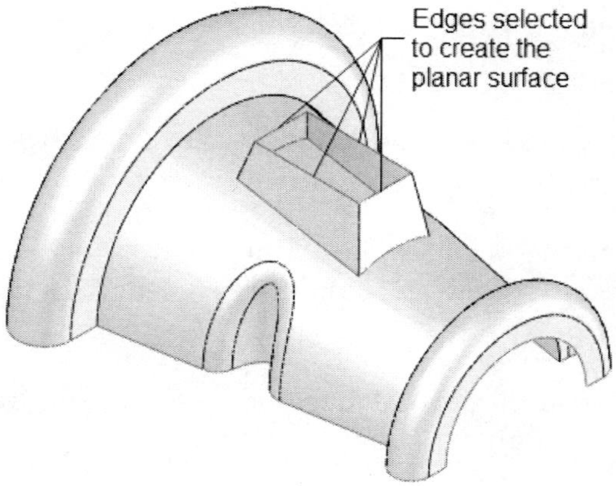

Figure 16-16 Edges selected to create the planar surface

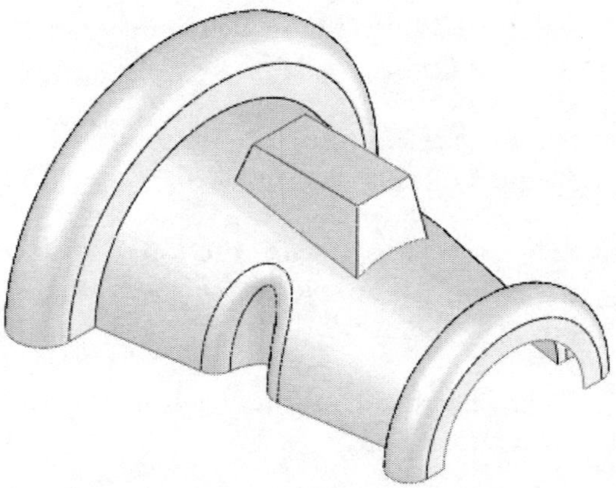

Figure 16-17 Resultant planar surface

Trimming the Base Surface Using the Lofted Surface

Next, you need to trim the base surface by using the lofted surface.

1. Choose the **Trim Surface** button from the **Surfaces CommandManager** to invoke the **Trim Surface PropertyManager**.

2. Select the **Standard** radio button from the **Trim Type** rollout. Next, select the lofted surface as the trim tool and then select the base surface as the piece to be kept.

3. Choose the **OK** button from the **Trim Surface PropertyManager**.

Figure 16-18 shows the model after the base feature has been trimmed by using the lofted surface.

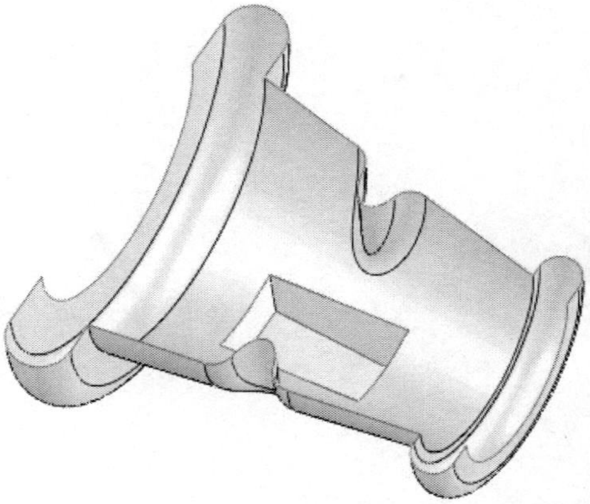

Figure 16-18 Model after trimming the base surface

Knitting all Surfaces

Next, you need to knit the base surface, the lofted surface, and the planar surface together by using the **Knit Surface** tool to create fillets on the edges. Note that, if you want to add wall thickness to the surface model created using multiple surfaces, you need to knit all the surfaces first.

1. Choose the **Knit Surface** button from the **Surfaces CommandManager**; the **Knit Surface PropertyManager** is invoked.
2. Select the base surface, the lofted surface, and the planar surface from the drawing area; the names of the selected surfaces are displayed in the **Surfaces and Faces to Knit** selection box.
3. Next, choose the **OK** button from the **Surface Knit PropertyManager**; the selected surfaces are knitted.
4. Add required fillets to the model. For dimensions of the fillets, refer to Figure 16-2. Final surface model after adding the fillets is shown in Figure 16-19.

Adding Thickness to the Surface Model

Next, you need to add thickness to the surface model.

1. Choose the **Thicken** button from the **Surfaces CommandManager** to invoke the **Thicken PropertyManager**.
2. Select the surface model and then set **1** in the **Thickness** spinner. Also select the **Thicken Side 2** button.
3. Choose the **OK** button from the **Thicken PropertyManager**.

Figure 16-20 shows the rotated view of the final model after thickening the surface.

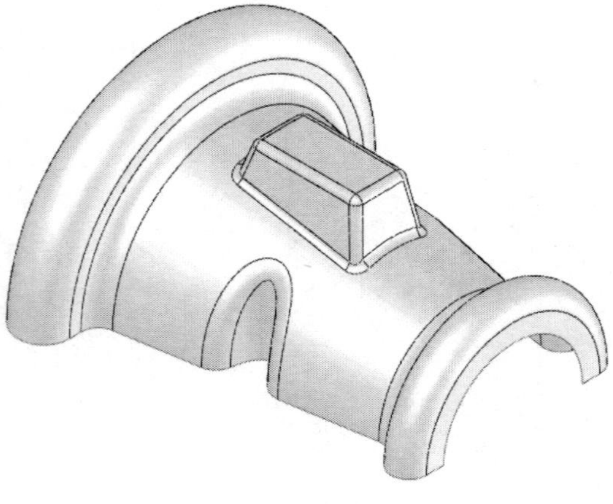

Figure 16-19 *Final surface model*

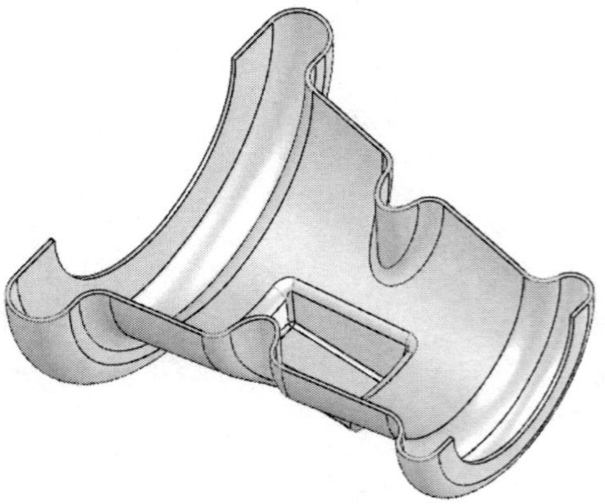

Figure 16-20 *The rotated view of the final model after thickening the surface*

Saving the Model

1. Choose the **Save** button from the Menu Bar and save the drawing with the name *c16_tut01* at the following location and close the file.

   \Documents\SOLIDWORKS\c16

Tutorial 2

In this tutorial, you will create the cover of a hair dryer, as shown in Figure 16-21. First, you will create the model by using surfaces and then thicken it. The views and dimensions of the model are shown in Figure 16-22. **(Expected time: 1.5 hr)**

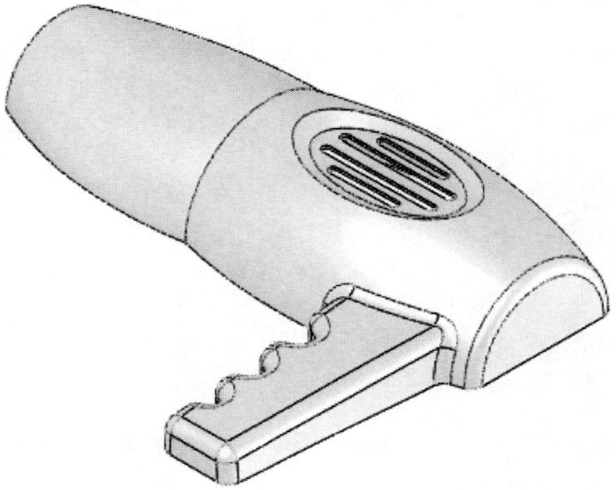

Figure 16-21 *Cover of hair dryer*

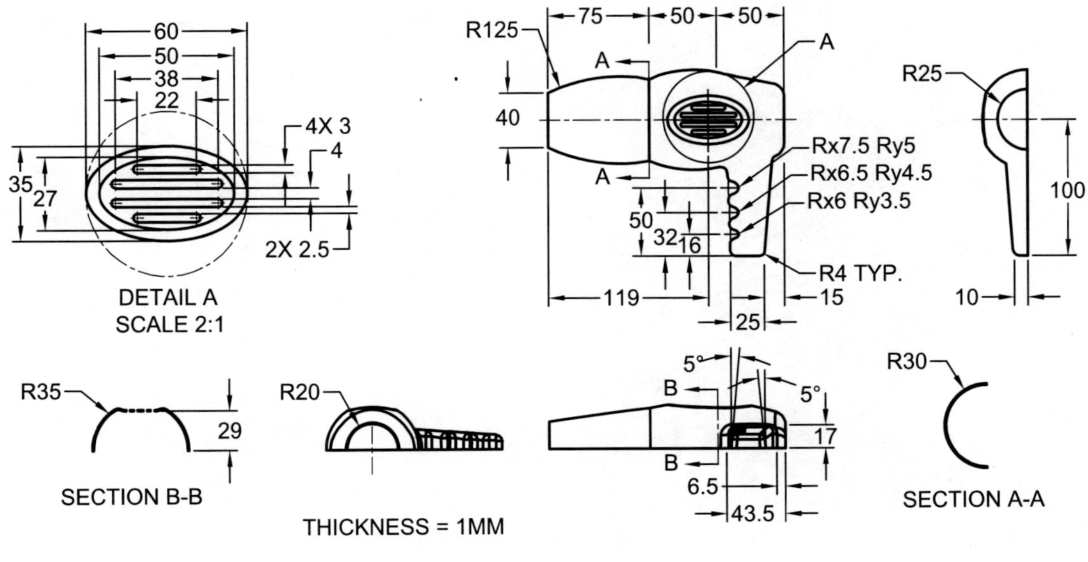

Figure 16-22 *Views and dimensions of the model for Tutorial 2*

The following steps are required to complete this tutorial:

a. First, create the base surface. The base surface is created by lofting the open sections along the guide curves, refer to Figures 16-23 and 16-24.
b. Create a planar surface to close the right face of the base surface, refer to Figures 16-25 and 16-26.
c. Create the basic structure of the handle of the hair dryer cover by creating a lofted surface between two open sections, refer to Figures 16-27 through 16-29.
d. Trim the unwanted portion of the lofted surface that is used to create the handle, refer to Figures 16-30 and 16-31.
e. Create a planar surface to close the front face of the handle, refer to Figure 16-32.
f. Extrude the elliptical sketches to create the grips of the handle and then trim the unwanted surfaces, refer to Figures 16-33 through 16-36.
g. Create a dip on the top surface of the hair dryer, refer to Figures 16-37 through 16-42.
h. Trim the surface to create air vents, refer to Figure 16-43.
i. Knit all surfaces together and add fillets to the model, refer to Figure 16-44.
j. Thicken the surface, refer to Figure 16-45.

Creating the Base Surface

To create the hair dryer cover, you first need to create the base surface of the model. The base surface will be created by lofting semicircular sections along the guide curves. These sections will be created on different planes. Therefore, you first need to create three planes at an offset distance from the Right plane.

1. Start SOLIDWORKS part document using the **New SOLIDWORKS Document** dialog box.

2. Create three planes at an offset distance from the Right plane, as shown in Figure 16-23. For the offset distance of planes, refer to Figure 16-22.

3. Create sections and guide curves to create a lofted surface, as shown in Figure 16-23. For dimensions, refer to Figure 16-22.

4. Invoke the **Lofted Surface** tool and create lofted surface, as shown in Figure 16-24.

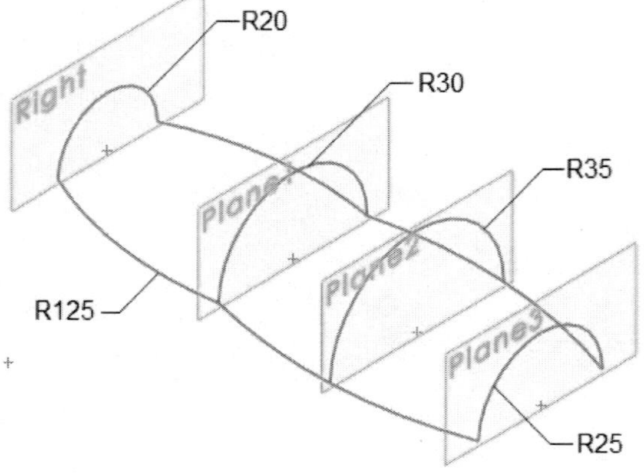

Figure 16-23 *Sections and guide curves to create a lofted surface*

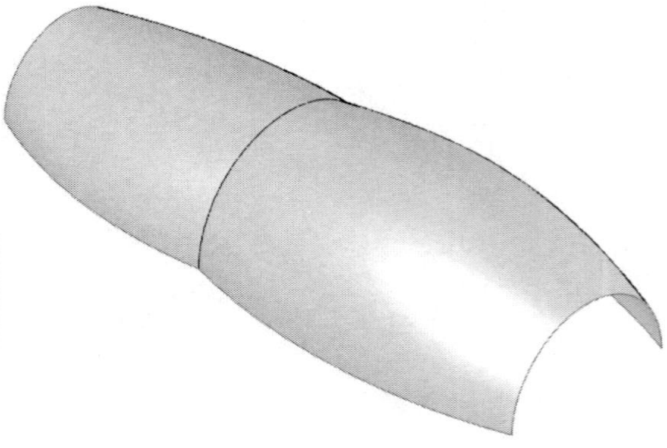

Figure 16-24 *Resultant lofted surface*

5. Invoke the sketching environment by selecting **Plane3** as the sketching plane.

6. Create a closed sketch to create a planar surface, as shown in Figure 16-25.

7. Invoke the **Planar Surface PropertyManager** and then select the closed sketch from the drawing area. Next, choose the **OK** button from the **Planar Surface PropertyManager**; the planar surface is created, as shown in Figure 16-26.

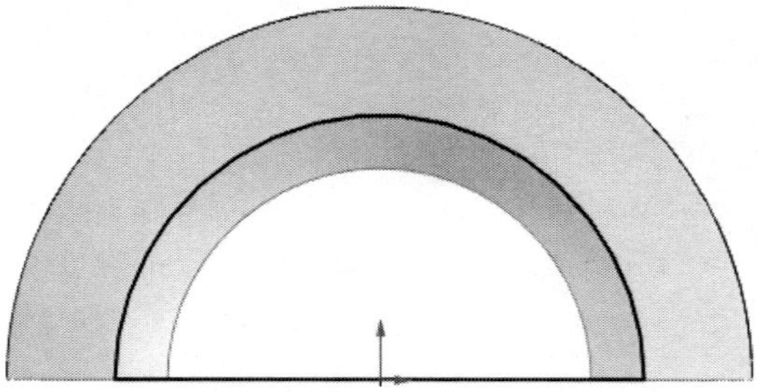

Figure 16-25 *Sketch for creating the planar surface*

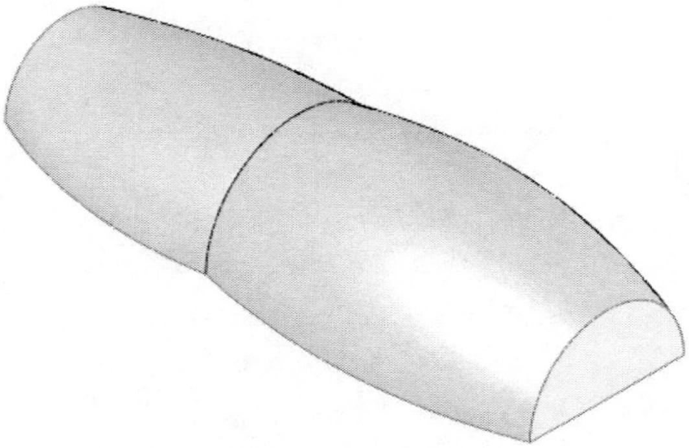

Figure 16-26 *Resultant planar surface*

Creating the Base Surface for the Handle

Next, you need to create the base surface for the handle. The base surface for the handle will be created by lofting two open sections. The first section for the lofted surface will be created on a plane at an offset distance from the Front plane and the second section will be created on the Front plane. Therefore, you first need to create a plane at an offset distance from the Front plane.

1. Create a plane at an offset distance of **100** mm from the Front plane.

2. Invoke the sketching environment using the newly created plane as the sketching plane.

3. Create an open sketch, as shown in Figure 16-27, and exit the sketching environment.

4. Next, invoke the sketching environment by using the Front plane as the sketching plane.

5. Create an open sketch, as shown in Figure 16-28, and exit the sketching environment.

6. Create the lofted surface by using the **Lofted Surface** tool, as shown in Figure 16-29.

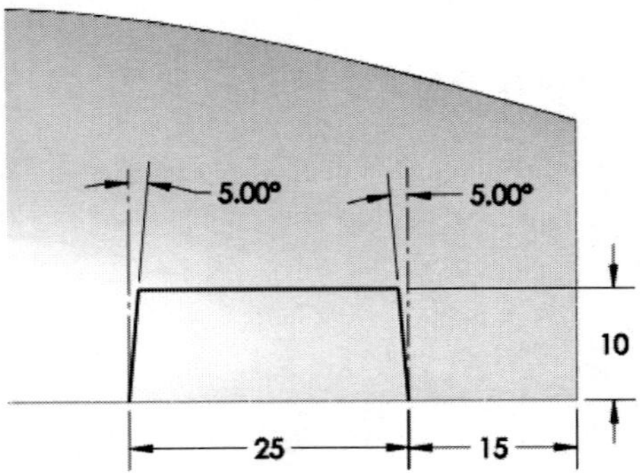

Figure 16-27 *Sketch of the first section for creating the lofted surface of the handle*

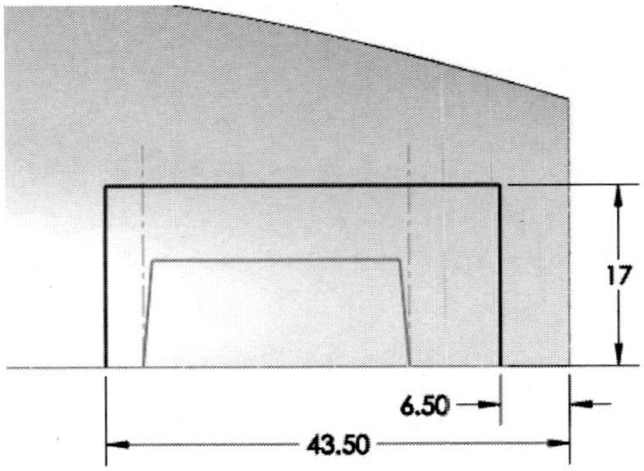

Figure 16-28 *Sketch of the second section for creating the lofted surface of the handle*

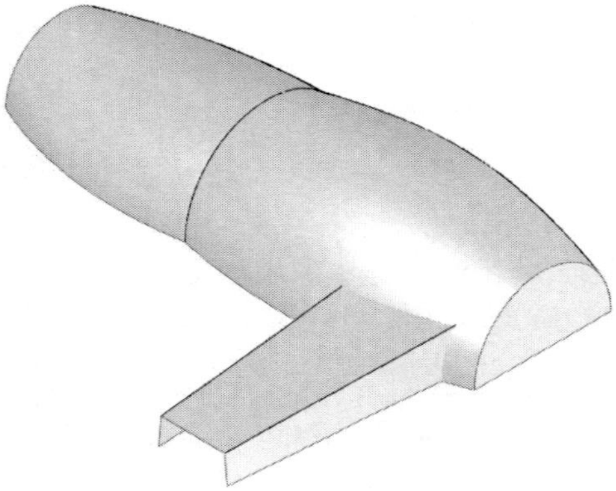

Figure 16-29 *Resultant lofted surface*

Trimming the Unwanted Portion from the Lofted Surface of the Handle

If you rotate the model after creating the lofted surface for the handle, you will observe that a portion of the lofted surface needs to be trimmed. The method for trimming the unwanted portion of the lofted surface is discussed next.

1. Invoke the **Trim Surface PropertyManager** and then select the **Mutual** radio button from the **Trim Type** rollout.

2. Select the trimming surfaces and the pieces to be kept, as shown in Figure 16-30. Next, choose the **OK** button from the **Trim Surface PropertyManager**. Figure 16-31 shows the resultant trimmed surface.

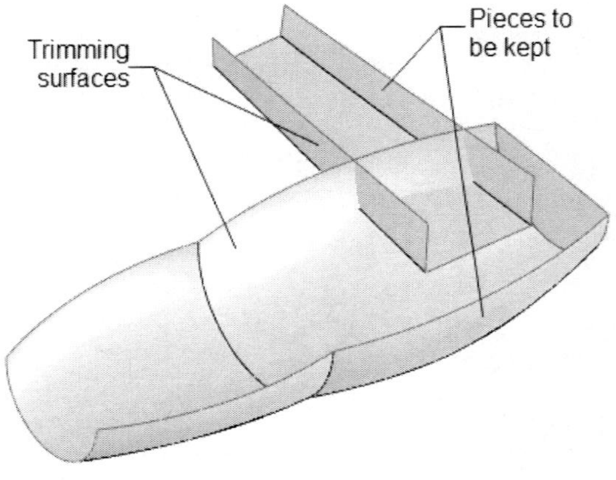

Figure 16-30 *Trimming surfaces and pieces to be kept*

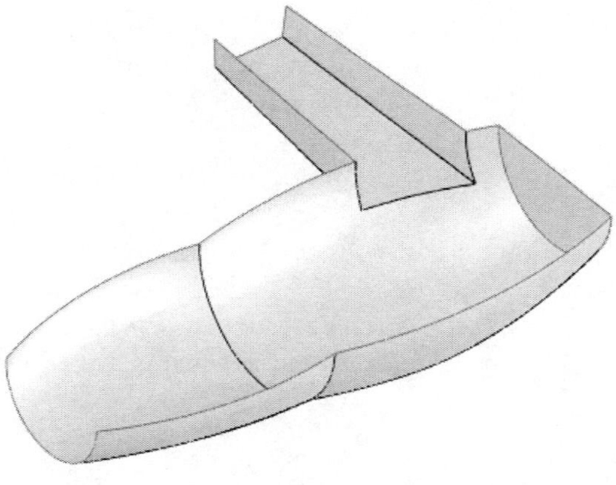

Figure 16-31 *Resultant trimmed surface*

Creating the Planar Surface

Next, you need to create a planar surface to close the front face of the handle.

1. Select the plane created at an offset distance from the Front plane and invoke the sketching environment.

2. Create a closed sketch to create the planar surface and then exit the sketching environment.

3. Choose the **Planar Surface** button from the **Surfaces CommandManager**; the **Planar Surface PropertyManager** is displayed. Next, select the closed sketch from the drawing area; the name of the selected sketch is displayed in the **Bounding Entities** selection box of the **Planar Surface PropertyManager**. Also, preview of the planar surface is displayed in the drawing area. Next, choose the **OK** button from the PropertyManager.

   The model after creating the planar surface is shown in Figure 16-32.

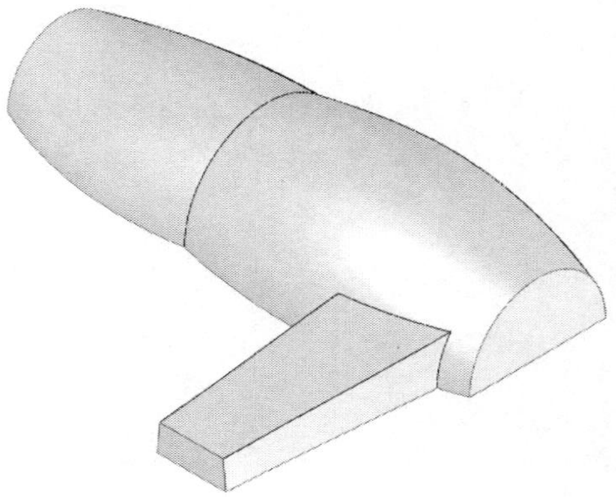

Figure 16-32 *Model after creating the planar surface*

Creating Grips on the Handle

Next, you need to create grips on the handle of the hair dryer. The grip will be created by extruding the elliptical surface and then trimming the unwanted portion of the surfaces.

1. Invoke the sketching environment using the Top plane as the sketching plane.
2. Create a sketch for extruding the surface, as shown in Figure 16-33.
3. Invoke the **Surface-Extrude PropertyManager** and extrude the sketch up to a depth of **25** mm. The extruded surface is displayed, as shown in Figure 16-34.

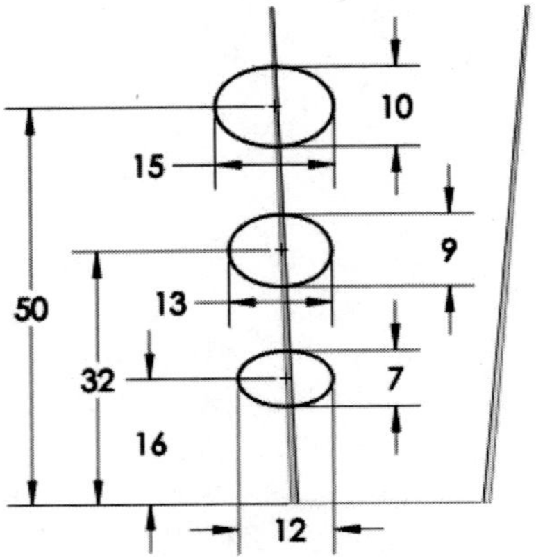

Figure 16-33 *Sketch created for extruding the surface*

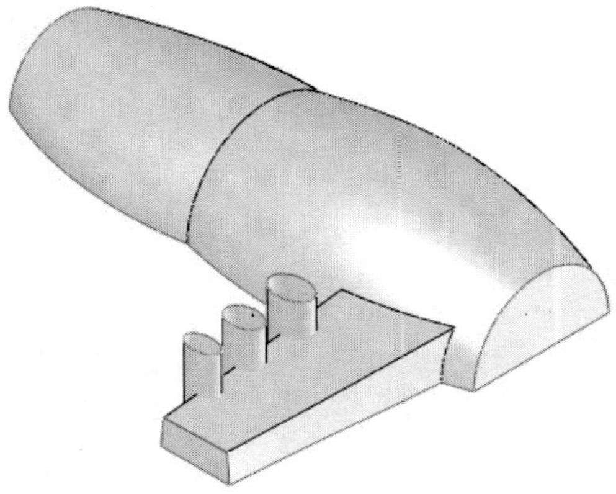

Figure 16-34 *Resultant extruded surface*

Next, you need to trim the unwanted portions of the extruded surface and the handle to achieve the desired shape of the grips.

4. Invoke the **Trim Surface PropertyManager** and then select the **Mutual** radio button from the **Trim Type** rollout.
5. Select the trimming surfaces and the pieces to be kept, as shown in Figure 16-35.
6. Choose the **OK** button from the **Trim Surface PropertyManager**; the surface is trimmed. Figure 16-36 shows the resultant trimmed surface.

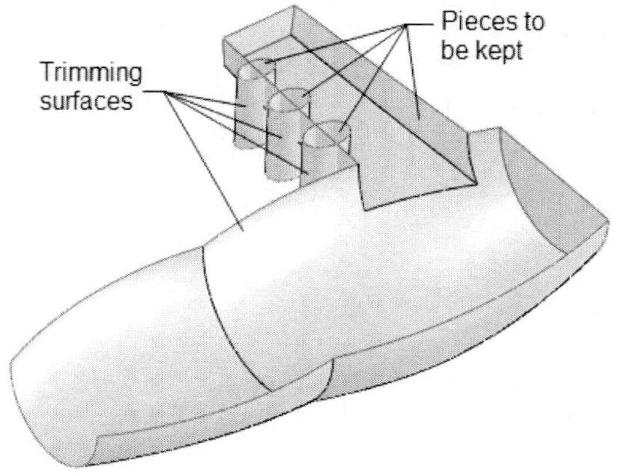

Figure 16-35 Trimming surfaces and the pieces to be kept

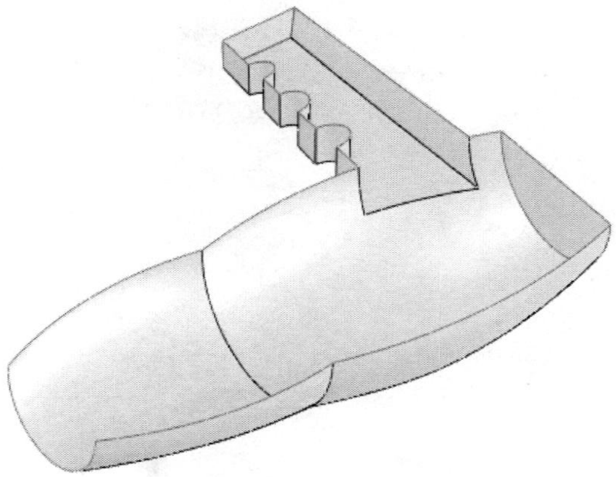

Figure 16-36 Resultant trimmed surface

Creating a Dip on the Upper Surface of the Base Surface

Next, you need to create a dip on the base surface. To do so, you need to use various tools for offsetting planes, creating lofted surface, trimming unwanted surfaces, and creating a planar surface.

1. Invoke the sketching environment using the Top plane as the sketching plane.
2. Create a sketch, as shown in Figure 16-37, and then exit the sketching environment.
3. Next, choose **Curves > Project Curve** from the **Surfaces CommandManager** and project the newly created sketch on the base surface. The model after projecting the sketch is displayed, as shown in Figure 16-38.

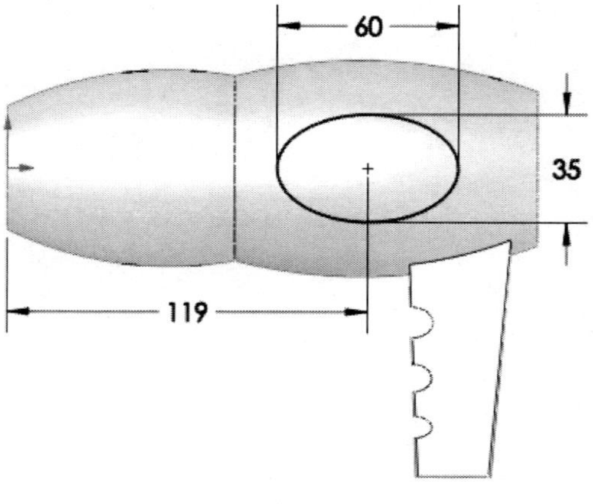

Figure 16-37 Sketch created

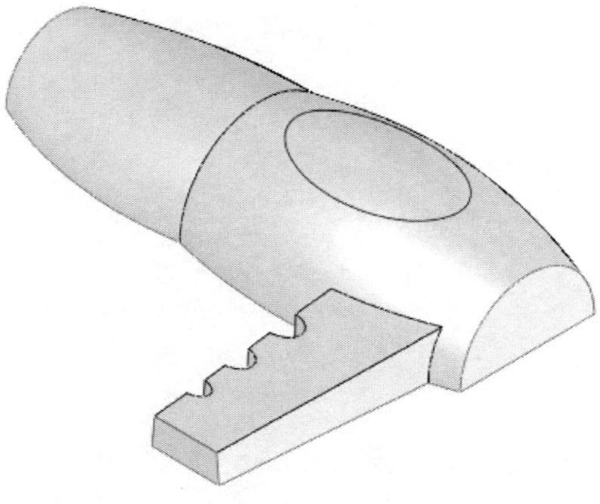

Figure 16-38 Resultant projected curve

4. Create a plane at an offset distance of **29** mm from the Top plane in the upward direction.
5. Next, invoke the sketching environment by using the newly created plane as the sketching plane and create a sketch, as shown in Figure 16-39. Next, exit the sketching environment.

6. Invoke the **Surface-Loft PropertyManager** by choosing the **Lofted Surface** button. Next, create a lofted surface by using the sketch and the projected curve created earlier. The lofted surface after hiding the base surface is shown in Figure 16-40.

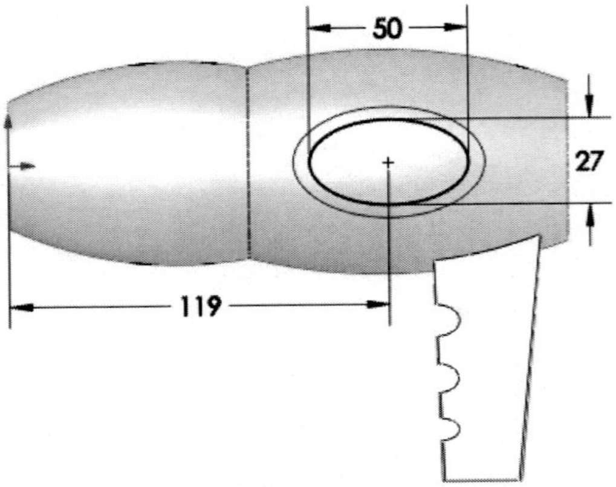

Figure 16-39 *Sketch to be created*

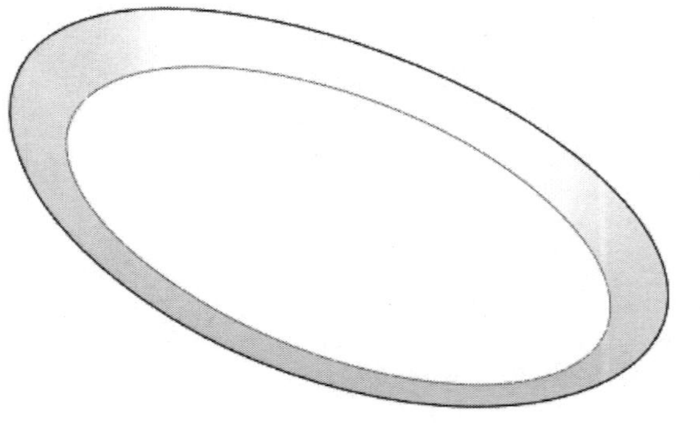

Figure 16-40 *Lofted surface*

7. To hide/show surface bodies, expand the **Surface Bodies** folder in the **FeatureManager Design Tree**. Next, select the surface and right-click on it to invoke the shortcut menu. Next, choose **Hide/Show** from the shortcut menu; the view of the selected surface gets modified.

8. Invoke the **Trim Surface** tool and trim the base surface using the newly created lofted surface. The model after trimming the surface is shown in Figure 16-41. Next, create the planar surface by using the **Planar Surface** tool, as shown in Figure 16-42.

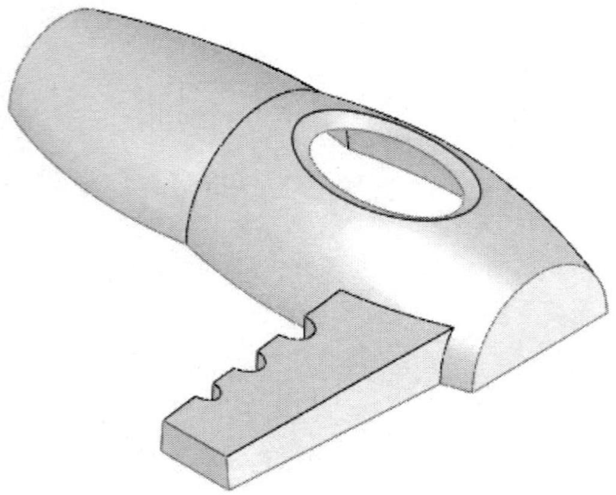

Figure 16-41 *Model after trimming the surface*

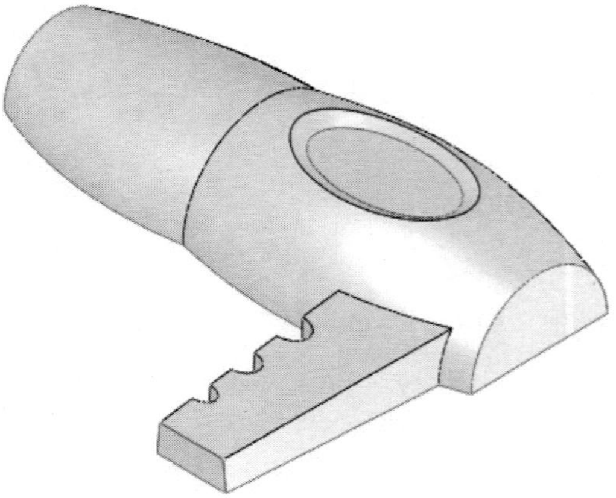

Figure 16-42 *Planar surface*

Creating Air Vents

Next, you need to create air vents on the newly created planar surface. Air vents are created by drawing the sketch on the planar surface and then trimming the surface.

1. Select the newly created planar surface as the sketching plane and then invoke the sketching environment.

2. Create the sketch of air vents. For dimensions, refer to Figure 16-22.

3. Invoke the **Trim Surface** tool by choosing the **Trim Surface** button from the **Surfaces CommandManager**. Then, select the pieces to keep or remove from the planar surface for creating air vents.

   The surface model after creating air vents is shown in Figure 16-43.

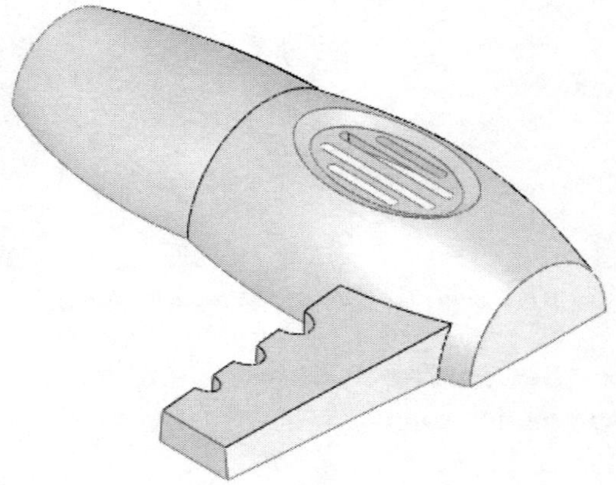

Figure 16-43 *Surface model after trimming the planar surface*

Knitting all Surfaces

After creating all the surfaces, you need to knit all the surfaces together and then add fillets to surfaces and thicken the model.

1. Choose the **Knit Surface** button from the **Surfaces CommandManager**; the **Knit Surface PropertyManager** is displayed and you are prompted to select the surfaces to be knit.

2. Expand the **Surface Bodies** folder in the design area of the **FeatureManager Design Tree** and then select all surface bodies from it.

3. Next, choose the **OK** button from the **Knit Surface PropertyManager**. If the surface do not knit properly then adjust the gaps using the list box provided at the bottom of the **Knit Surface PropertyManager**.

4. Add fillets at the required places of the surface model. The model after adding fillets is displayed in Figure 16-44.

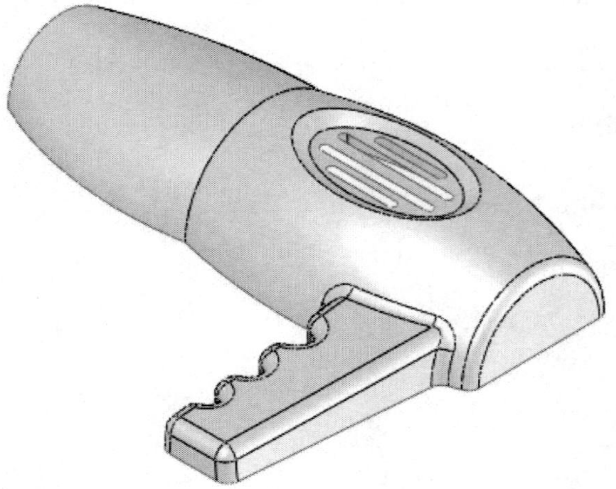

Figure 16-44 *Surface model after adding fillets*

Adding Thickness to the Surface Model

After creating the entire model, you need to add thickness to the surface model.

1. Choose the **Thicken** button from the **Surfaces CommandManager**; the **Thicken PropertyManager** is invoked and you are prompted to select the surface to thicken.

2. Set the value as **1** in the **Thickness** spinner and also select the **Thicken Side 2** button. Select the surface model from the drawing area; preview of the thickened model is displayed in the drawing area.

3. Next, choose the **OK** button from the **Thicken PropertyManager**; the final model is displayed, as shown in Figure 16-45.

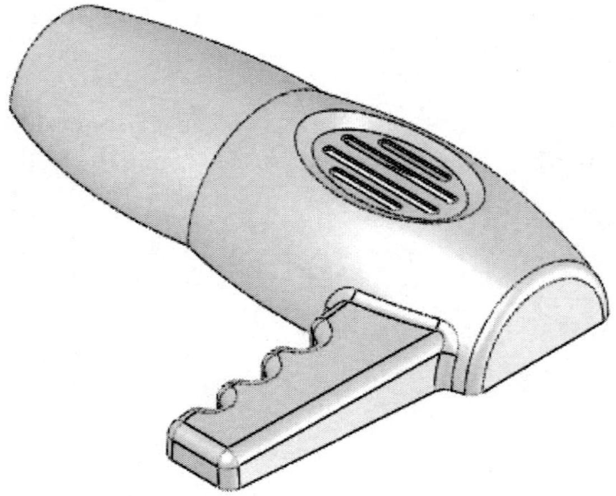

Figure 16-45 *Final model*

Saving the Model

1. Choose the **Save** button from the Menu Bar and save the drawing with the name *c16_tut02* at the following location and close the file.

   \Documents\SOLIDWORKS\c16

EXERCISES

Exercise 1

In this exercise, create the model shown in Figure 16-46 by using surfaces. After creating and knitting all surfaces, add thickness to the model. The views and dimensions of the model are shown in the same figure. **(Expected time: 1hr)**

Hint:
In this model, first you need to knit the surfaces and then fillet the edges of the resultant knit surface.

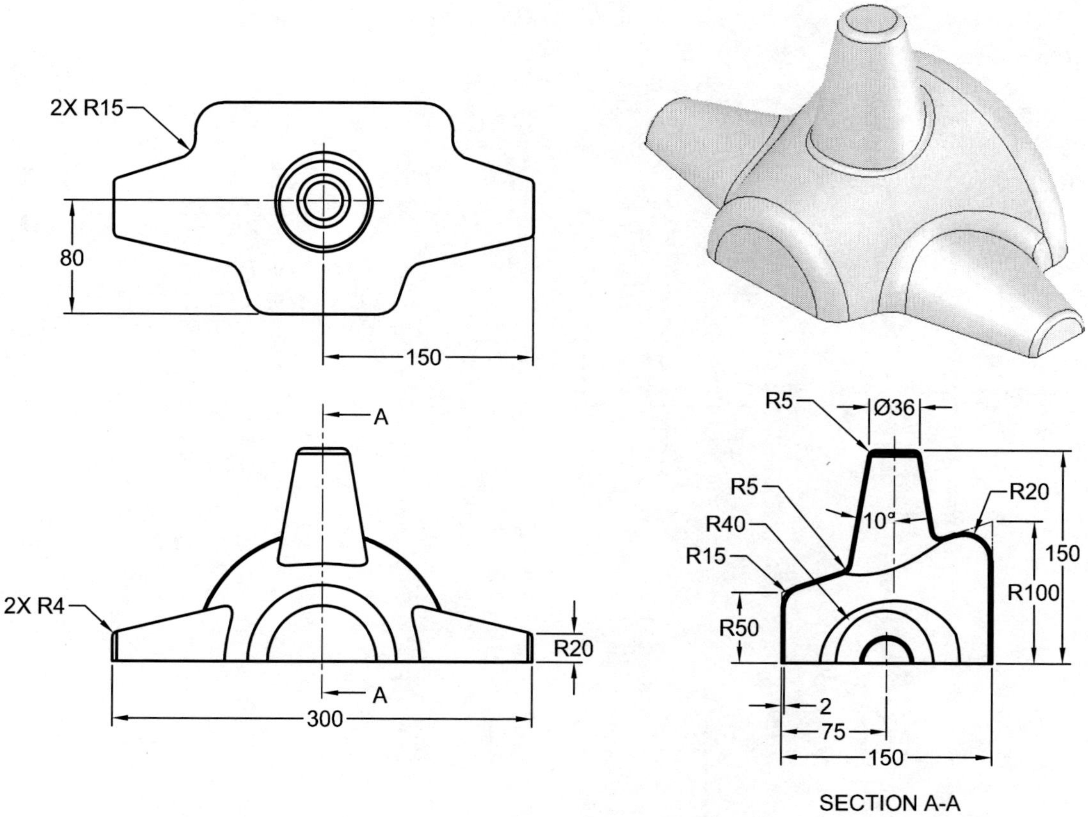

Figure 16-46 *Views and dimensions of the model for Exercise 1*

Exercise 2

In this exercise, create the model of binoculars shown in Figure 16-47 by using surfaces. First, create a closed surface model and knit all the surfaces together and then solidify it. The views and dimensions of the model are shown in Figure 16-48. **(Expected time: 1hr)**

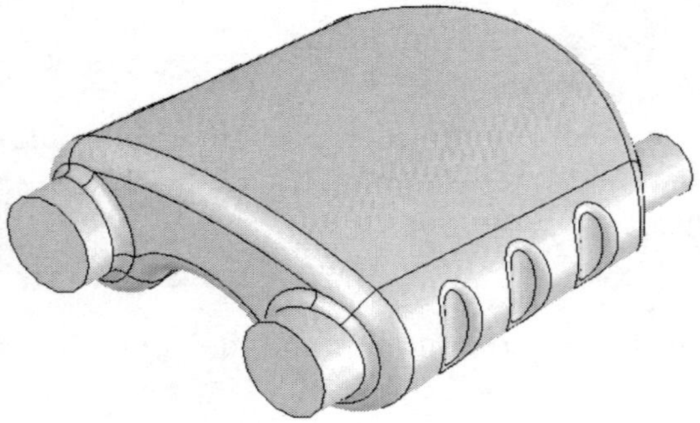

Figure 16-47 *Model for Exercise 2*

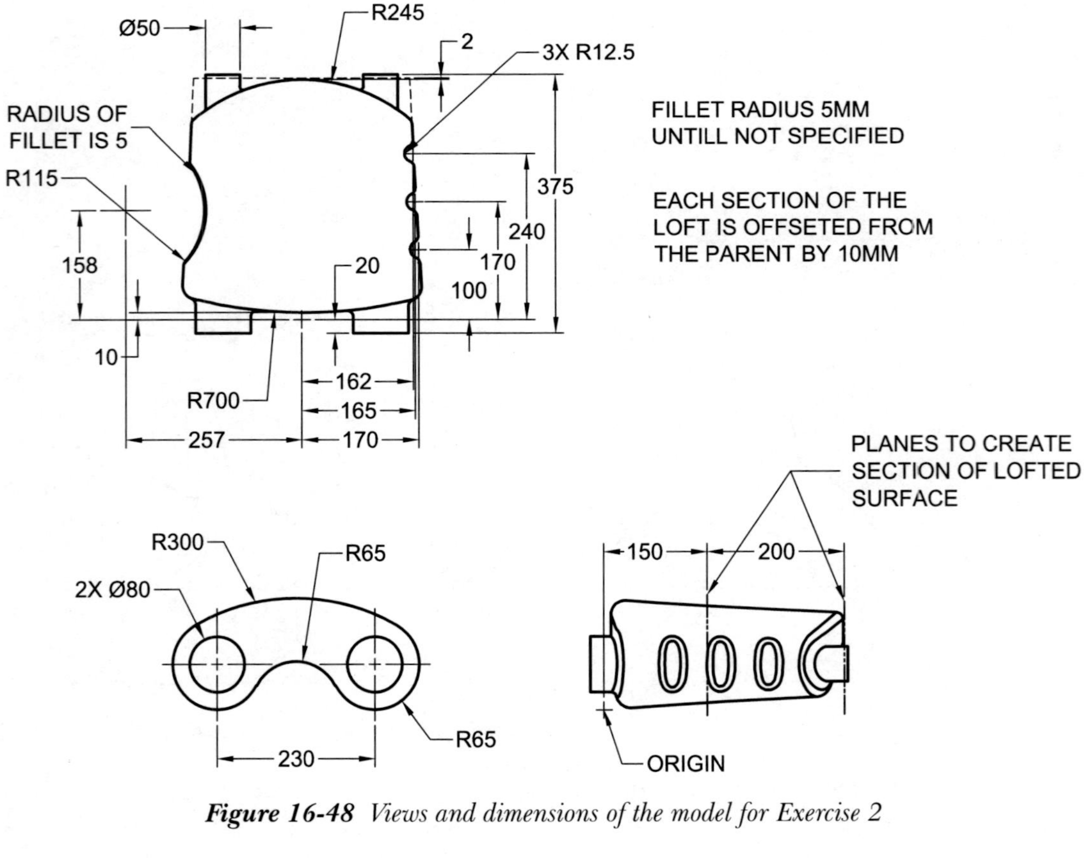

Figure 16-48 *Views and dimensions of the model for Exercise 2*

Chapter 17

Sheet Metal Design

Learning Objectives

After completing this chapter, you will be able to:

- *Create base, edge, and miter flanges*
- *Create tabs, closed corners, and hems*
- *Break corners of sheet metal components*
- *Create cuts on the flat faces of sheet metal components*
- *Create the flat pattern of sheet metal components*
- *Generate the drawing views of the flat pattern of the sheet metal components*

SHEET METAL DESIGN

In SOLIDWORKS, you can design the sheet metal components using various tools available for manipulating the sheet metal components in the **Part** mode. Generally, the solid models of the sheet metal components are created to generate the flat pattern of the sheet, study the design of the dies and punches, and study the process plan for designing the tools needed for manufacturing the sheet metal components. In a tool room or a machine shop, the most important thing that you need before designing the press tool, bending tool or any other tool for creating a sheet metal component is the flat pattern layout of a component. Figure 17-1 shows the model of a sheet metal component and Figure 17-2 shows its flat pattern layout. A flat pattern layout displays the flattened view of the sheet metal component, refer to Figure 17-2.

Figure 17-1 *Solid model of a sheet metal component*

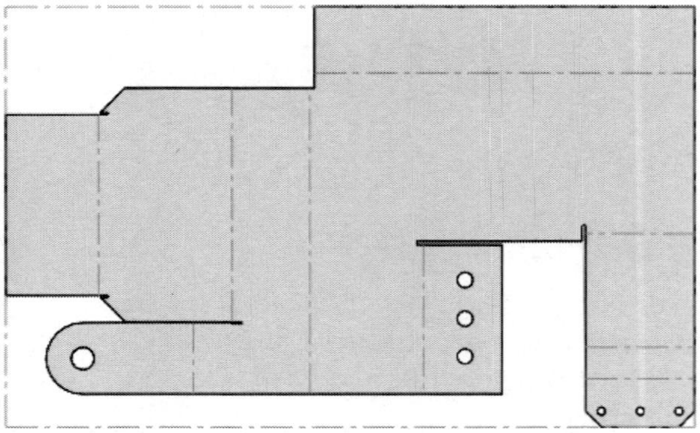

Figure 17-2 *Flat pattern layout of the sheet metal component*

As discussed earlier, the sheet metal components can also be created in the **Part** mode of SOLIDWORKS. To create a sheet metal component, start a new SOLIDWORKS document in the **Part** mode and then invoke the **Sheet Metal CommandManager**. If this CommandManager is not available by default, invoke it by right-clicking on the tab of a CommandManager and choosing **Sheet Metal** from the shortcut menu. All tools that are used to design a sheet metal component are available in this CommandManager. You can also invoke these tools from the **Sheet Metal** toolbar.

DESIGNING THE SHEET METAL COMPONENTS BY CREATING THE BASE FLANGE

The most widely used method of designing a sheet metal component is by first creating the base flange. In this method, first you will create the base flange and then add the sheet metal feature on the base flange to obtain the required sheet metal component. In this method, all the parameters related to the sheet metal such as the bending radius, the bend allowance, and the relief are defined while creating the base flange. Various tools used to create the sheet metal components are discussed next.

Creating the Base Flange

CommandManager:	Sheet Metal > Base Flange/Tab
SOLIDWORKS menus:	Insert > Sheet Metal > Base Flange

To create a sheet metal component, you first need to create a base feature or a base sheet. This base sheet is known as the base flange. You can create a base flange from a closed sketch or an open sketch.

Creating the Edge Flange

CommandManager:	Sheet Metal > Edge Flange
SOLIDWORKS menus:	Insert > Sheet Metal > Edge Flange

Edge flange is a bent sheet metal wall created at an angle on the edge of an existing base flange or an existing flange.

Creating a Tab and Slot

CommandManager:	Sheet Metal > Tab and Slot (Customize to add)
SOLIDWORKS menus:	Insert > Sheet Metal > Tab and Slot

The **Tab and Slot** tool is used to create tabs and slots (holes) respectively on the bodies to be interlocked. This feature helps in minimizing the use of welding fixtures. This tool can be used on parts other than sheet metal parts.

Creating the Miter Flange

CommandManager:	Sheet Metal > Miter Flange
SOLIDWORKS menus:	Insert > Sheet Metal > Miter Flange

The **Miter Flange** tool is used to create a series of flanges along the edges of a sheet metal component. The profile of the miter flange is defined by the sketch created on a sketching plane normal to the direction of extrusion of the flange.

Creating Closed Corners

CommandManager:	Sheet Metal > Corners flyout > Closed Corner
SOLIDWORKS menus:	Insert > Sheet Metal > Closed Corner

When you create walls using the **Edge Flange** tool, there may be a gap between the corners due to relief. You can close this gap and create a closed corner.

Creating Welded Corners

CommandManager:	Sheet Metal > Corners flyout > Welded Corner
SOLIDWORKS menus:	Insert > Sheet Metal > Welded Corner

The **Welded Corner** tool allows you to create a weld bead along the corners of a folded sheet metal part. You can create a weld bead between miter flanges, edge flanges, and closed corners. The welded corners are supressed in flattened state.

Breaking the Corners

CommandManager:	Sheet Metal > Corners flyout > Break-Corner/Corner-Trim
SOLIDWORKS menus:	Insert > Sheet Metal > Break-Corner

In SOLIDWORKS, you are provided with an option to break the edges of the sheet metal components to create chamfer or fillet. The edges of the sheet metal components are chamfered or filleted using the **Break-Corner/Corner-Trim** tool.

Creating Corner Relief

CommandManager:	Sheet Metal > Corners flyout > Corner Relief
SOLIDWORKS menus:	Insert > Sheet Metal > Corner Relief

The **Corner Relief** tool is used to modify sheet metal corners so that they can be bended or welded easily in the manufacturing process.

Creating Hems

CommandManager:	Sheet Metal > Hem
SOLIDWORKS menus:	Insert > Sheet Metal > Hem

Hems are generally used to bend a small area of sheet in order to eliminate the sharp edges in a sheet metal component. Hems are also used to connect two walls by interlocking their edges.

Creating the Swept Flange

CommandManager:	Sheet Metal > Swept Flange (Customize to add)
SOLIDWORKS menus:	Insert > Sheet Metal > Swept Flange

The **Swept Flange** tool is used to create a series of flanges along a sketched path or edges of a sheet metal component. The profile of the swept flange is defined by the sketch created on a sketching plane normal to the direction of extrusion of the flange.

Creating a Flat Pattern View of the Sheet Metal Components

CommandManager:	Sheet Metal > Flatten
Toolbar:	Sheet Metal > Flatten

The flat pattern view of a sheet metal component is extensively used in the tool room or the machine shop to define the size of the raw sheet, and also the shape of the sheet that you need before bending. It is also used for process planning to start the manufacturing of the tool that will create the sheet metal component.

TUTORIALS

Tutorial 1

In this tutorial, you will create the sheet metal component shown in Figure 17-3. Flat pattern of the sheet metal component and its views and dimensions are shown in Figures 17-4 and 17-5, respectively. You will first create the base flange and then the other features. After creating the model, you will create its flat pattern. **(Expected time: 45 min)**

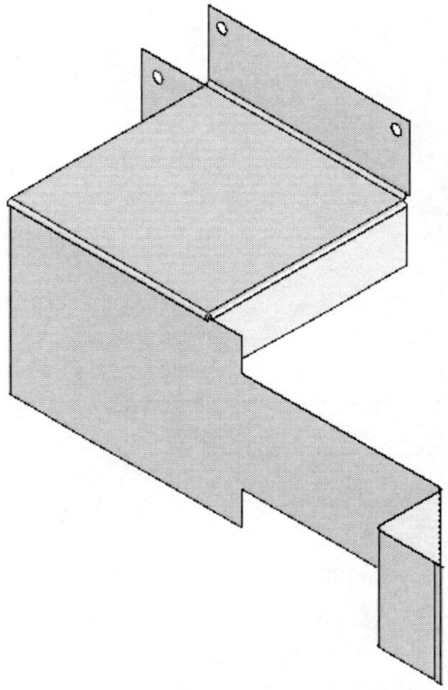

Figure 17-3 *Sheet metal component*

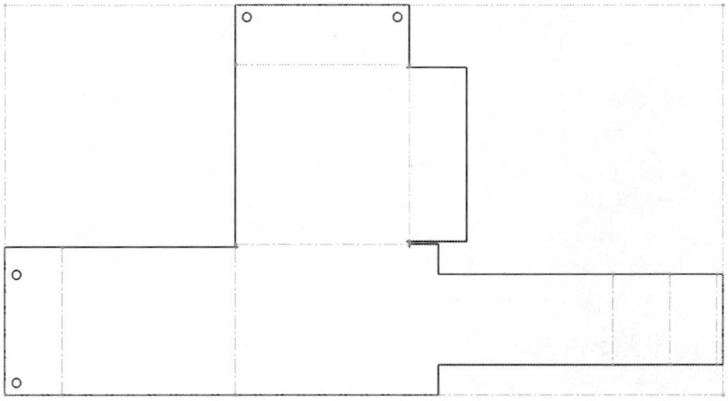

Figure 17-4 *Flat pattern of the sheet metal component*

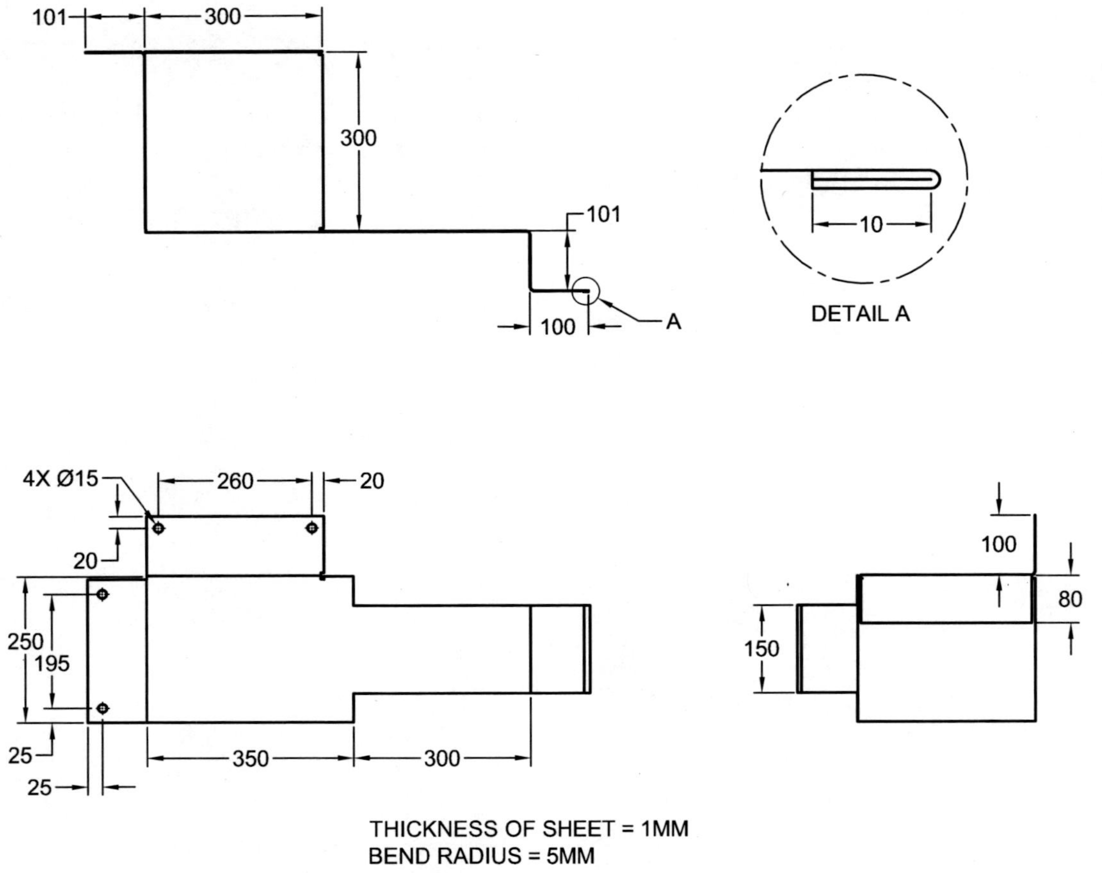

Figure 17-5 *Drawing views and dimensions for Tutorial 1*

The following steps are required to complete this tutorial:

a. Create the base flange of the sheet metal component, as shown in Figure 17-6.
b. Add other required flanges to the sheet metal component, as shown in Figure 17-12.
c. Create the tab feature, as shown in Figure 17-14.
d. Add a hem to the right most flange, refer to Figure 17-17.
e. Create the flat pattern of the sheet metal component.
f. Save the model.

Creating the Base Flange

For creating this sheet metal component, you first need to create the base flange. The base flange will be created by using a rectangular sketch drawn on the Front Plane.

1. Start a new SOLIDWORKS document in the **Part** mode.

2. Invoke the sketching environment using the **Front Plane** as the sketching plane.

3. Create a rectangle of 350x250 mm as the sketch of the base flange, refer to Figure 17-5.

4. Choose the **Base Flange/Tab** tool from the **Sheet Metal CommandManager**; the **Base Flange PropertyManager** is displayed.

5. Set the value of the following parameters and choose the **OK** button from the **Base Flange PropertyManager**.

   Thickness: **1** K-Factor: **1** Auto Relief Type: **Rectangular** Ratio: **0.5**

   Figure 17-6 shows the base flange created on the Front Plane.

Creating the First Edge Flange

After creating the base flange, you need to create the first edge flange using the top edge of the base flange as the reference. You will observe that the **Sheet-Metal1** feature is displayed in the **FeatureManager Design Tree**. Modify the default bend radius using the **Sheet-Metal1** feature before creating the first edge flange.

1. Select the **Sheet-Metal** node from the **FeatureManager Design Tree** to invoke the pop-up toolbar. Next, choose the **Edit Feature** option from the pop-up toolbar.

2. Set the value in the **Bend Radius** spinner to **5** and choose the **OK** button from the **Sheet-Metal PropertyManager**.

3. Choose the **Edge Flange** tool from the **Sheet Metal CommandManager**; the **Edge-Flange1 PropertyManager** is displayed. Also, you are prompted to select the linear edge of a planar face to create the edge flange.

4. Select the edge of the base flange, as shown in Figure 17-7.

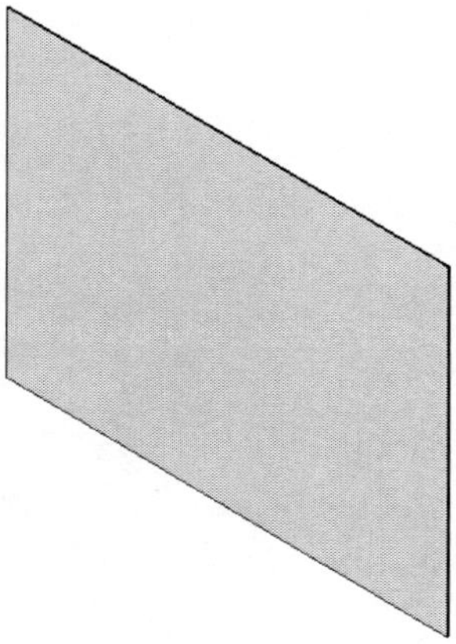

Figure 17-6 *Base flange*

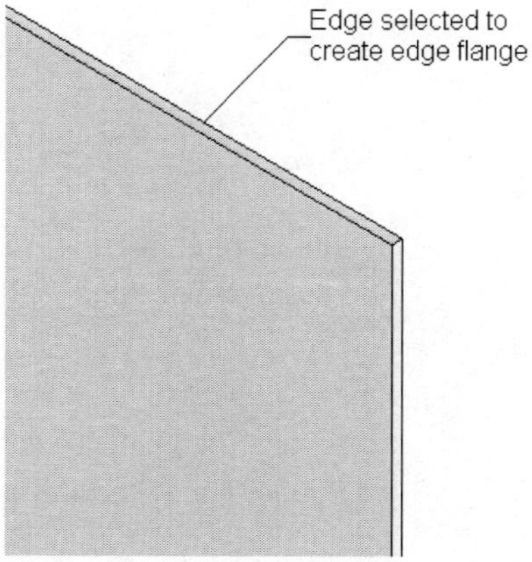

Figure 17-7 *Edge selected to create edge flange*

5. In the **Flange Length** rollout, set the value in the **Length** spinner to **300**. Choose the **Material Outside** button from the **Flange Position** rollout. Make sure that the flange is being created in the backward direction.

   Next, you need to edit the profile of the edge flange.

6. Choose the **Edit Flange Profile** button from the **Flange Parameters** rollout. As soon as you choose this button, the sketching environment is invoked and the **Profile Sketch** dialog box is displayed.

7. Edit the sketch, refer to Figure 17-8. Choose the **Finish** button from the **Profile Sketch** dialog box. The model after creating the first flange is as shown in Figure 17-9.

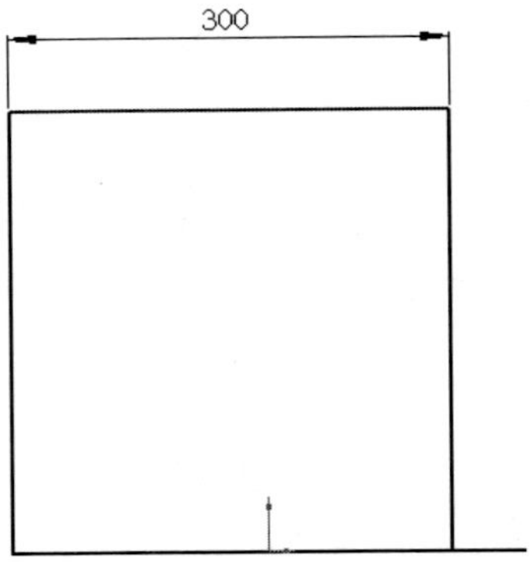

Figure 17-8 Modified sketch of the edge flange

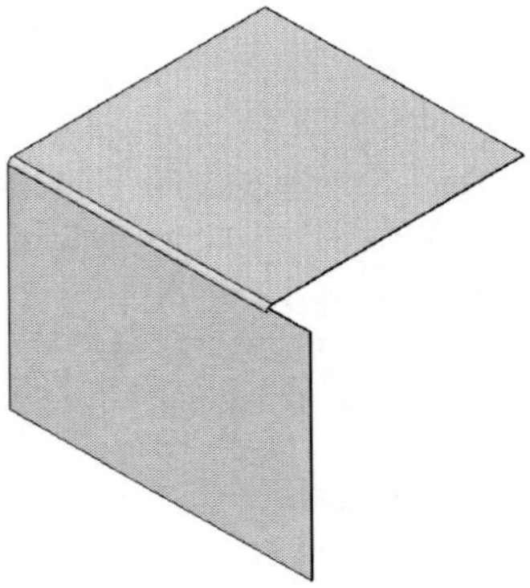

Figure 17-9 Model after creating the edge flange

Creating the Second Edge Flange

Next, you need to create the second edge flange with the holes.

1. Choose the **Edge Flange** tool from the **Sheet Metal CommandManager** to invoke the **Edge-Flange PropertyManager**.

2. Select the edge, as shown in Figure 17-10; a preview of the edge flange is displayed.

3. Set the value in the **Length** spinner to **100** and choose the **OK** button from the **Edge-Flange PropertyManager**.

4. Next, create two circular holes of diameter **15** mm each, refer to Figure 17-11.

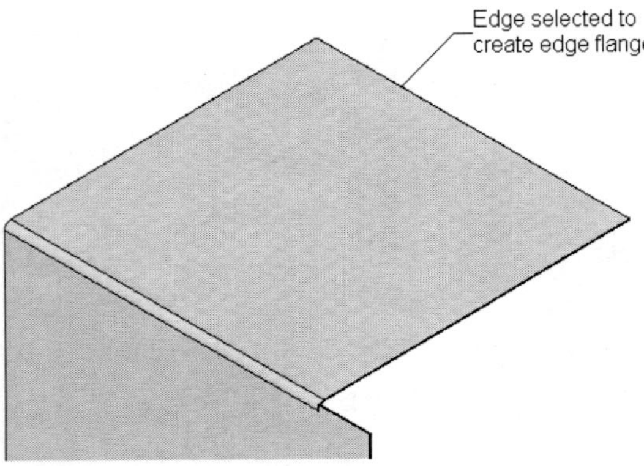

Figure 17-10 Edge selected to create the edge flange

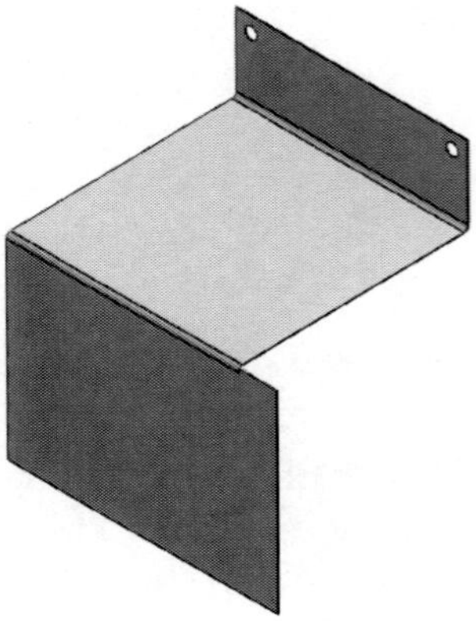

Figure 17-11 Model after creating the second edge flange

5. Similarly, create the edge flanges on the left side of the sheet metal component.

6. Edit the sketch of the edge flanges and draw the sketch for the holes. Refer to Figure 17-5 for dimensions. Edge-flanges with the holes are shown in Figure 17-12.

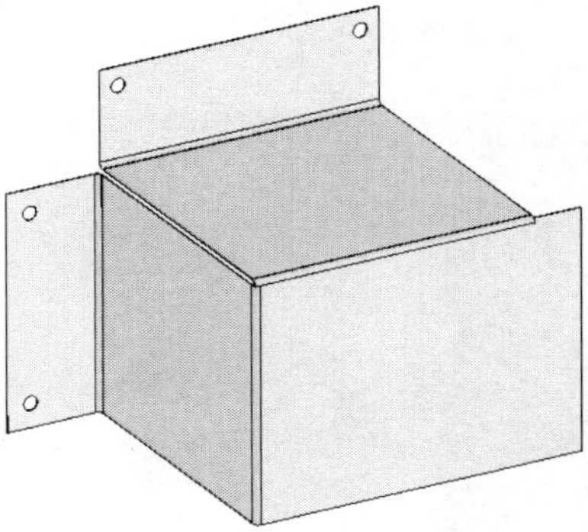

Figure 17-12 *Model after creating edge flanges and holes on the left of the model*

Creating the Tab and Flange Features

Next, you need to add a tab feature to the sheet metal component. A tab feature is used to add material to the base flange or any other flange feature.

1. Select the front face of the base flange as the sketching plane and invoke the sketching environment.

2. Draw the sketch of the tab feature, as shown in Figure 17-13.

3. Choose the **Base-Flange/Tab** tool from the **Sheet Metal CommandManager**; the **Base Flange PropertyManager** is displayed. Select the **Merge Result** check box if it is not selected by default, and choose the **OK** button; the tab feature is created, as shown in Figure 17-14.

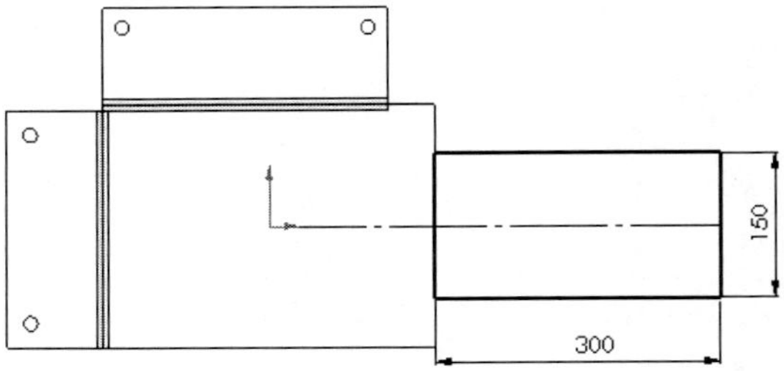

Figure 17-13 *Sketch of the tab feature*

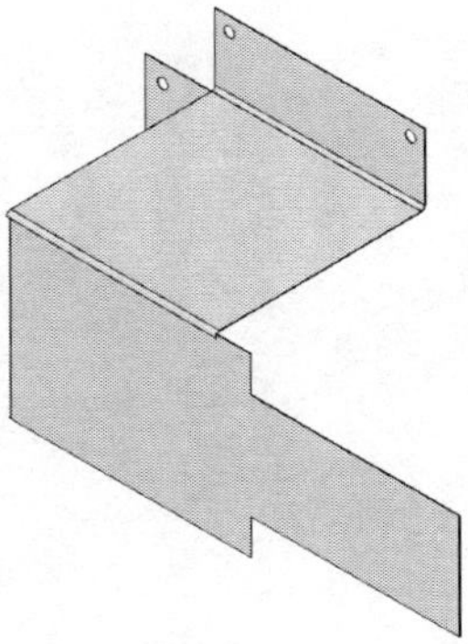

Figure 17-14 *Model after creating the tab feature*

4. Create three more edge flanges, two of length 100 mmand one of length 80 mm, refer to Figure 17-5. The model after creating all edge flanges is shown in Figure 17-15.

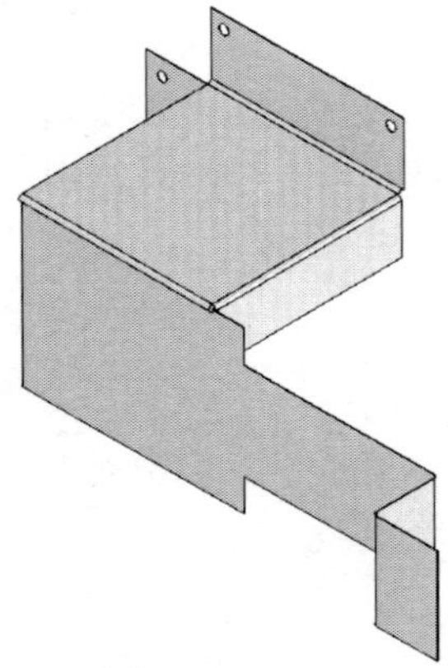

***Figure 17-15** Model after creating all edge flanges*

Creating the Hem Feature

After creating the other sheet metal features, you need to create the hem on the right most edge flange.

1. Choose the **Hem** tool from the **Sheet Metal CommandManager**; the **Hem PropertyManager** is displayed.

2. Choose the **Bend Outside** button from the **Edges** rollout and the **Closed** button from the **Type and Size** rollout. Next, set the value in the **Length** spinner to **10**.

3. Select the edge to create the hem, as shown in Figure 17-16; the hem is created, as shown in Figure 17-17.

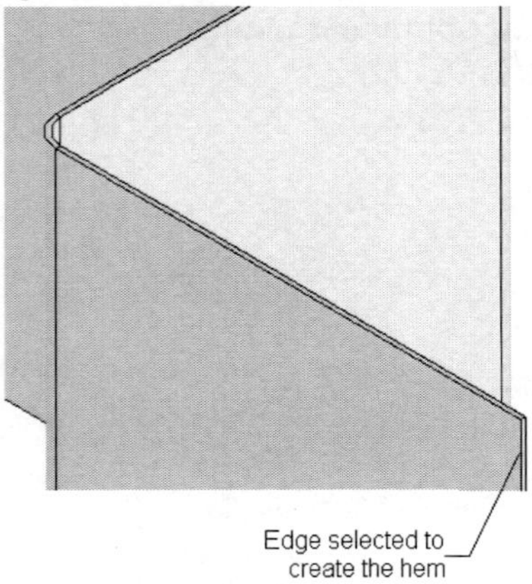

***Figure 17-16** Edge selected to create the hem*

***Figure 17-17** Resultant hem*

Creating the Flat Pattern

After creating the sheet metal component, you need to create its flat pattern.

1. Choose the **Flatten** tool from the **Sheet Metal CommandManager** to create the flat pattern of the sheet metal component.

2. Orient the flattened model parallel to the screen.

The flattened sheet metal component is displayed in Figure 17-18.

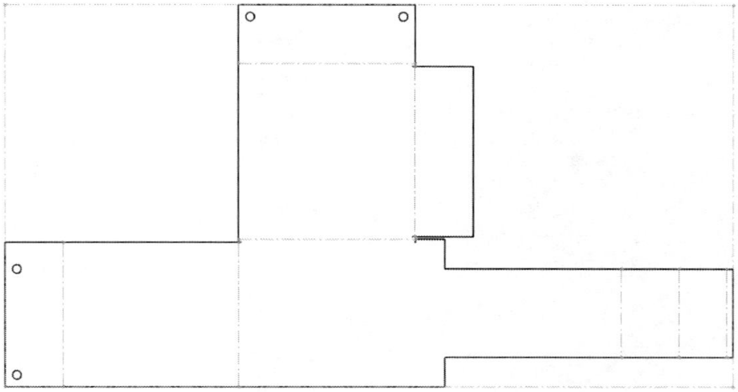

Figure 17-18 *Flattened sheet metal component*

Saving the Model

Next, you need to save the model.

1. Choose the **Save** button from the Menu Bar and save the drawing with the name *c17_tut01* at the location given below and then close the file.

\Documents\SOLIDWORKS\c17

Tutorial 2

In this tutorial, you will create the sheet metal component shown in Figure 17-19. The flat pattern of the sheet metal component is displayed in Figure 17-20. After creating the model, you need to generate its drawing views, as shown in Figure 17-21. The drawing views and the dimensions of the model are displayed in Figure 17-22. **(Expected time: 1 hr)**

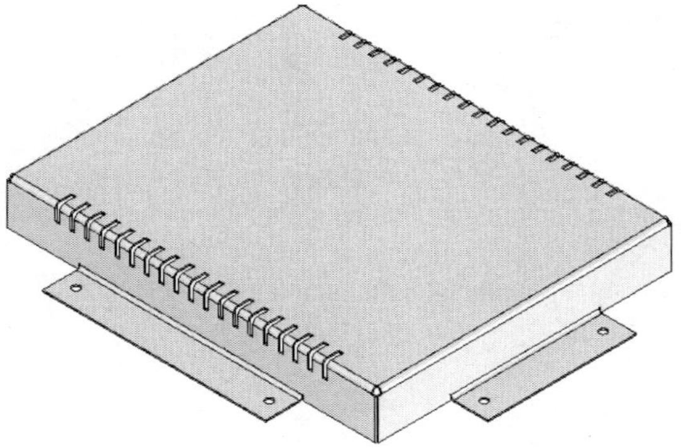

Figure 17-19 *Sheet metal component*

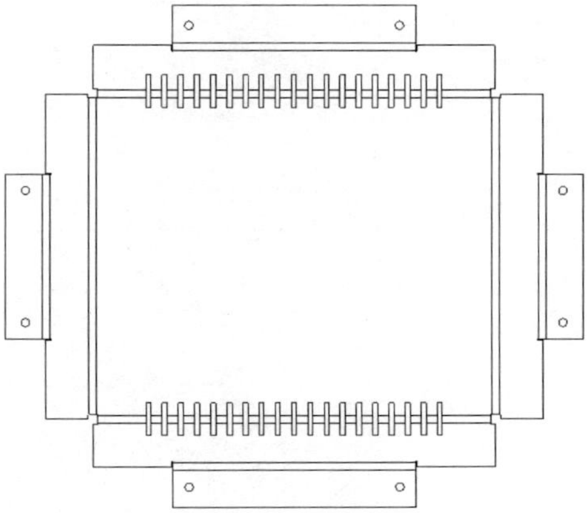

Figure 17-20 *Flattened sheet metal component*

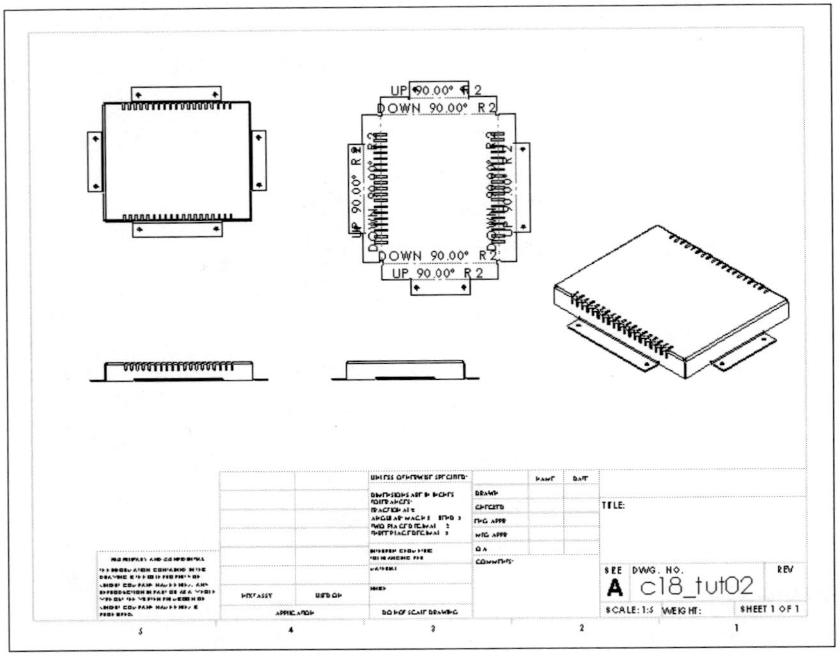

Figure 17-21 *Drawing views of the model*

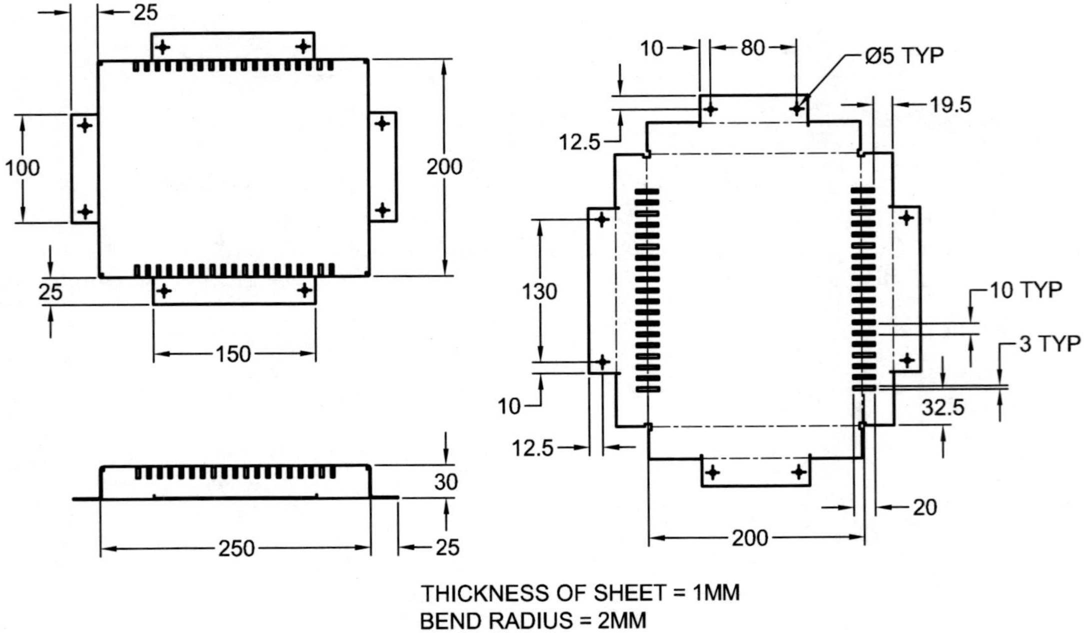

Figure 17-22 *Views and dimensions for Tutorial 2*

The following steps are required to complete this tutorial:

a. Create the base feature by extruding the sketch created on the Top Plane.
b. Shell the base feature.
c. Convert the shelled solid model into sheet metal component.
d. Roll back the model to the state where no bends were added to the sheet metal component.
e. Add flanges to the sheet metal component.
f. Create the flat pattern of the sheet metal component.
g. Create slots using the cut feature and pattern them on the sides of the flattened sheet metal component.
h. Refold the sheet metal component.
i. Generate the drawing views of the sheet metal component.
j. Save the model.

Creating the Base Feature

You will create this model by converting a shelled part into a sheet metal component. Therefore, you first need to create the base feature of the model by extruding the sketch created on the Top Plane.

1. Start a new SOLIDWORKS document in the **Part** mode.

2. Invoke the sketching environment by selecting the Top Plane as the sketching plane.

3. Draw the sketch of the base feature that consists of a rectangle of **250**x**200**.

4. Extrude the sketch upto a distance of **30** mm; the base feature of the model is displayed, as shown in Figure 17-23.

Shelling the Base Feature

Next, you need to add the shell feature to the model by removing the bottom face of the base feature.

1. Rotate the model so that its bottom face is clearly visible.

2. Invoke the **Shell** tool and select the bottom face of the base feature as the face to remove.

3. Set the value in the **Thickness** spinner to **1** and choose the **OK** button from the **Shell PropertyManager**; the shell feature is created, as shown in Figure 17-24.

> **Tip**
> *You can also convert a solid model into a sheet metal component by using the* ***Convert to Sheet Metal*** *tool.*

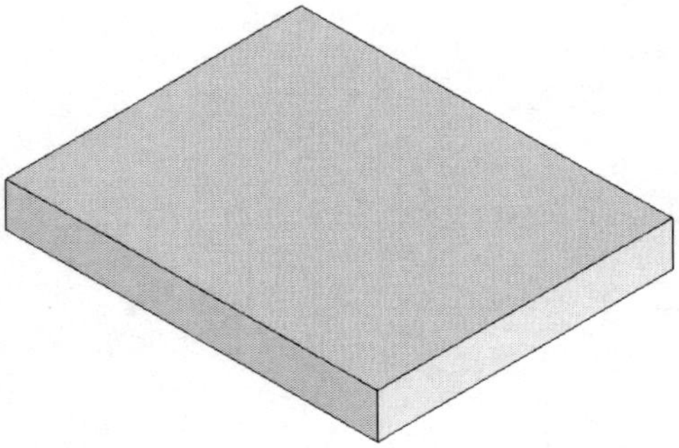

Figure 17-23 Base feature

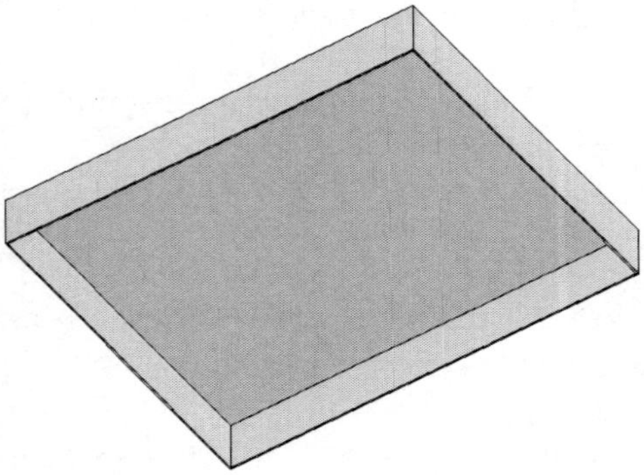

Figure 17-24 Model after shelling

Converting the Shelled Model into Sheet Metal Component

After creating the base feature and shelling the model, you need to convert it into a sheet metal component using the **Insert Bends** tool.

1. Choose the **Insert Bends** tool from the **Sheet Metal CommandManager** to invoke the **Bends PropertyManager**.

2. Select the top face of the model to fix.

3. Click once in the **Edges to Rip** selection box in the **Rip Parameters** rollout to activate the selection mode.

4. Select all the inner vertical edges of the base feature as the edges to rip.

5. Set the value **2** in the **Bend Radius** spinner.

6. Choose the **OK** button from the **Bends PropertyManager**; the **SOLIDWORKS** message box is displayed which informs you that the Auto relief cuts were made for one or more bends. Choose the **OK** button from this message box.

   The model after converting the solid shelled model into the sheet metal component is shown in Figure 17-25.

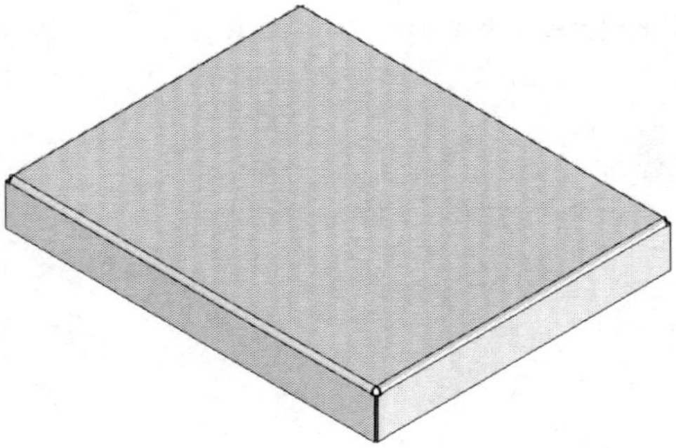

Figure 17-25 Solid shelled model after converting into sheet metal component

Creating the Edge Flanges

Next, you need to add the edge flanges to the sheet metal component. Before adding the edge flanges, you need to roll back the model to the stage when there were no bends in the sheet metal component.

1. Choose the **No Bends** button from the **Sheet Metal CommandManager** to roll back the model to the stage where it had no bends. Note that the four walls of the model are not unbent.

2. Now, select the bottom face of one of the four walls of the model as the sketching plane and invoke the sketching environment.

3. Draw the sketch of the flanges, as shown in Figure 17-26.

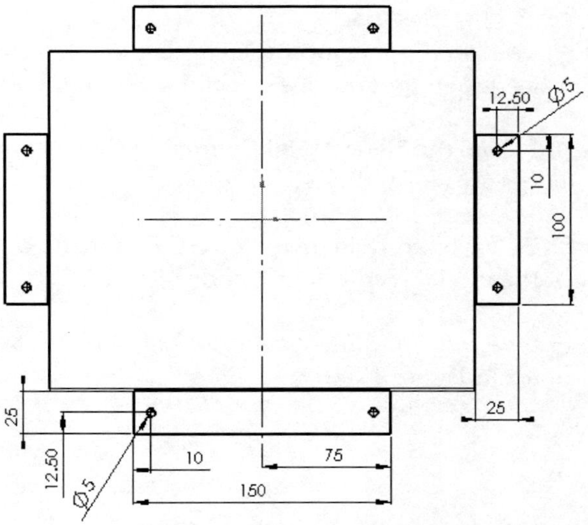

Figure 17-26 *Sketch of the flanges*

4. Invoke the **Boss-Extrude PropertyManager**. Select the **Link to thickness** check box and choose the **Reverse Direction** button. Make sure that the **Merge result** check box is selected in the **Direction 1** rollout.

5. Choose the **OK** button from the **Boss-Extrude PropertyManager**. Figure 17-27 shows the model after extruding the flanges.

Now, you need to roll the model back to the bending stage.

6. Again, choose the **No Bends** tool from the **Sheet Metal CommandManager** to roll the model to the bending stage.

Figure 17-28 shows the model after having been rolled back to the bending stage.

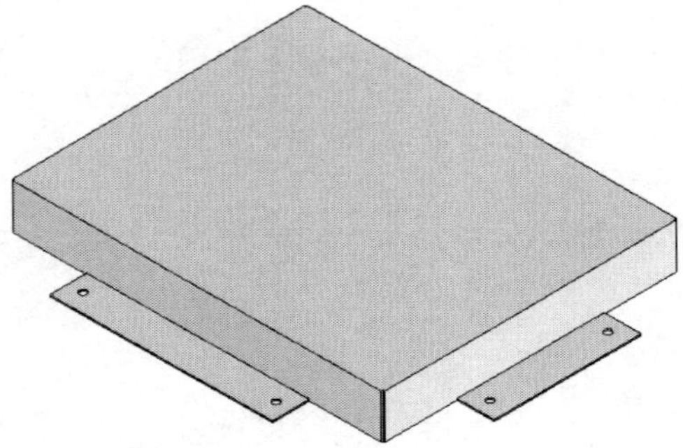

Figure 17-27 *Model after extruding the flanges*

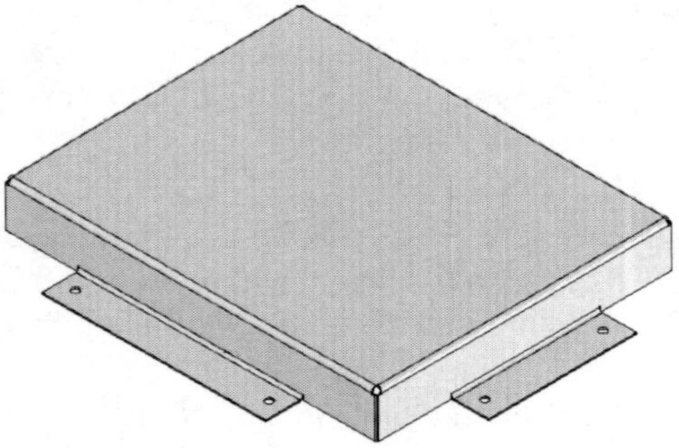

Figure 17-28 *Model rolled back to the bending stage*

Creating Cuts across the Bends

The next feature that you need to create is the cut feature across the bends. For creating this feature, you first need to unfold the sheet metal component.

1. Choose the **Unfold** tool from the **Sheet Metal CommandManager** and select the top face of the model as the fixed face.

2. Choose the **Collect All Bends** button and then choose **OK** from the **Unfold PropertyManager**. The unfolded view of the model is shown in Figure 17-29.

3. Select the top face of the model and invoke the sketching plane. Next, draw the sketch of the cut feature, as shown in Figure 17-30.

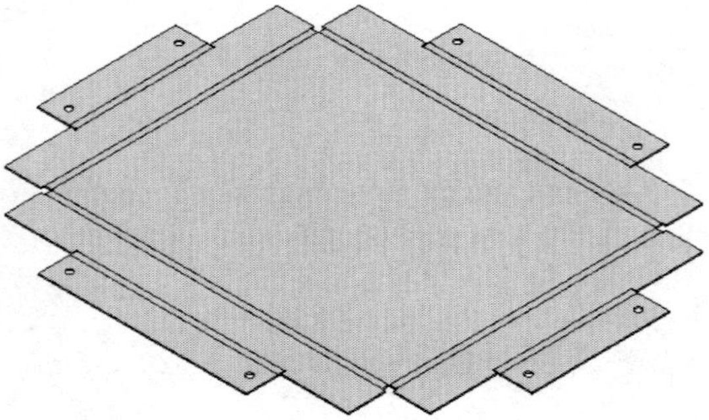

Figure 17-29 *Unfolded view of the model*

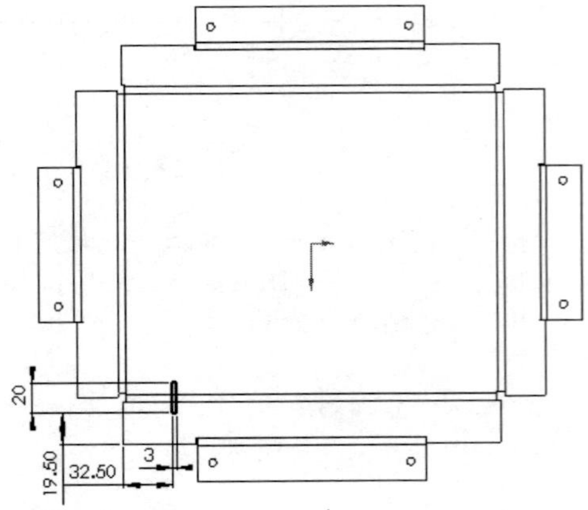

Figure 17-30 *Sketch of the cut feature*

4. Invoke the **Cut-Extrude PropertyManager** and create the cut feature. The model after adding the cut feature is shown in Figure 17-31.

5. Pattern the cut feature. The model after patterning the cut feature is shown in Figure 17-32.

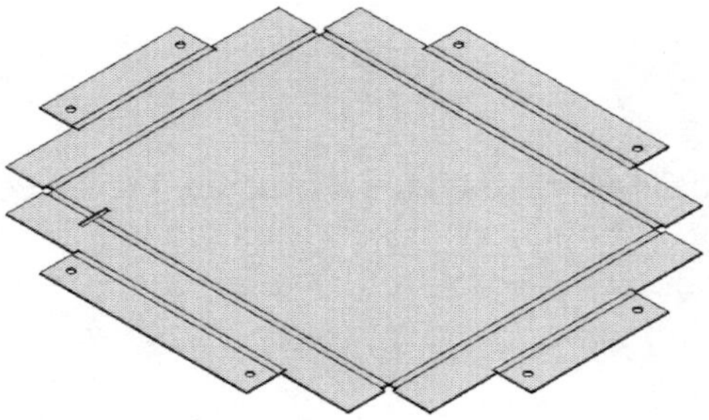

Figure 17-31 Model after adding the cut feature

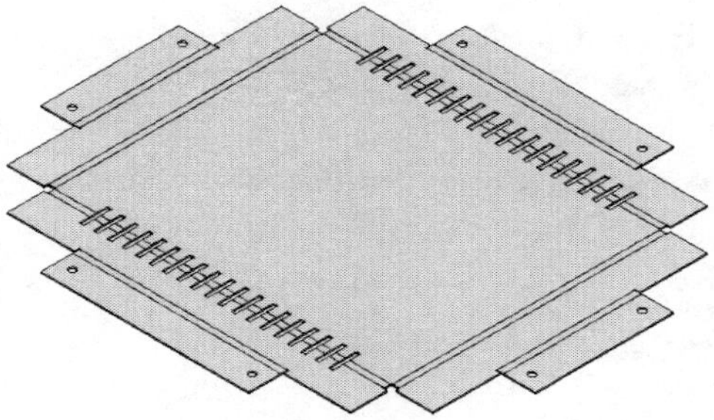

Figure 17-32 Model after patterning the cut feature

Refolding the Sheet Metal Part

After creating the sheet metal component, you need to refold the sheet.

1. Choose the **Fold** button from the **Sheet Metal CommandManager**; the **Fold PropertyManager** is displayed and the top face is selected by default.

2. Choose the **Collect All Bends** button and then choose **OK**. The sheet metal component after refolding is shown in Figure 17-33.

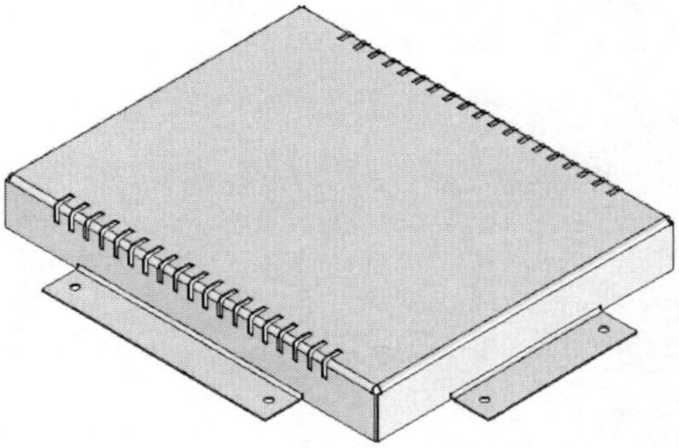

Figure 17-33 Sheet metal component after refolding

3. Save the model with the name *c17_tut02* at the following location:

   \Documents\SOLIDWORKS\c17

Generating the Drawing Views of the Sheet Metal Component

You will generate the drawing views of the sheet metal component on an A4 size drawing sheet with third angle projection.

1. Choose **New > Make Drawing from Part/Assembly** from the Menu Bar; the **Sheet Format/Size** dialog box is displayed.

2. Select the **A4 (ANSI) Landscape** option from the list box in the **Sheet Format/Size** message box and then choose **OK**.

3. If the default projection type is in the first angle, change it to the third angle using the **Sheet Properties** dialog box.

4. Generate the three default standard views of the model by dragging it from the **View Palette** task pane. Figure 17-34 shows the drawing sheet after generating the three standard views.

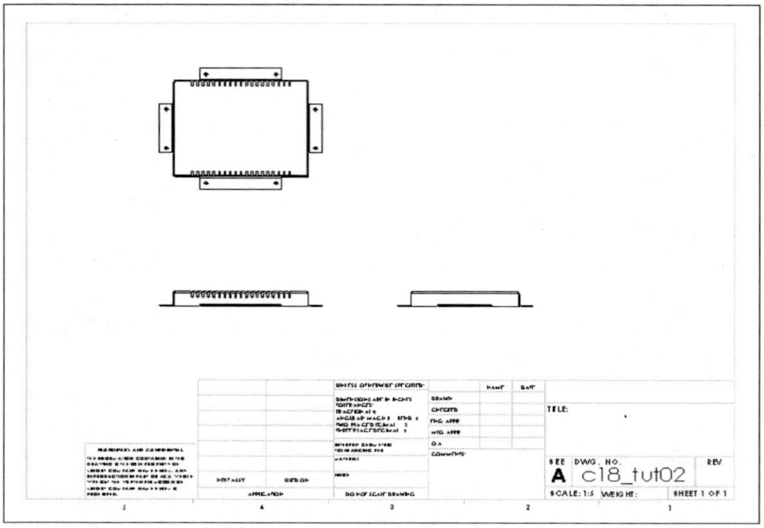

Figure 17-34 *Drawing sheet after generating the three standard views*

After generating the three standard views, you need to generate the drawing view of the flat pattern.

5. Invoke the **Model View PropertyManager**; the *c17_tut02* file is selected by default and its name is displayed in the **Open documents** list box. Choose the **Next** button from the **Model View PropertyManager**.

6. Select the **Flat pattern** check box in the **More views** list box of the **Orientation** rollout. Make sure that the **DEFAULT-FLAT-PATTERN** option is selected in the drop-down list of the **Reference Configuration** rollout of the PropertyManager.

7. Move the cursor to the top-right corner of the drawing sheet and left-click to place the flat pattern of the drawing view. Next, choose the **OK** button from the PropertyManager.

   Figure 17-35 shows the drawing sheet after generating the drawing view of the flat pattern.

   Next, you need to generate the isometric view.

8. Generate an isometric view and place the view close to the right corner of the drawing sheet. Change the scale of the isometric view to 1:4.

   The drawing sheet after generating all drawing views is displayed in Figure 17-36.

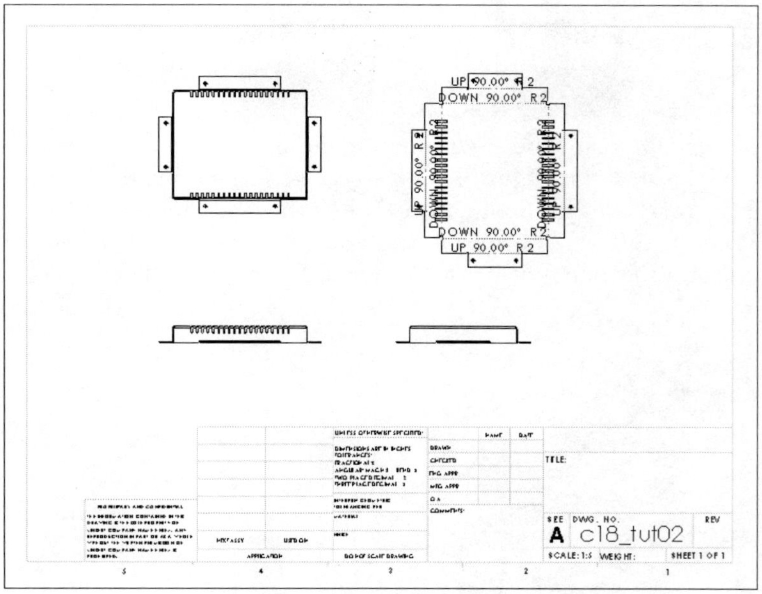

Figure 17-35 Drawing sheet after generating the flat pattern view

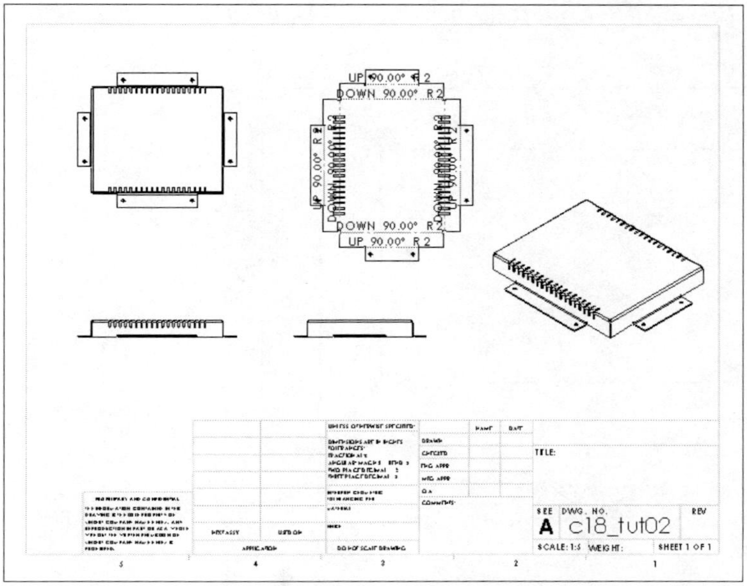

Figure 17-36 Final drawing sheet

Saving the Model

Next, you need to save the model after generating the drawing views.

1. Choose the **Save** button from the **Menu Bar** and save the file with the name *c17_tut02* at the following location:

\Documents\SOLIDWORKS\c17

EXERCISE

Exercise 1

In this exercise, you will create the sheet metal component shown in Figure 17-37. The flat pattern of this model is shown in Figure 17-38. To create this model, first create the base flange. Then, you need to create a miter flange to complete the sheet metal component. The default bend radius is 2 mm, K-Factor is 0.5, and Rectangular Relief ratio is 0.5. Thickness of the sheet is 1 mm. Rip gap for miter flange is 2 mm. Views and dimensions for this model are shown in Figure 17-39. **(Expected time: 20 min)**

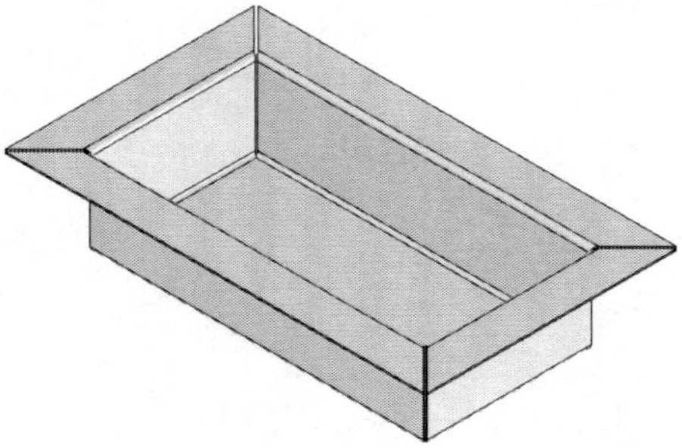

Figure 17-37 *Sheet metal component*

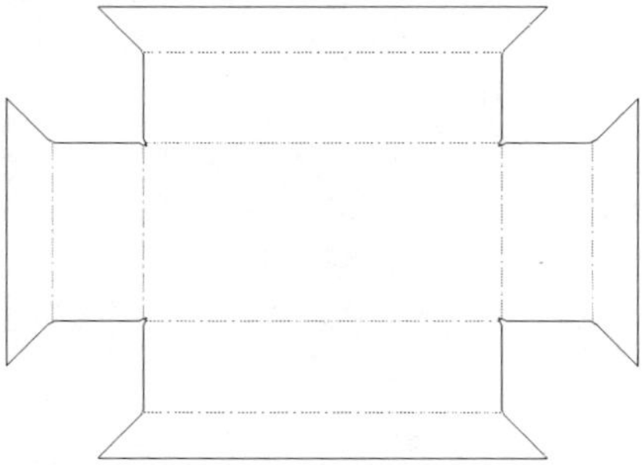

Figure 17-38 *Flat pattern of the sheet metal component*

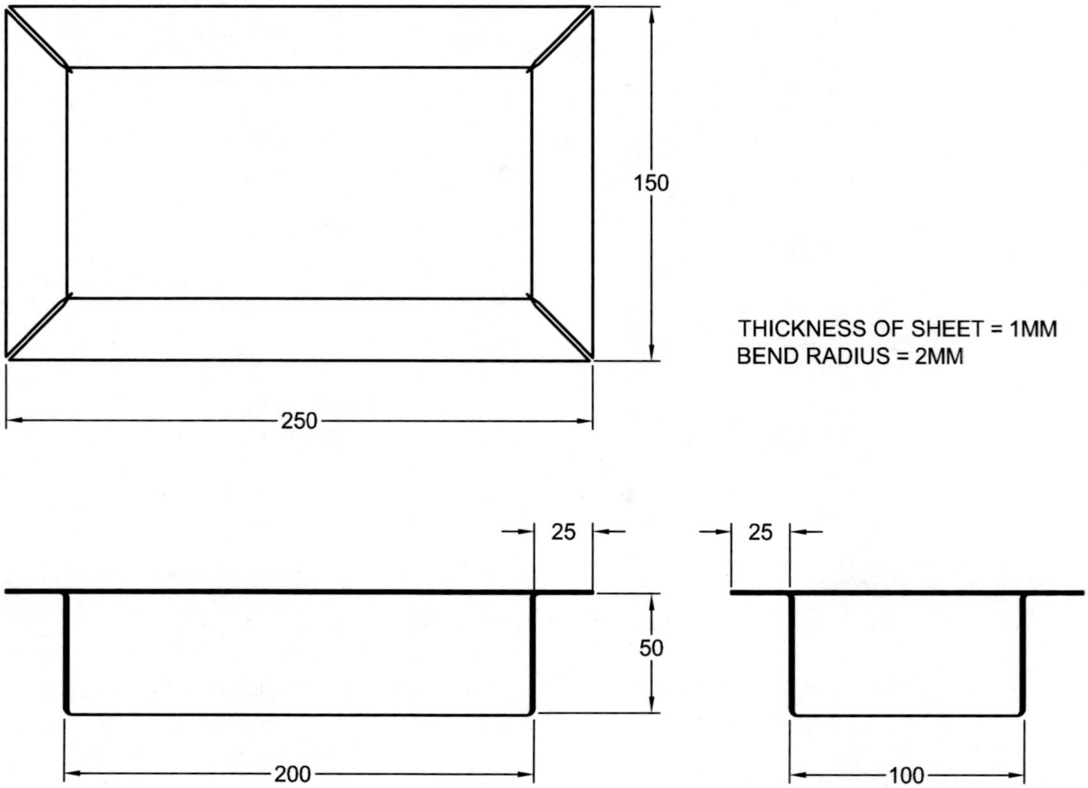

Figure 17-39 *Views and dimensions of the sheet metal component for Exercise 1*

Index

Symbols

A

B

C

D

E

Made in United States
Cleveland, OH
05 November 2026

47763598R00172